EDEXCEL A LEVEL

D1188965

BIOLOGY

Ed Lees
Martin Rowland
C. J. Clegg

NEW COLLEGE NOTTINGHAM
265378

DYNAMIC

HODDER
EDUCATION
AN HACHETTE UK COMPANY.

In order to ensure that this resource offers high-quality support for the associated Pearson qualification, it has been through a review process by the awarding body. This process confirms that this resource fully covers the teaching and learning content of the specification or part of a specification at which it is aimed. It also confirms that it demonstrates an appropriate balance between the development of subject skills, knowledge and understanding, in addition to preparation for assessment.

Endorsement does not cover any guidance on assessment activities or processes (e.g. practice questions or advice on how to answer assessment questions), included in the resource nor does it prescribe any particular approach to the teaching or delivery of a related course.

While the publishers have made every attempt to ensure that advice on the qualification and its assessment is accurate, the official specification and associated assessment guidance materials are the only authoritative source of information and should always be referred to for definitive guidance.

Pearson examiners have not contributed to any sections in this resource relevant to examination papers for which they have responsibility.

Examiners will not use endorsed resources as a source of material for any assessment set by Pearson.

Endorsement of a resource does not mean that the resource is required to achieve this Pearson qualification, nor does it mean that it is the only suitable material available to support the qualification, and any resource lists produced by the awarding body shall include this and other appropriate resources.

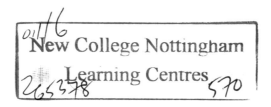
New College Nottingham
Learning Centres

Although every effort has been made to ensure that website addresses are correct at time of going to press, Hodder Education cannot be held responsible for the content of any website mentioned in this book. It is sometimes possible to find a relocated web page by typing in the address of the home page for a website in the URL window of your browser.

Hachette UK's policy is to use papers that are natural, renewable and recyclable products and made from wood grown in sustainable forests. The logging and manufacturing processes are expected to conform to the environmental regulations of the country of origin.

Orders: please contact Bookpoint Ltd, 130 Milton Park, Abingdon, Oxon OX14 4SB. Telephone: +44 (0)1235 827720. Fax: +44 (0)1235 400454. Lines are open 9.00a.m.–5.00p.m., Monday to Saturday, with a 24-hour message answering service. Visit our website at www.hoddereducation. co.uk

© C. J. Clegg, Ed Lees, Martin Rowland 2015

First published in 2015 by
Hodder Education,
An Hachette UK Company
Carmelite House, 50 Victoria Embankment
London, EC4Y 0DZ

Impression number   10 9 8 7 6 5 4 3 2 1

Year   2019 2018 2017 2016 2015

All rights reserved. Apart from any use permitted under UK copyright law, no part of this publication may be reproduced or transmitted in any form or by any means, electronic or mechanical, including photocopying and recording, or held within any information storage and retrieval system, without permission in writing from the publisher or under licence from the Copyright Licensing Agency Limited. Further details of such licences (for reprographic reproduction) may be obtained from the Copyright Licensing Agency Limited, Saffron House, 6–10 Kirby Street, London EC1N 8TS.

Cover photo © maw89 – Fotolia

Typeset in BemboStd, 11/13 pt by Aptara, Inc.

Printed in Italy

A catalogue record for this title is available from the British Library.

ISBN 9781471807343

Contents

Get the most from this book

Welcome to the **Edexcel A level Biology 1 Student's Book**. This book covers Year 1 of the Edexcel A level Biology specification and all content for the Edexcel AS Biology specification.

The following features have been included to help you get the most from this book.

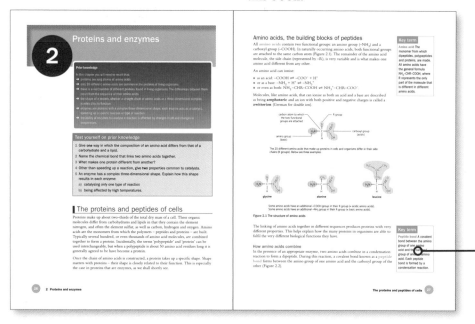

Key terms and formulae

These are highlighted in the text and definitions are given in the margin to help you pick out and learn these important concepts.

Examples

Examples of questions and calculations feature full workings and sample answers.

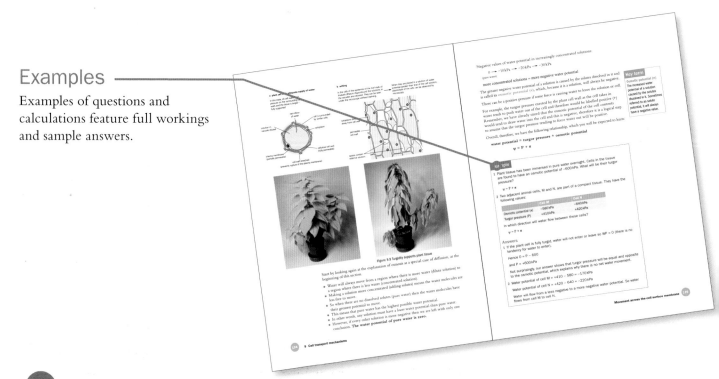

Test yourself questions

These short questions, found throughout each chapter, are useful for checking your understanding as you progress through a topic.

Activities and Core practicals

These practical-based activities will help consolidate your learning and test your practical skills. Edexcel's Core practicals are clearly highlighted.

In this edition the authors describe many important experimental procedures to conform to recent changes in the A level curriculum.

Teachers should be aware that, although there is enough information to inform students of techniques and many observations for exam purposes, there is not enough information for teachers to replicate the experiments themselves, or with students, without recourse to CLEAPSS Hazcards or Laboratory worksheets which have undergone a risk assessment procedure.

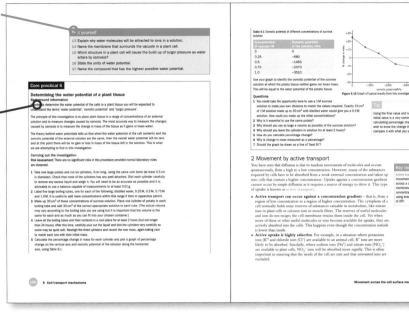

Exam practice questions

You will find Exam practice questions at the end of every chapter. These follow the style of the different types of questions you might see in your examination and are colour coded to highlight the level of difficulty. Test your understanding even further with Maths questions and Stretch and challenge questions.

Tips

These highlight important facts, common misconceptions and signpost you towards other relevant topics.

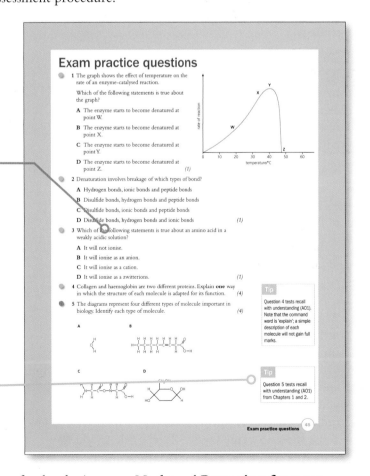

Dedicated chapters for developing your **Maths** and **Preparing for your exam** are also included in this book.

Introduction

Welcome to *Edexcel A level Biology 1*. This book has been written to cover the content of the first year of the Pearson Edexcel Level 3 Advanced GCE in Biology B (9BI0). Since the subject content is the same, it also covers the content of the Pearson Edexcel Level 3 Advanced Subsidiary GCE in Biology B (8BI0).

The first twelve chapters of this book cover the essential concepts, facts, principles and terminology of the four Biology topics in the AS qualification and the first year of A level course. The final chapters contain advice about developing skills that are assessed in your final examinations. Since you will need to put into practice the principles they contain, you are advised to read these final chapters at the start of your course. You will also find it useful to dip into them during the course, especially when preparing your examination strategy.

The biological content of each chapter has been written to ensure you cover everything that you can be expected to recall with understanding in your examinations. To help you and to encourage you to take an active part in your learning, the chapters contain the following features.

- Clear definitions of technical terms that you will be expected to understand and use correctly.
- Tips that offer reminders, hints or warnings.
- Regularly spaced Test yourself questions that encourage recall and understanding.
- Activities that encourage you to apply your knowledge, analyse information and make judgements.

At the end of each chapter you will find a series of Exam Practice Questions. By simulating the types of questions that you will encounter in your examinations, these questions will help you to develop, and maintain, the skills that will be assessed. To help you further, these questions have been graded to indicate their accessibility level (⬤ = AS/A level grade E–C, ⬤ = AS grade C–A/A level grade C and ⬤ = AS grade A/A level grade C–A). In addition, each chapter contains Stretch and challenge questions. Some of these are similar to questions in an A level paper that are targeted at the more able students. Others encourage you to use the learning resource centre in your college or school to carry out further research. Don't feel you have to do this alone; group research will develop your skills and can be more satisfying than working alone.

Practical work is an essential part of science. The Edexcel Biology specification contains core practicals that you are expected to carry out and on which you can be tested in examinations. This book covers these core practicals in a way that will encourage you to think about what you are doing, make comments on, or develop practical procedures, analyse results and make judgements. Again, these are all skills that can be assessed in your examinations.

Above all else, Biology is a fascinating subject which we hope you enjoy. We hope that some of our own enjoyment of Biology is reflected in this book and in the supporting material.

Acknowledgements

This book is an extensively revised, restructured and updated version of Edexcel Biology for AS by C J Clegg. We have relied heavily on the original book and are most grateful that C J Clegg has encouraged us to build on his work. We would also like to acknowledge the value of the detailed comments and suggestions from Liz Jones. The team at Hodder Education, led initially by Hanneke Remsing and then by Emma Braithwaite, has made an extremely valuable contribution to the development of the book and the website resources. In particular, we would like to thank Abigail Woodman, the project manager, for her expert advice and encouragement. We are also grateful for the skilful work on the print and electronic resources by Lydia Young.

Ed Lees and Martin Rowland

January 2015

Introducing the chemistry of life

1

Prior knowledge

In this chapter you will need to recall that:

→ carbohydrates contain the chemical elements carbon, hydrogen and oxygen in the ratio $C_x(H_2O)_y$
→ carbohydrates include simple sugars, such as glucose, and complex carbohydrates, such as cellulose and starch
→ simple sugars are used in respiration; complex carbohydrates might be glucose stores (starch) or structural components of cells (cellulose)
→ lipids also contain the chemical elements carbon, hydrogen and oxygen, but there is much less oxygen in lipids than in carbohydrates
→ lipids can be useful storage compounds and can also be structural components of cells
→ inorganic ions are present in the cytoplasm of cells and in body fluids; each type of ion has a specific function
→ water is essential for life and most of the mass of an organism comprises water
→ water is a reactant in many cell reactions; these reactions also occur in solution in water.

Test yourself on prior knowledge

1 Give **one** way in which the composition of a carbohydrate molecule is similar to that of a lipid and **one** way in which it is different from that of a lipid.

2 Name **one** type of carbohydrate used for energy storage in an animal and **one** used for energy storage in a plant.

3 How do animal lipids differ from plant lipids at room temperature?

4 Give **two** functions of lipids.

5 Calcium ions are essential for healthy growth in animals and plants. Give **one** function of calcium ions in humans and **one** function of calcium ions in plants.

6 In cells, water is a reactant in hydrolysis reactions and in condensation reactions. Describe the difference between these two types of reaction.

Some basic concepts

Chemical elements are the units of pure substance that make up our world. The Earth is composed of about 92 stable elements; living things are built from some of them. Table 1.1 shows a comparison between the most common elements in the Earth's crust and in us. You can see that the bulk of the Earth is composed of the elements oxygen, silicon, aluminium and iron. Of these, only oxygen is a major component of our cells.

Table 1.1 Most common elements

Earth's crust		Human body	
Element	% of atoms	Element	% of atoms
Oxygen	47.0	Hydrogen	63.0
Silicon	28.0	Oxygen	25.5
Aluminium	7.9	Carbon	9.5
Iron	4.5	Nitrogen	1.4
Calcium	3.5	Calcium	0.3
Sodium	2.5	Phosphorus	0.2

In fact, about 16 elements are required to build up all the molecules of the cell, and are therefore essential for life. Consequently, the full list of essential elements is a relatively short one. Furthermore, about 99 per cent of living matter consists of just four elements: carbon, hydrogen, oxygen and nitrogen.

The elements carbon, hydrogen and oxygen predominate because living things contain large quantities of water, and also because most other molecules present in cells and organisms are compounds of carbon combined with hydrogen and oxygen, including the carbohydrates and lipids. We will examine the structures and roles of carbohydrates and lipids shortly.

The element nitrogen is combined with carbon, hydrogen and oxygen in compounds called amino acids, from which proteins are constructed (Chapter 2). First, we will introduce some inorganic ions essential for organisms, and then discuss water.

Atoms, molecules and ions

The fundamental unit of chemical structure is the atom. Atoms group together to form molecules and molecules are the smallest part of most elements or compounds that can exist alone under normal conditions. For example, both oxygen and nitrogen naturally combine with another atom of the same type to form a molecule (O_2 and N_2, respectively).

If an atom gains or loses an electron, an ion is formed. Depending on their charge, ions migrate to the poles of an electric field. Positively charged ions migrate to the negative pole (cathode) and so are called cations. In contrast, negatively charged ions migrate to the positive pole (anode) and so are called anions.

Acids and bases

An acid is a compound that releases hydrogen ions in solution. We are familiar with the sharp taste that acids such as lemon juice or vinegar give to the tongue. These are relatively weak acids, weak enough to use on foods. The stronger the acid the more dangerous and corrosive it is, and the more hydrogen ions it releases. An example of a strong acid is hydrochloric acid. In water, this acid dissociates completely. The word dissociate means 'separates into its constituent ions':

$$HCl \rightarrow H^+ + Cl^-$$

hydrochloric acid   hydrogen ion (proton)   chloride ion

Key terms

Atom The smallest part of an element that can take part in a chemical change.

Ions Charged particles formed when atoms gain or lose electrons. **Cations** are positively charged, whereas **anions** are negatively charged.

Acid A compound that releases hydrogen ions in solution. Acidic solutions have a pH value below 7.

With organic acids such as citric acid (present in lemon juice) and ethanoic acid (found in vinegar), which we recognise as weak acids, relatively few molecules dissociate, and few hydrogen ions are present:

$$C_3H_3COOH \quad \rightarrow \quad C_2H_5COO^- \quad + \quad H^+$$

| ethanoic acid | ethanoate ion | hydrogen ion (proton) |

A **base** is a compound that can take up hydrogen ions in solution. In doing so it can neutralise an acid, forming a salt and water in the process. Many bases are insoluble in water. Those that are soluble in water are called alkalis. Examples of strong bases (that are also alkalis) are sodium hydroxide and potassium hydroxide. Strong alkalis, like strong acids, are completely dissociated in water:

$$NaOH \quad \rightarrow \quad Na^+ \quad + \quad OH^-$$

| sodium hydroxide | sodium ion | hydroxide ion |

Key term

Base A compound that can take up hydrogen ions in solution. Basic solutions have a pH value above 7.

pH and buffers

pH is a measure of the acidity or alkalinity of a solution. Strictly, pH is a measure of the hydrogen ion concentration. Since these concentrations involve a very large range of numbers, the pH scale uses logarithms:

$$pH = -\log_{10} H^+ \text{ concentration}$$

The pH value of pure water is 7. A solution with a pH value less than 7 is acidic; strong acids have a pH value of 0 to 2. A solution with a pH value more than 7 is alkaline; strong alkalis have a pH value of 12 to 14.

pH can be measured experimentally, either using an indicator solution or a pH meter. For example, universal pH indicator is a mixture of several different indicators, and changes colour with the pH, as shown in Figure 1.1.

Tip

Remember when dealing with logarithmic values that a value of 2 ($\log_{10} 100$) is ten times greater than a value of 1 ($\log_{10} 10$), not two times greater.

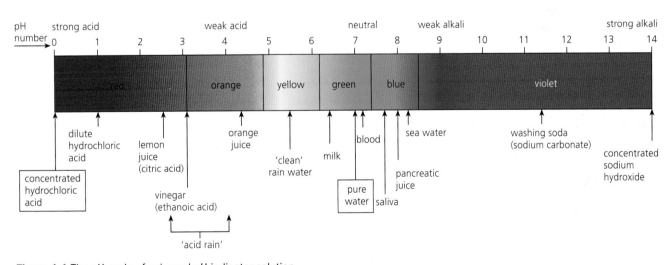

Figure 1.1 The pH scale of universal pH indicator solution

pH is very important in living organisms, largely because pH affects the shape of enzymes, almost all of which are proteins (page 34). In a mammal's body there are mechanisms that stabilise pH at a value just slightly above pH 7.0. If the pH varies much from this value this lack of stabilisation is quickly fatal. For plants that obtain essential mineral ions from the soil solution, the pH of the soil affects the availability of the ions for absorption.

Key term

Buffer solution
A solution that resists changes in pH; usually a mixture of a weak acid and one of its soluble salts.

A buffer solution is one that will resist pH change when diluted, or if a little acid or alkali is added. Many buffers used in laboratory experiments contain a weak acid (such as ethanoic acid and one of its soluble salts, for example sodium ethanoate). In this case, if acid is added, the excess hydrogen ions are immediately removed by being combined with ethanoate ions to form undissociated ethanoic acid. Alternatively, if alkali is added, the excess hydroxyl ions immediately combine with hydrogen ions, forming water. At the same time, more of the ethanoic acid dissociates, adding more hydrogen ions to the solution. The pH does not change in either case.

In the body of a mammal, the blood is very powerfully buffered by the presence of a mixture of phosphate ions, hydrogencarbonate ions and blood proteins (page 226). The blood is held between pH 7.35 and 7.45.

Test yourself

1 Distinguish between a sodium atom and a sodium ion.

2 A pH value is calculated as $-\log_{10}$ hydrogen ion concentration. By how many times is the concentration of hydrogen ions in a solution with a pH value of 2 greater than one with a pH value of 8?

3 Explain the importance of using buffer solutions during investigations into the rate of enzyme-controlled reactions.

4 Explain the meaning of the term *dissociation*.

5 Explain why a positively charged ion is called a cation.

Inorganic ions used by plants

Metabolism involves a range of inorganic ions, in addition to those mentioned above. Table 1.2 shows four inorganic ions whose roles in plants you are required to know.

Key term

Metabolism All the chemical reactions that occur within an organism.

Table 1.2 The role of selected ions in plants

Inorganic ion	Role in plants
Nitrate (NO_3^-)	Used to synthesise the nitrogenous bases in DNA and RNA nucleotides and to synthesise the amino groups of amino acids.
Calcium (Ca^{2+})	Used to synthesise calcium pectate, which exists as a layer, called the middle lamella, between the walls of adjacent plant cells.
Magnesium (Mg^{2+})	Used to synthesise the photosynthetic pigment, chlorophyll.
Phosphate (PO_4^{3-})	Used to synthesise adenosine triphosphate (ATP) from adenosine diphosphate (ADP) and to synthesise DNA and RNA

Water

Living things are typically solid, substantial objects, yet water forms the bulk of their structures – between 65 and 95 per cent by mass of most multicellular plants and animals (about 80 per cent of a human cell consists of water). Despite this, and the fact that water has some unusual properties, it is a substance that is often taken for granted.

Water is composed of atoms of the elements hydrogen and oxygen. One atom of oxygen and two atoms of hydrogen combine by sharing of electrons in an arrangement known as a **covalent bond** (see Figure 1.2). The large nucleus of the oxygen atom draws electrons (negatively charged) away from the smaller hydrogen nuclei (positively charged) with an interesting consequence. Although overall the water molecule is electrically neutral, there is a weak negative charge (represented by δ^-) on the oxygen atom and a weak positive charge (represented by δ^+) on each hydrogen atom. In other words, the water molecule carries an unequal distribution of electrical charge within it. This arrangement is known as a **polar molecule**.

Key terms

Covalent bond
A relatively strong chemical link between two atoms in which electrons are shared between them.

Polar molecule
A molecule that contains weak positive charges (represented by δ^+) and weak negative charges (represented by δ^-)

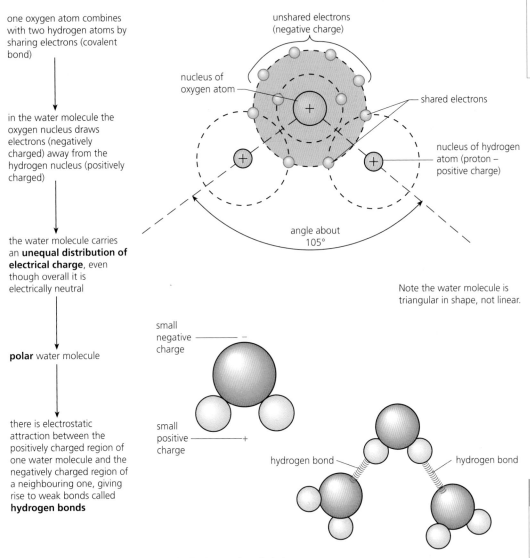

one oxygen atom combines with two hydrogen atoms by sharing electrons (covalent bond)

↓

in the water molecule the oxygen nucleus draws electrons (negatively charged) away from the hydrogen nucleus (positively charged)

↓

the water molecule carries an **unequal distribution of electrical charge**, even though overall it is electrically neutral

↓

polar water molecule

↓

there is electrostatic attraction between the positively charged region of one water molecule and the negatively charged region of a neighbouring one, giving rise to weak bonds called **hydrogen bonds**

unshared electrons (negative charge)

nucleus of oxygen atom

shared electrons

nucleus of hydrogen atom (proton – positive charge)

angle about 105°

Note the water molecule is triangular in shape, not linear.

small negative charge

small positive charge

hydrogen bond   hydrogen bond

Figure 1.2 A water molecule and the hydrogen bonds it forms

Hydrogen bonds

The positively charged hydrogen atoms of one molecule are attracted to negatively charged oxygen atoms of nearby water molecules by forces called **hydrogen bonds**. These are weak bonds compared with covalent bonds, yet they are strong enough to hold water molecules together. This is called **cohesion**; it not only attracts water molecules to each other but also to another charged particle or charged surface. In fact, hydrogen bonds largely account for the unique properties of water, which are examined next.

Key terms

Hydrogen bond A relatively weak link between two atoms in which a weakly negative atom attracts another weakly positive atom.

Cohesion The force by which hydrogen bonds hold polar molecules together, or to a charged surface.

Solvent properties of water

Because water molecules are polar, water is a powerful solvent for other polar substances (Figure 1.3). These include:

- ionic substances like sodium chloride (Na^+ and Cl^-). All ions become surrounded by a shell of orientated water molecules (Figure 1.3)

- carbon-containing (organic) molecules with ionised groups, such as the carboxyl group ($-COO^-$) and amino group ($-NH_3^+$). Soluble organic molecules like sugars dissolve in water due to the formation of hydrogen bonds with their slightly charged hydroxyl groups ($-OH^-$).

Once they have dissolved, the **solute** molecules are free to move around in water (the **solvent**) and, as a result, are more chemically reactive than when in the undissolved solid state.

Polar substances that can dissolve in, or mix in, water are termed hydrophilic (water-loving). On the other hand, non-polar substances are repelled by water, as in the case of oil on the surface of water. Non-polar substances are hydrophobic (water-hating).

Key terms

Hydrophilic Refers to substances that will mix with water.

Hydrophobic Refers to substances that will not mix with water.

Ionic compounds like NaCl dissolve in water:

$NaCl \rightleftharpoons Na^+ + Cl^-$

with a group of orientated water molecules around each ion.

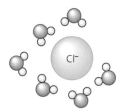

Sugars and alcohols dissolve due to hydrogen bonding between polar groups in their molecules (e.g. –OH) and the polar water molecules.

Figure 1.3 Water as universal solvent

High specific heat capacity of water

A lot of heat is required to raise the temperature of water. This is because heat is needed to break the hydrogen bonds between water molecules. This property of water is its **specific heat capacity**. The specific heat capacity of water is extremely high ($4.184\,kJ\,kg^{-1}\,°C^{-1}$). Consequently, the temperature of aquatic environments like streams and rivers, ponds, lakes and seas is very slow to change when the surrounding air temperature changes. Aquatic environments have much more stable temperatures than do terrestrial (land) environments.

Another consequence is that the temperature of cells and the bodies of organisms does not change readily. Bulky organisms, particularly, tend to have a stable temperature in the face of a fluctuating surrounding temperature, whether in extremes of heat or cold.

Tip

The specific heat capacity of water given in the text is $4.184\,kJ\,kg^{-1}\,°C^{-1}$. You need to be confident in using compound units. In this case, the unit means that it takes 4.184 kJ of heat to increase the temperature of 1 kg of water by 1 degree Celsius.

Surface tension of water

Compared with other liquids, water has extremely strong adhesive and cohesive properties.

As we saw earlier, cohesion is the force by which charged molecules stick together. Water molecules are held together by hydrogen bonding. In practice, these bonds continually break and reform with other surrounding water molecules but, at any one moment, a large number are held together by their hydrogen bonds.

At an air–water interface, cohesion between water molecules results in **surface tension**. The outermost molecules of water form hydrogen bonds with water molecules below them. This gives a very high surface tension to water, which you can see being exploited by the pond skater in Figure 1.4. This insect has a waxy cuticle that prevents wetting of its body and its mass is not great enough to break the surface tension of the water.

Figure 1.4 A pond skater moving over the water surface

Incompressibility of water

Water is essentially incompressible. Incompressibility is a common property of liquids but water is especially so. There is much less distance between the molecules in a liquid than in a gas, and the intermolecular force of the hydrogen bonds aids this property. Because of this incompressibility, a water-filled cavity within an organism can act as a hydrostatic skeleton.

Maximum density of water at 4 °C

Most liquids contract on cooling, reaching maximum density at their freezing point. Water is unusual in reaching its maximum density at 4 °C (Figure 1.5). So as water freezes, the ice formed is less dense than the cold water around it. As a consequence, ice floats on top of very cold water. The floating layer of ice insulates the water below. The consequence is that lakes rarely freeze solid; aquatic life can generally survive freezing temperatures.

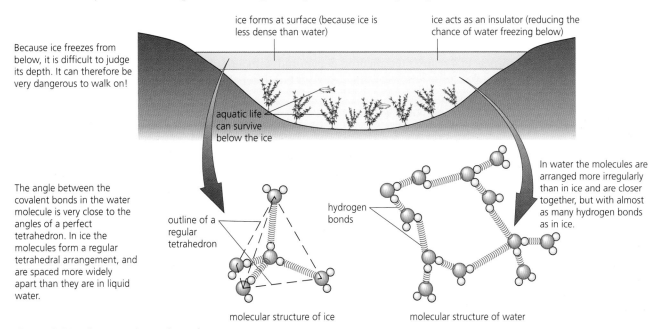

Figure 1.5 Ice forms on the surface of water

Test yourself

6 Suggest what symptoms would be shown by a plant growing in magnesium-deficient soil. Explain your answer.

7 What is meant by the term *metabolism*?

8 The text includes five properties of water that make it essential in biology. Use your knowledge of metabolism to suggest a sixth property.

9 Water that evaporates from the leaves of a flowering plant is replaced when a water column is pulled up the plant in xylem tissue. This water column is under negative pressure (tension). Explain why this does *not* cause the water column to break.

10 Explain why the pond skater shown in Figure 1.4 can walk on water but you cannot.

Introducing the carbon of organic compounds

Key term

Organic compound
A compound in which carbon atoms are linked by covalent bonds to each other and to hydrogen molecules. The molecules of organic compounds can be very large and can exist as chains or rings of carbon atoms.

Carbon is a relatively uncommon element of the Earth's crust but, as Table 1.1 showed, in cells and organisms it is the third most abundant element. The majority of the carbon compounds found in living organisms are relatively large molecules in which many carbon atoms are linked together and to hydrogen and oxygen atoms by covalent bonds. They are known as **organic compounds**.

Some carbon-containing compounds are not like this – for example, the gas carbon dioxide (CO_2) and hydrogencarbonate ions (HCO_3^-) are not organic forms of carbon.

The properties of carbon

Carbon has remarkable properties. It has a relatively small atom, but it is able to form four strong, stable, covalent bonds. As you can see in Figure 1.6, these bonds point to the corners of a regular tetrahedron (a pyramid with a triangular base). This is because the four pairs of electrons that form the covalent bonds repel each other and so position themselves as far away from each other as possible.

Carbon atoms are able to react with each other to form extended chains. The resulting carbon 'skeletons' can be straight chains, branched chains or rings. Carbon also bonds covalently with other atoms, such as oxygen, hydrogen, nitrogen and sulfur, forming different groups of organic molecules with distinctive properties.

Covalent bonds are formed by sharing of electrons, one from the carbon atom and one from the neighbouring atom it reacts with:

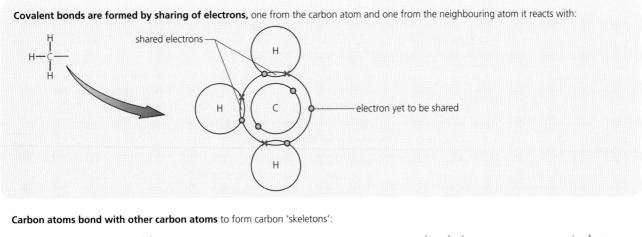

Carbon atoms bond with other carbon atoms to form carbon 'skeletons':

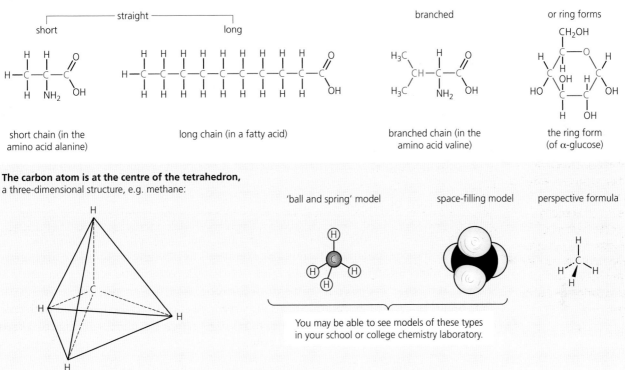

Figure 1.6 A tetrahedral carbon atom, its covalent bonds and the carbon 'skeletons' it can form

One inevitable outcome of these features is that there are vast numbers of organic compounds – more than the total of known compounds made from other elements, in fact. Biologists think the diversity of organic compounds has made possible the diversity of life. Fortunately, very many of the organic chemicals of living things fall into one of four discrete groups or 'families' of chemicals with many common properties, one of which is the carbohydrates. We will consider this family of molecules first before looking at a second family, the lipids.

Carbohydrates

Carbohydrates include sugars, starch, glycogen and cellulose. They contain only three elements: carbon, hydrogen and oxygen, in the ratio $C_x(H_2O)_y$. Table 1.3 summarises features of the three types of carbohydrates you should recognise.

Table 1.3 Carbohydrates of cells and organisms.

Type of carbohydrate	Features
Monosaccharides	Simple sugars, including: • trioses ($C_3H_6O_3$), which you will learn more about in Year 2 • pentoses ($C_5H_{10}O_5$), e.g. ribose and deoxyribose (see Chapter 3) • hexoses ($C_6H_{12}O_6$), e.g. glucose, fructose, galactose.
Disaccharides	Two simple sugars chemically linked by a glycosidic bond during a condensation reaction, e.g. • sucrose = glucose + fructose • lactose = glucose + galactose • maltose = glucose + glucose.
Polysaccharides	Very many simple sugars chemically linked by glycosidic bonds, e.g. • starch (a fuel store in plants) • glycogen (a fuel store in animals) • cellulose (a major component of plant cell walls).

Monosaccharides – the simple sugars

Monosaccharides are carbohydrates with relatively small molecules. They are soluble in water and taste sweet. In biology, glucose is an especially important monosaccharide because:

● all green leaves manufacture glucose using light
● all cells use glucose in respiration – we call it one of the respiratory substrates.

The structure of glucose

Glucose is a hexose, i.e. it has a **molecular formula** of $C_6H_{12}O_6$. This type of formula tells us what the component atoms are, and the numbers of each in the molecule. But the molecular formula does not tell us how these atoms are arranged within a molecule. You can see in Figure 1.7 that glucose can be written on paper as a linear molecule. It does not exist in this form because, as you saw in Figure 1.6, the four bonds in each of its carbon atoms are arranged into a tetrahedron; the molecule cannot be 'flat'. Rather, glucose is folded, taking a ring or cyclic form. Figure 1.7 also shows the **structural formula** of glucose.

The carbon atoms of an organic molecule can be numbered. This allows us to identify which atoms are affected when the molecule reacts and changes shape. For example, as the glucose ring forms, the oxygen on carbon atom 5 (carbon-5) becomes linked to that on carbon atom 1 (carbon-1). As a result, the glucose ring contains five carbon atoms and an oxygen atom; again, you can see this in Figure 1.7.

Isomers of glucose

Molecules with the same molecular formula but different structural formulae are known as **isomers**. Many organic compounds exist in isomeric forms, and so it is often important to know the structure of an organic compound as well as its composition.

In the ring structure of glucose the positions of −H and −OH that are attached to carbon-1 can lie in one of two directions, giving rise to two isomers, known as **alpha-glucose** (α-glucose) and **beta-glucose** (β-glucose). You can see these isomers in Figure 1.8. The significance of the differences between them will become apparent when we compare the structures of starch, glycogen and cellulose (pages 13–16).

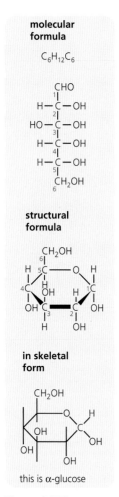

molecular formula

$C_6H_{12}C_6$

structural formula

in skeletal form

this is α-glucose

Figure 1.7 The structure of alpha glucose

Key terms

Molecular formula
Shows the nature and number of atoms in a molecule, for example $C_5H_{10}O_5$.

Structural formula
Shows the way in which the atoms within a molecule are arranged in space.

Isomers Two or more different structural formulae of the same molecular formula.

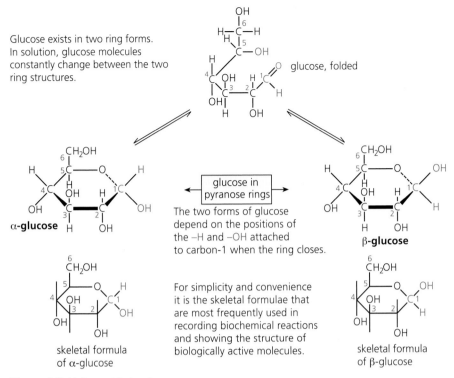

Glucose exists in two ring forms. In solution, glucose molecules constantly change between the two ring structures.

glucose, folded

glucose in pyranose rings

The two forms of glucose depend on the positions of the –H and –OH attached to carbon-1 when the ring closes.

α-glucose

β-glucose

For simplicity and convenience it is the skeletal formulae that are most frequently used in recording biochemical reactions and showing the structure of biologically active molecules.

skeletal formula of α-glucose

skeletal formula of β-glucose

Figure 1.8 Alpha and beta glucose

A test for the presence of glucose, a 'reducing sugar'

Glucose and some other sugars are known as 'reducing sugars'. This is because, when they are heated with an alkaline solution of copper(II) sulfate (a blue solution, called Benedict's solution), the carboxyl group (–COOH) that their molecule contains (known as an aldehyde group) reduces Cu^{2+} ions of copper(II) sulfate to Cu^{+} ions, which then form a brick-red precipitate of copper(I) oxide. In the process, the aldehyde group is oxidised to a carbonyl group (–C=O).

This reaction is used to test for reducing sugar, and is known as **Benedict's test** (Figure 1.9). If no reducing sugar is present the solution remains blue after heating. The colour change observed depends on the concentration of reducing sugar. The greater the concentration the more precipitate is formed, and the more the colour changes:

blue → green → yellow → brown → red

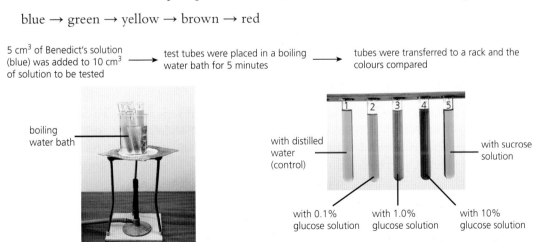

5 cm^3 of Benedict's solution (blue) was added to 10 cm^3 of solution to be tested → test tubes were placed in a boiling water bath for 5 minutes → tubes were transferred to a rack and the colours compared

boiling water bath

with distilled water (control)

with sucrose solution

with 0.1% glucose solution

with 1.0% glucose solution

with 10% glucose solution

Figure 1.9 The test for reducing sugar. Wear eye protection when performing this test

Other monosaccharides of importance in living cells

Glucose, fructose and galactose are examples of hexose sugars commonly occurring in cells and organisms, but it is only the structure of α- and β-glucose that you need to know. Other monosaccharide sugars produced by cells and used in metabolism include a 3-carbon sugar (Table 1.4), and two 5-carbon sugars (**pentoses**), namely ribose and deoxyribose. These pentoses are components of the nucleic acids and you will learn about their structure in Chapter 3.

Table 1.4 Other monosaccharides important in cell chemistry

Length of carbon chain	Name of sugar	Molecular formula	Formula	Roles
3C = triose	glyceraldehyde	$C_2H_6O_2$		intermediate in respiration and photosynthesis
5C = pentoses	ribose	$C_5H_{10}O_5$		in RNA, ATP and hydrogen acceptors NAD and NADP
	deoxyribose	$C_5H_{10}O_4$		in DNA

Disaccharides

A disaccharide is a carbohydrate made of two monosaccharides linked together. For example, sucrose is formed from a molecule of glucose and a molecule of fructose chemically linked together.

Condensation and hydrolysis reactions

When two monosaccharide molecules are combined to form a disaccharide, a molecule of water is also formed as a product, and so this type of reaction is known as a condensation reaction. The bond between monosaccharide residues, after the removal of H–O–H between them, is called a glycosidic bond (Figure 1.10). This is a strong, covalent bond. The condensation reaction is catalysed by an enzyme (Chapter 2).

In the reverse process, disaccharides are 'digested' to their component monosaccharides in a hydrolysis reaction. Of course this reaction involves adding a molecule of water ('hydro-') as splitting ('-lysis') of the glycosidic bond occurs. It is catalysed by an enzyme, too, but it is a different enzyme from the one that brings about the condensation reaction.

Key terms

Condensation reaction
A reaction in which two molecules are chemically linked together with the elimination of a molecule of water.

Glycosidic bond
A covalent bond between two monosaccharides.

Hydrolysis reaction
A reaction in which a molecule of water is used in breaking a chemical bond (the reverse of a condensation reaction).

Apart from sucrose, other disaccharide sugars produced by cells and used in metabolism include:

- maltose, formed by a condensation reaction of two molecules of glucose
- lactose, formed by a condensation reaction of galactose and glucose.

sucrose + water $\underset{\text{condensation}}{\overset{\text{hydrolysis}}{\rightleftharpoons}}$ glucose + fructose

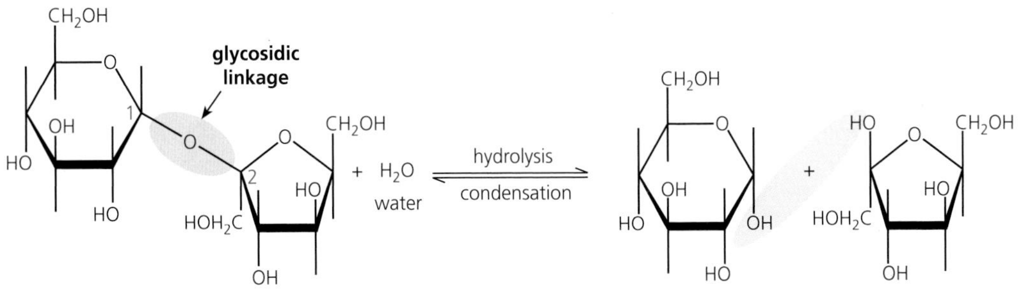

This structural formula shows us how the glycosidic linkage forms/breaks.

maltose + water $\underset{\text{condensation}}{\overset{\text{hydrolysis}}{\rightleftharpoons}}$ glucose + glucose

lactose + water $\underset{\text{condensation}}{\overset{\text{hydrolysis}}{\rightleftharpoons}}$ galactose + glucose

Figure 1.10 Sucrose, a disaccharide, and the monosaccharides that form it

Polysaccharides

A polysaccharide is built from many monosaccharides linked by glycosidic bonds formed during condensation reactions. '*Poly*' means many, and in fact thousands of saccharide (sugar) units make up a polysaccharide. So a polysaccharide is a giant molecule, a macromolecule. Normally each polysaccharide contains only one type of **monomer**. A chemist calls this a **polymer** because it is constructed from a huge number of *identical* monomers.

Some polysaccharides function as fuel stores. Both glycogen and starch are examples, as we shall shortly see. On the other hand, some polysaccharides, such as cellulose, have a structural role. Cellulose has huge molecules that are not so easily hydrolysed by enzyme action.

Key terms

Polymer A large molecule comprising repeated, identical smaller molecules (**monomers**) linked together by chemical bonds.

Residues When monomers are linked together in a polymer, we can no longer refer to them as molecules. Instead, we can call them residues.

Starch

Starch is a mixture of two polysaccharides, both of which are polymers of α-glucose:

- **amylose** – an unbranched chain of α-glucose residues
- **amylopectin** – branched chains of α-glucose residues.

The glycosidic bonds between α-glucose residues in starch bring the molecules together in such a way that a helix forms. The whole starch molecule is then stabilised by countless hydrogen bonds between parts of the component glucose **residues**.

Starch is the major storage carbohydrate of most plants. It is laid down as compact grains. It is useful because its molecules are both compact and insoluble, but are readily hydrolysed to form sugar when required. Of course, enzymes are involved in this reaction, too.

We sometimes see 'soluble starch' as an ingredient of manufactured foods. Here the starch molecules have been broken down into short lengths, making them dissolve more easily.

A test for the presence of starch

We test for starch by adding a solution of iodine in potassium iodide. Iodine molecules fit neatly into the centre of a starch helix, creating a blue-black colour (Figure 1.11).

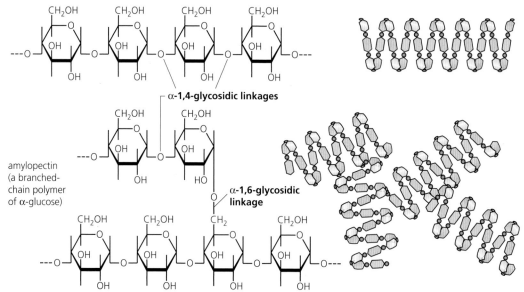

amylose (a straight-chain polymer of α-glucose)

α-1,4-glycosidic linkages

amylopectin (a branched-chain polymer of α-glucose)

α-1,6-glycosidic linkage

In the test for starch with iodine in potassium iodide solution, the blue-black colour comes from a starch/iodine complex:

a) on a potato tuber cut surface

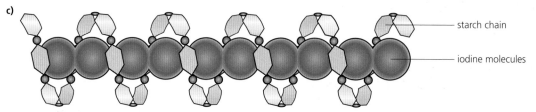

1% starch solution

0.1% starch solution

0.01% starch solution

b) on starch solutions of a range of concentrations

c)

starch chain

iodine molecules

Figure 1.11 Starch

1 Introducing the chemistry of life

Glycogen

Glycogen is also a polymer of α-glucose. It is chemically very similar to amylopectin, although larger and more highly branched. Granules of glycogen are seen in liver cells and muscle fibres when observed by the electron microscope, but they occur throughout the human body, except in the brain cells (where there are virtually no carbohydrate reserves). During prolonged and vigorous exercise we draw on our glycogen reserves first. Only when these are exhausted does the body start to metabolise stored fat.

Structural formula

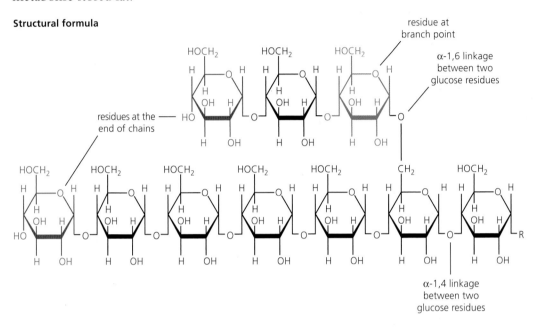

Diagram to show the branching pattern of a glycogen molecule

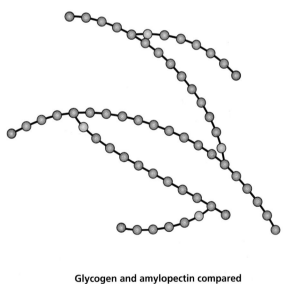

TEM of a liver cell (x7000)

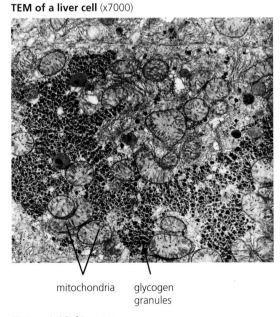

mitochondria   glycogen granules

Figure 1.12 Glycogen

Glycogen and amylopectin compared

glycogen	amylopectin
branch point every 10 glucose residues	branch point every 30 glucose residues

Cellulose

Cellulose is a polymer of around 2000 to 3000 units of beta-glucose (β-glucose). Look back to Figure 1.8, which shows the difference in structure between α-glucose and β-glucose.

Can you spot what it is?

The only difference between the two molecules is the way in which the −H and −OH groups are bonded to carbon-1. In α-glucose the −H group is uppermost whereas in β-glucose the −OH group is uppermost. Although this might seem trivial, it has a big effect when molecules of β-glucose become linked together. As Figure 1.13 shows, the way glycosidic bonds form causes adjacent β-glucose units to be upside down with respect to each other. These glycosidic bonds are referred to as β-1,4 glycosidic bonds. This arrangement leads to cellulose molecules being long, straight-chains.

About 200 of these chains naturally become packed into fibres, held together by hydrogen bonds (Figure 1.13). The strength of plant cell walls results from the combined effect of the bonds between β-glucose monomers, the hydrogen bonds within and between these chains of β-glucose and the way in which the fibres are arranged in different directions.

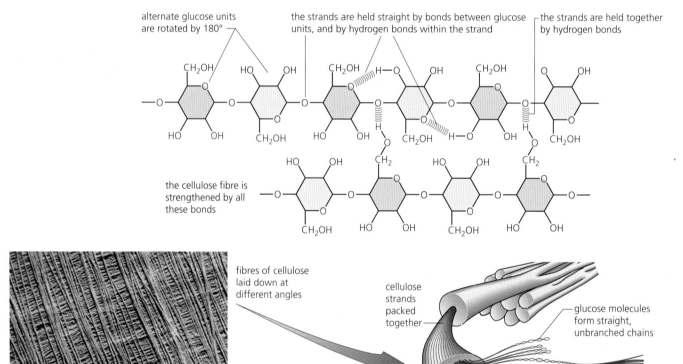

Figure 1.13 The chemistry and structure of cellulose

11 Explain the difference between a molecular formula and a structural formula.

12 α-glucose and β-glucose are isomers. Explain what this means.

13 What is the difference between a pentose and a hexose sugar?

14 Lactose is a disaccharide found in milk. Into which monosaccharides is it broken down in your intestines?

15 Starch is a polymer. What is meant by a *polymer*?

16 Both starch and glycogen can be broken down to provide glucose, used in respiration. Name the type of reaction by which both are broken down.

17 The reaction by which amylose and amylopectin are hydrolysed produces disaccharides.

a) Name the disaccharide formed.

b) Which compound, amylase or amylopectin, would you expect to be hydrolysed faster? Explain your answer.

Lipids

The second 'family' of organic molecules to consider here is the lipids. These occur in mammals as fats and in plants as oils. Fats and oils appear to be rather different substances, but the basic difference between them is that, at about 20°C (room temperature), oils are liquid and fats are solid. Like the carbohydrates, lipids also contain the elements carbon, hydrogen and oxygen, but in lipids the proportion of oxygen is much less.

Lipids are insoluble in water, i.e. are hydrophobic. However, lipids can be dissolved in organic solvents such as alcohol (for example ethanol).

Here we will consider only two types of lipid: triglycerides and phospholipids.

Triglycerides

Triglycerides are formed during condensation reactions between glycerol (an alcohol) and three fatty acids. The bonds formed are known as ester bonds.

Fatty acids are long hydrocarbon chains, anything between 14 and 22 carbon atoms long.

The structures of a fatty acid commonly found in cells and that of glycerol are shown in Figure 1.14 and the steps to triglyceride formation in Figure 1.15. Enzymes catalyse the condensation reactions by which triglycerides are formed.

> **Key term**
>
> **Ester bond** The bond formed during a condensation reaction between a fatty acid and glycerol.

Fatty acid

this is palmitic acid, with 16 carbon atoms

molecular formula of palmitic acid:
$CH_3(CH_2)_{14}COOH$

the carboxyl group ionises to form hydrogen ions, i.e. it is a weak acid

Glycerol

molecular formula of glycerol:
$C_3H_5(OH)_3$

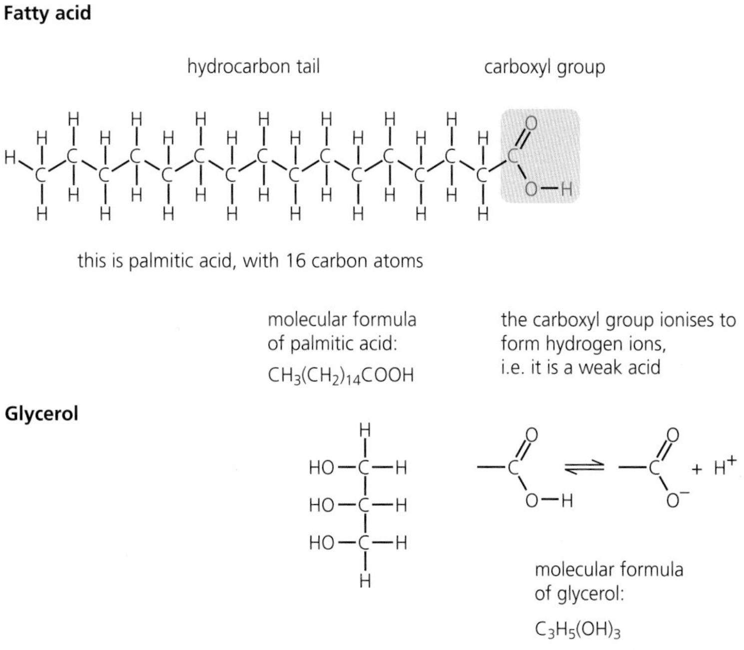

Figure 1.14 Fatty acids and glycerol, the building blocks of lipids

The hydrophobic properties of triglycerides are caused by the hydrocarbon chains of the component fatty acids. A molecule of triglyceride is quite large, but relatively small when compared with macromolecules such as starch. However, because of their hydrophobic properties, triglyceride molecules clump together (aggregate) into huge globules in the presence of water, making them appear to be macromolecules.

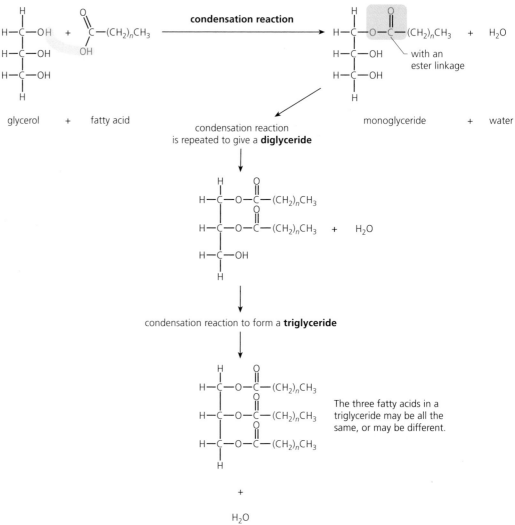

Figure 1.15 Formation of triglyceride

Saturated and unsaturated lipids

We have seen that the length of the hydrocarbon chains is different from fatty acid to fatty acid. These chains can differ in another way, too. To understand this latter difference, we need to note another property of carbon atoms and the ways they can combine together in chains. This concerns the existence of double covalent bonds (Figure 1.16).

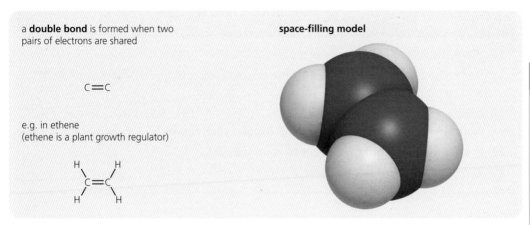

a **double bond** is formed when two pairs of electrons are shared

C=C

e.g. in ethene
(ethene is a plant growth regulator)

space-filling model

Figure 1.16 A carbon-carbon double covalent bond

A double covalent bond is formed when adjacent carbon atoms share *two pairs* of electrons, rather than the single electron pair shared in a single covalent bond. Carbon compounds that contain double carbon–carbon bonds are known to chemists as unsaturated compounds. On the other hand, when all the carbon atoms of the hydrocarbon tail of an organic molecule are combined together by single bonds, the compound is described as saturated. This difference is illustrated in Figure 1.17.

Key terms

Unsaturated fatty acid A fatty acid in which one or more pairs of adjacent carbon atoms in the hydrocarbon chain are linked by a double covalent bond (represented as C=C).

Saturated fatty acid One in which all the bonds between carbon atoms in the hydrocarbon chain are single covalent bonds (represented as C-C).

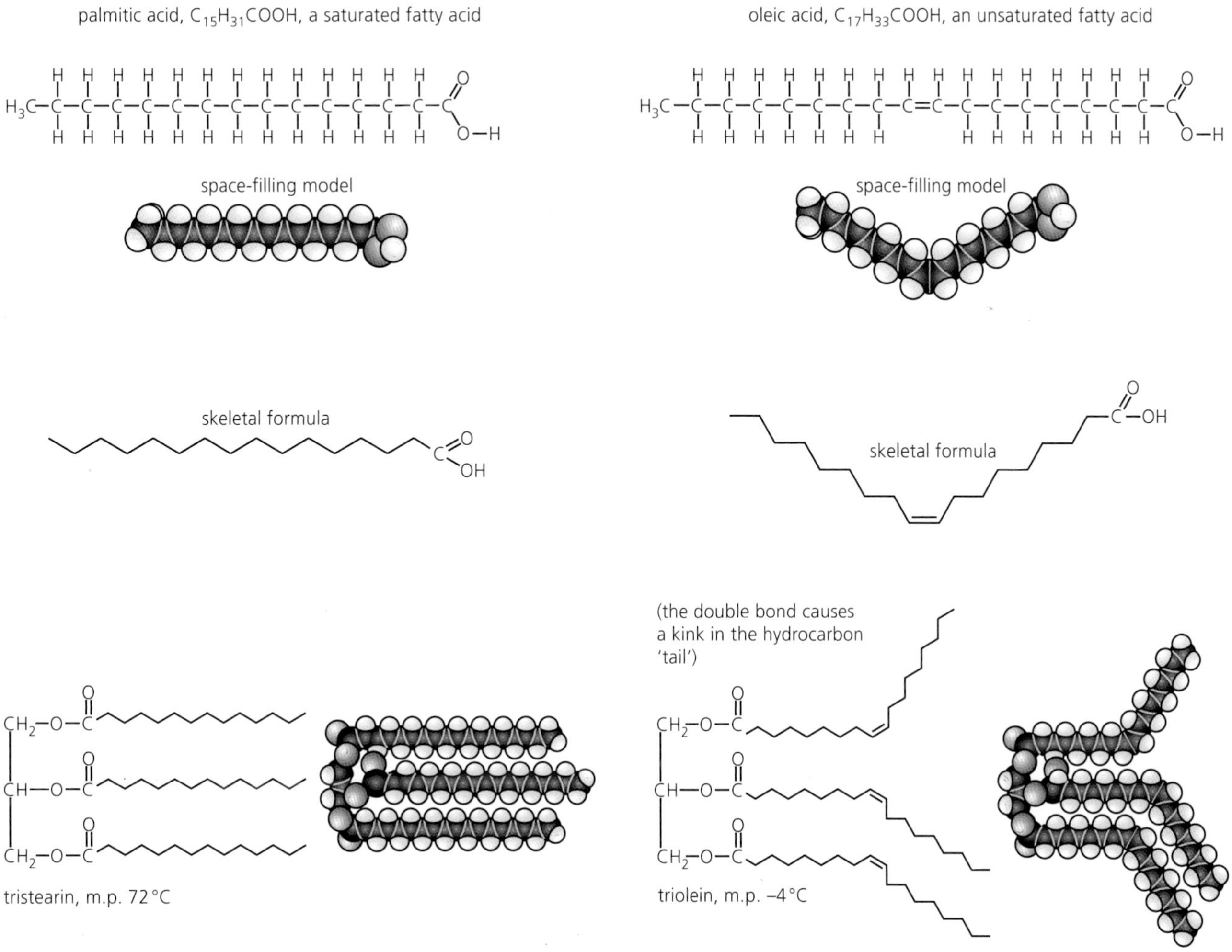

palmitic acid, $C_{15}H_{31}COOH$, a saturated fatty acid

space-filling model

skeletal formula

tristearin, m.p. 72 °C

oleic acid, $C_{17}H_{33}COOH$, an unsaturated fatty acid

space-filling model

skeletal formula

(the double bond causes a kink in the hydrocarbon 'tail')

triolein, m.p. −4 °C

Figure 1.17 Saturated and unsaturated fatty acids and triglycerides formed from them

18 State the molecular formula of glycerol.

19 Name the type of bond formed by the condensation of glycerol and a single fatty acid to produce a monoglyceride.

20 A fatty acid can be represented as $CH_3-(CH_2)_n-COOH$. Would this represent a saturated or unsaturated fatty acid? Explain your answer.

21 Is a triglyceride a polymer? Explain your answer.

22 What makes a triglyceride hydrophobic?

The roles of lipids in living organisms

You need to be familiar with three ways in which the structure of lipids relates to their role in living organisms.

Energy storage

When triglycerides are oxidised during respiration, energy is released. Some is lost to the environment as heat but some is used to make ATP – the energy currency of cells, introduced in Chapter 9. Mass for mass, when fully respired, lipids release more than twice as much energy as do carbohydrates (Table 1.5). Lipids, therefore, form a more 'concentrated' energy store than do carbohydrates.

A fat store is especially typical of animals that endure long unfavourable seasons in which they survive on reserves of food stored in the body. Oils are often a major energy store in the seeds and fruits of plants, and it is common for fruits and seeds to be used commercially as a source of edible oils for humans, for example maize, olives and sunflower.

Table 1.5 Lipids and carbohydrates as energy stores – a comparison

Feature	Lipids	Carbohydrates
Energy released on complete breakdown / $kJ\,g^{-1}$	~37	~17
Ease of breakdown	Not easily hydrolysed – energy released slowly	More easily hydrolysed – energy released quickly
Solubility	Hydrophobic, so do not cause osmotic water uptake by cells	Sugars are highly soluble in water, so can cause osmotic water uptake by cells
Production of metabolic water	A great deal of metabolic water produced on oxidation	Less metabolic water produced on oxidation

Waterproofing

Since lipids are hydrophobic, they repel water.

Oily secretions from the sebaceous glands, found in the skin of mammals, act as a water repellent, preventing fur and hair from becoming waterlogged when wet. Birds have a preen gland that fulfils the same function for feathers. You might have seen birds preening – they use their beaks to spread lipids from this gland over their feathers.

Insulation

Lipids are poor conductors of both heat and hydrophilic ions.

Triglycerides are stored in mammals as adipose tissue, typically under the skin, where it is known as subcutaneous fat. Fat reserves like these have a restricted blood supply (Figure 1.18) so little body heat is distributed to the fat under the skin. In these circumstances, the subcutaneous fat functions as a heat insulation layer.

Myelin is a lipid found in the surface membranes of cells that wrap around the long fibres of nerve cells in animals (*Edexcel A level Biology 2*, Chapter 12). Over much of its length, the many layers of myelin insulate the fibre, preventing the passage of sodium and potassium ions that are essential for the conduction of the nerve impulse. As a result, nerve impulses travel along nerve fibres surrounded by myelin much faster than along those that are not surrounded by myelin.

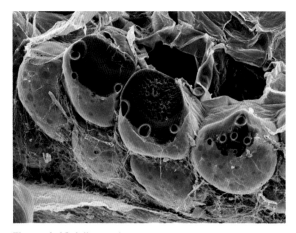

Figure 1.18 Adipose tissue

Phospholipids

A phospholipid has a similar chemical structure to a triglyceride, except one of the fatty acid groups is replaced by a phosphate group.

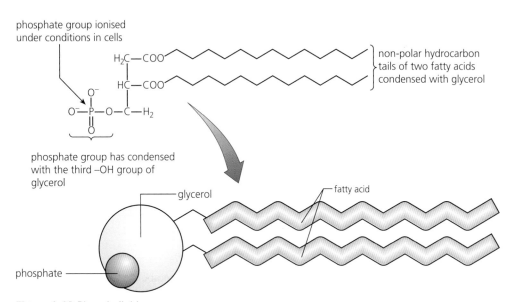

Figure 1.19 Phospholipid

You can see from Figure 1.19 that a phospholipid molecule has a 'head' composed of a glycerol to which is attached an ionised phosphate group. Since hydrogen bonds readily form between this phosphate group and water molecules, this part of the molecule has hydrophilic properties. The remainder of a phospholipid consists of two long, fatty acid residues, comprising hydrocarbon chains. As we have seen above, these 'tails' have hydrophobic properties.

So phospholipid molecules are unusual in being partly hydrophilic and partly hydrophobic. The effect of this is that a small quantity of phospholipid in contact with water will float, with the hydrocarbon tails exposed above the water. It forms a single layer (monolayer) of phospholipids (Figure 1.20).

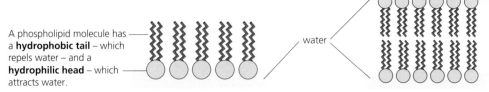

Phospholipid molecules **in contact with water** form a **monolayer**, with heads dissolved in the water and the tails sticking outwards.

When **mixed with water**, phospholipid molecules arrange themselves into a **bilayer**, in which the hydrophobic tails are attracted to each other.

A phospholipid molecule has a **hydrophobic tail** – which repels water – and a **hydrophilic head** – which attracts water.

water

Figure 1.20 Phospholipid molecules and water

Key term

Bilayer A single structure made of two layers of molecules, usually used to describe the arrangement of phospholipids in a cell membrane.

When slightly more phospholipid is added the molecules arrange themselves as a bilayer, with the hydrocarbon 'tails' facing together, away from the water, and the hydrophilic heads in the water (Figure 1.20). Their hydrophobic/hydrophilic nature and their ability to form a bilayer are two extremely important properties of phospholipids, as you will see when we consider cell surface membranes in Chapter 9.

Test yourself

23 Some people believe that a camel stores water in its hump. In fact, the hump is a lipid store. Use information in Table 1.5 to suggest how this lipid store is an adaptation to living in desert conditions.

24 Explain how the structure of triglycerides results in their waterproofing properties.

25 How does a molecule of triglyceride differ from a molecule of phospholipid?

26 When mixed with water, phospholipids often form micelles – small droplets with the fatty acid 'tails' on the inside and the 'heads' on the outside. Suggest why.

Exam practice questions

1 The general formula for a carbohydrate is:

   A $C_6H_{12}O_6$

   B $C_6(H_2O)_6$

   C $C_n(H_2O)_{2n}$

   D $C_x(H_2O)_y$

2 Sucrose is a disaccharide formed by the condensation of:

   A glucose and fructose

   B glucose and glucose

   C glucose and galactose

   D fructose and galactose

3 Phospholipids are mainly

   A used as an energy store

   B found in plasma membranes

   C used for waterproofing

   D hormones

4 Copy and complete the table to show similarities and differences between glycogen and starch. *(4)*

Feature	Glycogen	Starch
Found in cells of which group of organisms?		
Formed from which monomer?		
Name of polymer(s)		
Degree of branching		

5 The structure of biological molecules is related to their function. Explain **two** ways in which the structure of glycogen is related to its function and **two** ways in which the structure of a triglyceride is related to its function. *(4)*

6 Cellulose forms cell walls in plants. Describe the structure of cellulose and explain how its properties contribute to the functions of cell walls. *(6)*

7 The diagram shows a molecule of α-glucose.

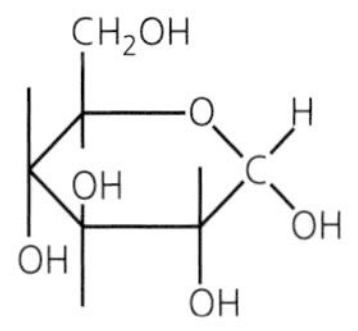

   a) How can you tell that this is α-glucose and **not** β-glucose. *(1)*

   b) i) Draw a diagram to show a condensation reaction between two molecules of α-glucose. *(2)*

      ii) Name the products of this reaction. *(1)*

      iii) Explain why the bond linking the two units of glucose is called an α-1,4 glycosidic bond. *(3)*

8 A group of scientists investigated milk production by dairy cattle. For 28 days, they fed one group of cows their normal diet and fed a second group of cows the same diet with added animal fat.

After 28 days, the scientists made several measurements on the cows and on the milk they produced. The table shows their results.

Result calculated by scientists	Cows with normal diet	Cows with normal diet plus added animal fat
Mean body mass/kg	547.9	552.0
Mean concentration of lipid in blood/mg per 100 cm³	469.1	605.7
Mean milk production/kg per cow per day	14.5	16.1
Mean content of butterfat in milk/%	4.6	4.7

a) Suggest a null hypothesis that the scientists were testing in their investigation. *(1)*

b) Explain the difference in the concentration of lipids in the blood of the two groups of cows. *(3)*

c) Calculate the percentage increase in milk production of the cows with added fat in their diet. *(1)*

d) Can you conclude from the data in the table that adding fat to the diet of the cows resulted in a change in the composition of the milk they produced? Explain your answer. *(4)*

9 Earthworms have a body that is divided into discrete segments. Figure 1 shows a longitudinal section through one segment. The outer region of the body has two layers of muscle surrounding an inner fluid-filled space, called the coelom. The fluid in this coelom (coelomic fluid) is mainly water.

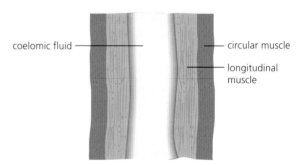

coelomic fluid — — circular muscle
— longitudinal muscle

Figure 1

The diagrams in Figure 2 represent different times as an earthworm moves forwards. Contraction of the circular muscle in one body segment causes that segment to become long and thin. Contraction of the longitudinal muscle in one body segment causes that segment to become short and fat.

Earthworms normally live in a burrow in the soil.

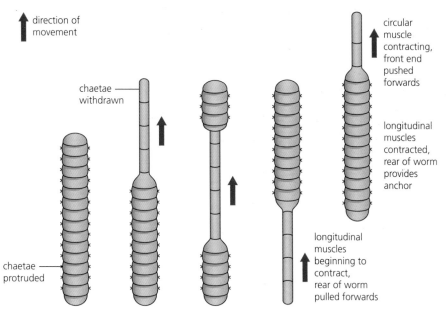

direction of movement

chaetae withdrawn

circular muscle contracting, front end pushed forwards

longitudinal muscles contracted, rear of worm provides anchor

chaetae protruded

longitudinal muscles beginning to contract, rear of worm pulled forwards

Figure 2

a) Use your knowledge of the properties of water to explain how the coelomic fluid enables the movement of the earthworm shown in Figure 2. *(4)*

b) The chaetae shown in Figure 2 are short, bristle-like structures. Suggest their role in the movement of the earthworm. *(4)*

Proteins and enzymes

Prior knowledge

In this chapter you will need to recall that:
- → proteins are long chains of amino acids
- → only 20 different amino acids are common in the proteins of living organisms
- → there is a vast number of different proteins found in living organisms. The differences between them result from the sequence of their amino acids
- → the shape of a protein, whether a straight chain of amino acids or a three-dimensional complex, is related to its function
- → enzymes are proteins with a complex three-dimensional shape; each enzyme acts as a catalyst, speeding up a specific reaction or type of reaction
- → the ability of enzymes to catalyse a reaction is affected by changes in pH and changes in temperature.

Test yourself on prior knowledge

1 Give **one** way in which the composition of an amino acid differs from that of a carbohydrate and a lipid.
2 Name the chemical bond that links two amino acids together.
3 What makes one protein different from another?
4 Other than speeding up a reaction, give **two** properties common to catalysts.
5 An enzyme has a complex three-dimensional shape. Explain how this shape results in each enzyme:
   a) catalysing only one type of reaction
   b) being affected by high temperatures.

The proteins and peptides of cells

Proteins make up about two-thirds of the total dry mass of a cell. These organic molecules differ from carbohydrates and lipids in that they contain the element nitrogen, and often the element sulfur, as well as carbon, hydrogen and oxygen. Amino acids are the monomers from which the polymers – peptides and proteins – are built. Typically several hundred, or even thousands of amino acid molecules, are combined together to form a protein. Incidentally, the terms 'polypeptide' and 'protein' can be used interchangeably, but when a polypeptide is about 50 amino acid residues long it is generally agreed to be have become a protein.

Once the chain of amino acids is constructed, a protein takes up a specific shape. Shape matters with proteins – their shape is closely related to their function. This is especially the case in proteins that are enzymes, as we shall shortly see.

Amino acids, the building blocks of peptides

All amino acids contain two functional groups: an amino group ($-NH_2$) and a carboxyl group ($-COOH$). In naturally occurring amino acids, both functional groups are attached to the same carbon atom (Figure 2.1). The remainder of the amino acid molecule, the side chain (represented by $-R$), is very variable and is what makes one amino acid different from any other.

An amino acid can ionise:

- as an acid: $-COOH \rightleftharpoons -COO^- + H^+$
- or as a base: $-NH_2 + H^+ \rightleftharpoons -NH_3^+$
- or even as both: $NH_2-CHR-COOH \rightleftharpoons NH_3^+-CHR-COO^-$.

Molecules, like amino acids, that can ionise as both an acid and a base are described as being **amphoteric** and an ion with both positive and negative charges is called a **zwitterion** (German for double ion).

Key term

Amino acid The monomer from which dipeptides, polypeptides and proteins, are made. All amino acids have the general formula $NH_2-CHR-COOH$, where R represents the only part of the molecule that is different in different amino acids.

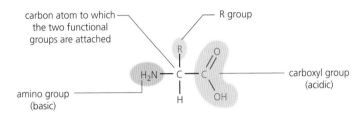

carbon atom to which the two functional groups are attached

R group

carboxyl group (acidic)

amino group (basic)

The 20 different amino acids that make up proteins in cells and organisms differ in their side chains (R groups). Below are three examples.

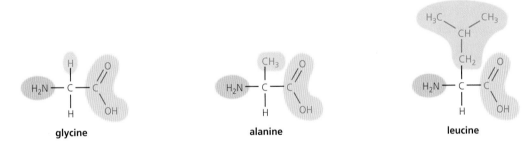

glycine

alanine

leucine

Some amino acids have an additional $-COOH$ group in their R group (= acidic amino acids).
Some amino acids have an additional $-NH_2$ group in their R group (= basic amino acids).

Figure 2.1 The structure of amino acids

The linking of amino acids together in different sequences produces proteins with very different properties. This helps explain how the many proteins in organisms are able to fulfil the very different biological functions they have.

How amino acids combine

In the presence of an appropriate enzyme, two amino acids combine in a condensation reaction to form a dipeptide. During this reaction, a covalent bond known as a peptide bond forms between the amino group of one amino acid and the carboxyl group of the other (Figure 2.2).

Key term

Peptide bond A covalent bond between the amino group of one amino acid and the carboxyl group of another amino acid. Each peptide bond is formed by a condensation reaction.

A further condensation reaction between the dipeptide and another amino acid results in a tripeptide. In this way, long strings of amino acid residues, linked by peptide bonds, are formed. As you will see in Chapter 3, polypeptides are assembled by adding one amino acid at a time.

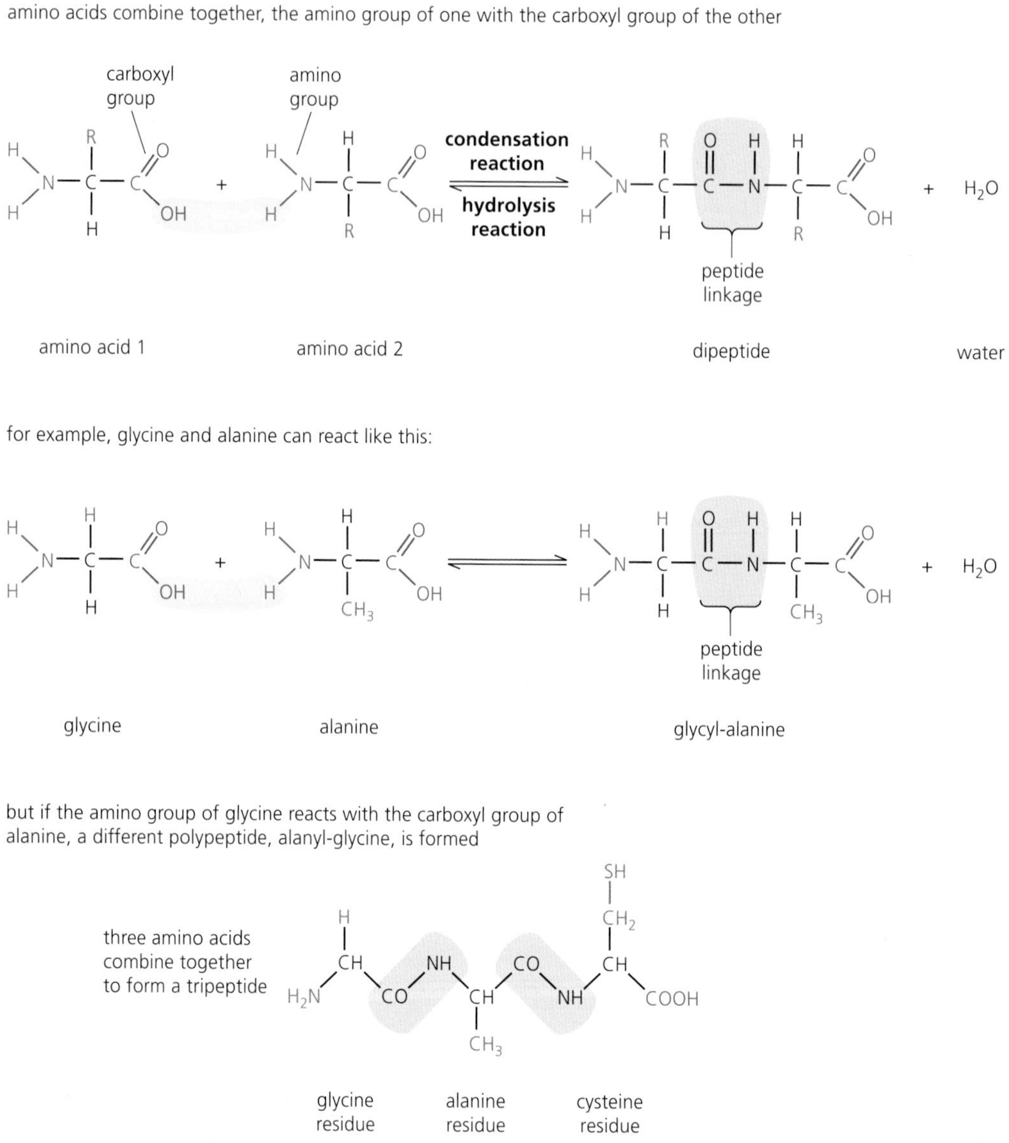
amino acids combine together, the amino group of one with the carboxyl group of the other
carboxyl group
amino group
condensation reaction
hydrolysis reaction
peptide linkage
amino acid 1
amino acid 2
dipeptide
water
for example, glycine and alanine can react like this:
peptide linkage
glycine
alanine
glycyl-alanine
but if the amino group of glycine reacts with the carboxyl group of alanine, a different polypeptide, alanyl-glycine, is formed
three amino acids combine together to form a tripeptide
glycine residue
alanine residue
cysteine residue

Figure 2.2 Peptide bond formation

Testing for the presence of protein: the biuret test

The biuret test is used as an indicator of the presence of protein because it gives a purple colour in the presence of peptide bonds (–CO–NH–).

We test for the presence of a protein by adding an equal quantity of sodium hydroxide solution to the test solution, followed by a few drops of 0.5% copper(II) sulfate solution. After gentle mixing, the appearance of a distinctive purple colour confirms that protein is present in the test solution (Figure 2.3).

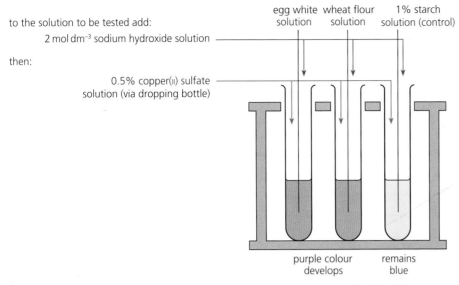

to the solution to be tested add:

2 mol dm^{-3} sodium hydroxide solution

then:

0.5% copper(II) sulfate solution (via dropping bottle)

egg white solution   wheat flour solution   1% starch solution (control)

purple colour develops   remains blue

Figure 2.3 The biuret test on protein solutions. Wear goggles when performing this test

The structure of proteins

We have already noted that the shape of a protein molecule is critical in determining the properties and the role that protein has in a cell. There are four levels of protein structure: primary, secondary, tertiary and quaternary.

The **primary structure** of a protein is the sequence of the amino acids in its molecule. As you have seen above, the amino acids are held together by peptide bonds. Proteins differ in the variety, number and order of their constituent amino acids. As you will see in Chapter 3, the order of amino acids in a polypeptide chain is controlled by the DNA of the cell producing it. Just changing one amino acid in the sequence might completely alter the properties of a protein.

The **secondary structure** of a protein develops immediately after its formation when parts of the polypeptide chain become folded or twisted, or both. The most common shapes are formed either by coiling, to produce an α helix, or folding into β sheets. These shapes are shown in Figure 2.4 and are permanent, being held in place by hydrogen bonds.

The **tertiary structure** of a protein is the compact structure, unique to that protein, that arises when the molecule is further folded and held in a particular complex three-dimensional shape. Figure 2.5 shows a polypeptide with a tertiary structure, together with the three types of bond that hold the shape in place – ionic bonds, hydrogen bonds and disulfide bonds.

The **quaternary structure** of a protein arises when two or more polypeptides become held together, forming a complex, biologically active molecule.

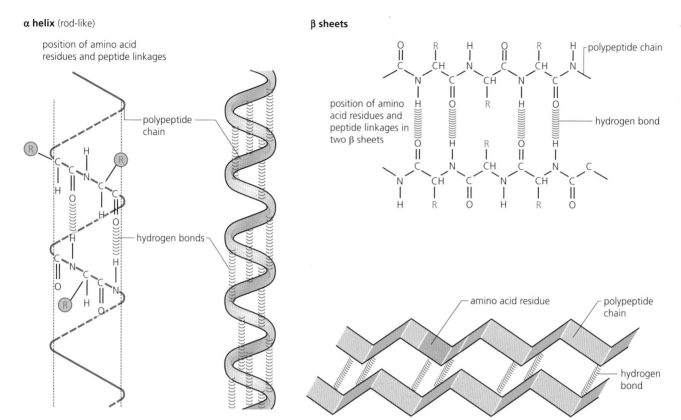

Figure 2.4 The secondary structure of proteins

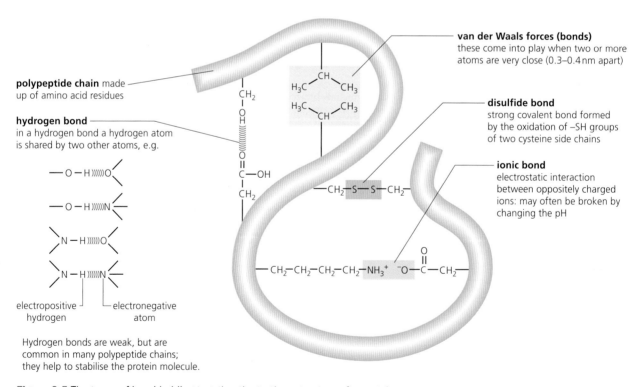

Figure 2.5 The types of bond holding together the tertiary structure of a protein

Fibrous and globular proteins

Collagen and haemoglobin are examples of proteins. Both consist of more than one polypeptide chain, i.e. they both have a quaternary structure. Their shape, however, is quite different. One is fibrous and the other is globular.

Fibrous proteins contain long, coiled polypeptide chains, shaped like a rod or wire. Collagen, a component of bone and tendons, is a fibrous protein. Figure 2.6 shows that a single collagen molecule has three polypeptide chains, held together by covalent bonds and hydrogen bonds, forming a triple helix. This structure makes collagen resistant to denaturing and provides its strength.

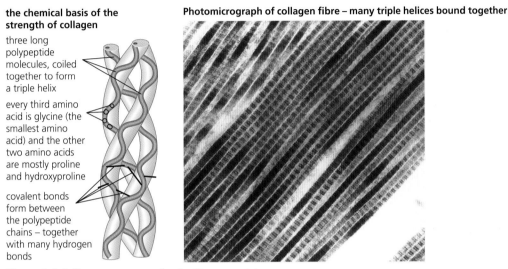

the chemical basis of the strength of collagen

three long polypeptide molecules, coiled together to form a triple helix

every third amino acid is glycine (the smallest amino acid) and the other two amino acids are mostly proline and hydroxyproline

covalent bonds form between the polypeptide chains – together with many hydrogen bonds

Photomicrograph of collagen fibre – many triple helices bound together

Figure 2.6 Collagen – an example of a fibrous protein

Globular proteins are more spherical. Their polypeptide chains wind in such a way that their hydrophilic amino acids are at the surface of the 'sphere' whilst their hydrophobic amino acids are at its core. Haemoglobin, the respiratory pigment found in our red blood cells, is an example of a globular protein. A haemoglobin molecule consists of four polypeptide chains, each bound to an iron-containing haem group (Figure 2.7). The iron in the haem groups binds to oxygen. As the first molecule of oxygen binds to a haem group, it causes a change in the shape of the haemoglobin molecule, making it easier for further oxygen molecules to bind. You will see the effect of this property when you learn about the transport of oxygen in Chapter 10.

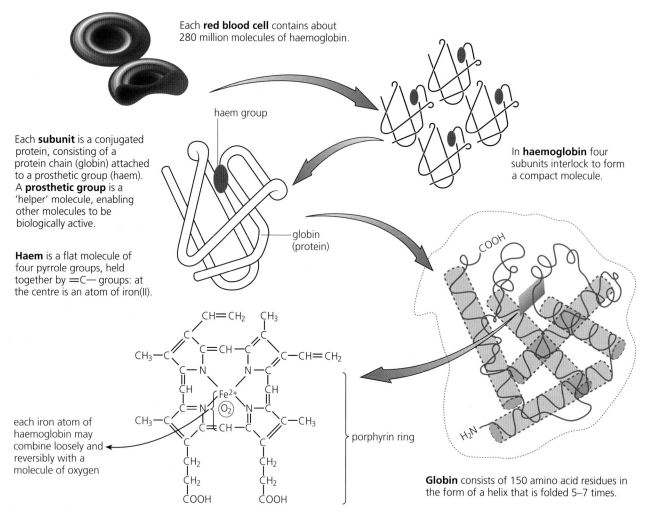

Each **red blood cell** contains about 280 million molecules of haemoglobin.

haem group

Each **subunit** is a conjugated protein, consisting of a protein chain (globin) attached to a prosthetic group (haem). A **prosthetic group** is a 'helper' molecule, enabling other molecules to be biologically active.

In **haemoglobin** four subunits interlock to form a compact molecule.

Haem is a flat molecule of four pyrrole groups, held together by $=C-$ groups: at the centre is an atom of iron(II).

globin (protein)

COOH

$CH=CH_2$ CH_3

CH_3-C $C=CH-C$ $C-CH=CH_2$

$C-N$ $N-C$

CH CH

each iron atom of haemoglobin may combine loosely and reversibly with a molecule of oxygen

$C=N$ Fe^{2+} $N-C$ O_2

CH_3-C $C-CH_3$

$C=CH-C$

CH_2 CH_2

CH_2 CH_2

$COOH$ $COOH$

porphyrin ring

H_2N

Globin consists of 150 amino acid residues in the form of a helix that is folded 5–7 times.

Figure 2.7 Haemoglobin: an example of a globular protein

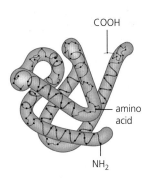

COOH

amino acid

NH$_2$

Figure 2.8 A molecule of myoglobin

Test yourself

1 In a chemical reaction, two amino acids join together.

   a) What *type* of chemical reaction occurs when two amino acids join together?

   b) Name the chemical bond that forms between these amino acids.

2 Amino acids can act as buffers, resisting changes in pH. Use your knowledge of amino acids to suggest how amino acids act as buffers.

3 How does the shape of a fibrous protein differ from that of a globular protein?

4 Myoglobin is an oxygen-carrying protein found in skeletal muscle cells. Figure 2.8 shows one molecule of myoglobin. How many levels of protein structure can you see in the diagram? Explain your answer.

5 Are enzyme molecules fibrous proteins or globular proteins?

Denaturation

We have seen the variety of shapes that different protein molecules might have. The different shapes are related to the functions these proteins have in cells and organisms. You will learn about the functions of several proteins throughout your course. For now, some of their important roles, together with their dynamic states, are reviewed in Figure 2.9.

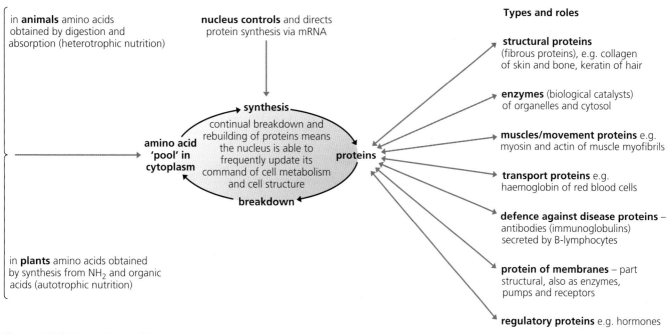

Figure 2.9 Cell proteins – origins, types and roles

Look again at the range of bonds that hold the 3-D shape of the tertiary structure of a protein in Figure 2.5. Their roles in maintaining the 3-D shape of the protein are confirmed when any protein loses its specific 3-D shape (we refer to this as denaturation). This happens only because these bonds that maintain the 3-D shape of the protein molecule are changed. This change can be caused by exposure to high temperature, to heavy metal ions, and to some organic solvents, but is also triggered by changes in pH and by certain other chemicals. When the shape of a protein changes, the protein may cease to be useful. The biochemistry of cells and organisms is extremely sensitive to conditions that alter proteins in this way.

Incidentally, when high temperature and heavy metal ions cause denaturation, the changes are irreversible, typically causing the protein to become elongated, disorganised strands, which are insoluble in water. This apart, however, it is often the case that proteins will revert back to their former shape, once the conditions that triggered denaturation are removed. This observation led to the idea that it is simply the amino acid sequence of a protein that decides its tertiary structure. This may well be true for many polypeptides and small proteins. However, in most proteins within the cell environment, folding is a speedy process in which some accessory proteins, including enzymes, are normally involved. These might determine the shape as much as, or more than, the primary structure does.

> **Key term**
>
> **Denaturation** A change in the shape of a protein that alters, or destroys, the ability of the protein to carry out its function. The change results from the breakage of bonds holding the shape of the protein together.

Test yourself

6   Is a protein a polymer? Explain your answer.

7   Name **three** types of chemical bond that might be present in a protein molecule.

8   Pickling is one form of food preservation. It involves immersing food in vinegar, i.e. weak ethanoic acid. Suggest how this slows decomposition of the food.

9   Muscle wastage is one of the symptoms of a protein-deficient diet. Explain why.

10   When a hen's egg is cooked its protein is denatured irreversibly. Is protein denaturation always irreversible?

Enzymes – biological catalysts

Key term

Metabolite The term used to describe any molecule involved in the reactions occurring in cells and organisms (in other words involved in metabolism).

There are literally many thousands of chemical reactions taking place within cells and organisms. We saw in Chapter 1 that metabolism is the name we give to these chemical reactions of life. The molecules involved are collectively called metabolites. Many of these are made within organisms. Other metabolites have been imported from the environment, for example water and oxygen.

Metabolism actually consists of chains (linear sequences) and cycles of enzyme-catalysed reactions, such as we see in protein synthesis (see Chapter 3), respiration (*Edexcel A level Biology 2*, Chapter 1) and photosynthesis (*Edexcel A level Biology 2*, Chapter 2). These reactions may be classified as one of just two types, according to whether they involve the build-up or breakdown of organic molecules.

● In **anabolic** reactions, larger molecules are built up from smaller molecules. Examples of anabolism are the synthesis of proteins from amino acids and the synthesis of polysaccharides from simple sugars.
● In **catabolic** reactions, larger molecules are broken down. Examples of catabolism are the digestion of complex foods and the breakdown of sugar in respiration.

Overall: **metabolism = anabolism + catabolism**

Introducing catalysis

For a reaction between two molecules to occur there must be successful collisions between them. The molecules must collide with each other at the right angle and with the right velocity. If the angle of collision is not correct, the molecules bounce apart. Only if the molecules are lined up and collide with the correct energies does a reaction occur.

Key term

Catalyst A substance that increases the rate of a chemical reaction. Only a small amount of the catalyst is required. Additionally, the catalyst is chemically unaffected by the reaction it catalyses and can be recovered once that reaction is complete.

Most chemical reactions do not occur spontaneously. In a laboratory or in an industrial process, chemical reactions can be made to occur by applying high temperatures, high pressures, extremes of pH, and by maintaining high concentrations of the reacting molecules. If these drastic conditions were not applied, very little of the chemical product would be formed quickly.

Alternatively, a catalyst can be used. You might recall from your GCSE science course the industrial use of a catalyst in the Haber process. If so, you will know that a catalyst (iron in the Haber process) is a substance that increases the rate of a reaction without itself being chemically changed. Chemical reactions in cells and organisms can occur at normal temperatures, under very mild, almost neutral, aqueous conditions and often at low concentrations of reactants, because cells contain many catalysts. We will now see how these catalysts enable reactions to occur in cells and organisms.

Enzymes as catalysts

The catalysts that are produced in cells are enzymes. Most are protein molecules. Like all catalysts, enzymes:

- do not change the nature of the reaction they catalyse
- are effective in small amounts
- remain chemically unchanged at the end of the reaction.

How enzymes work: the enzyme–substrate complex

An enzyme molecule (**E**) works by binding to a specific substance, known as its substrate molecule (**S**), at a specially formed pocket in the enzyme, called its active site. As the enzyme and substrate bind, they form an unstable enzyme–substrate complex (**ES**), which immediately breaks down to form the product(s) (**P**), plus the unchanged enzyme. Using these letters, this reaction can be shown as a simple equation:

$$\mathbf{E + S \rightleftharpoons ES \rightleftharpoons P + E}$$

In your GCSE science course, you probably learnt about the 'lock-and-key' model of enzyme action. In this model, the enzyme and substrate molecules have a fixed shape and the substrate fits into the active site of the enzyme just like a key fits into a lock. Further studies show that this model is too simplistic. In fact, an enzyme molecule *does* change shape as a substrate molecule binds to its active site, rather like a glove changes shape as you put your hand into it. Only during this binding of enzyme and substrate does the active site become truly complementary to the part of the substrate molecule to which it attaches. As often happens in science, new evidence disproves or modifies a previously held model or theory. In this case, the new model is called the induced-fit hypothesis. You can see how the substrate induces the enzyme to fit in Figure 2.10. Notice in this computer-generated model, how the active site of the enzyme is only a general fit until the substrates combine with it.

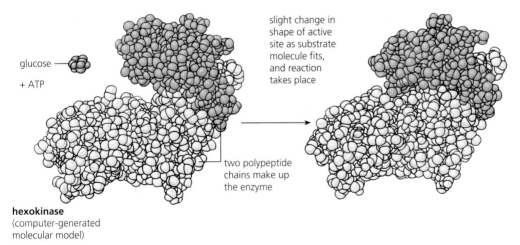

glucose
+ ATP

slight change in shape of active site as substrate molecule fits, and reaction takes place

two polypeptide chains make up the enzyme

hexokinase
(computer-generated molecular model)

Figure 2.10 The induced-fit hypothesis to explain enzyme action

Key terms

Enzyme An organic catalyst that speeds up a metabolic reaction in cells and organisms.

Substrate A molecule that binds to the active site of an enzyme to form an enzyme–substrate complex. As a result, the substrate is converted to product and the enzyme remains unchanged.

Active site Enzyme molecules are usually much bigger than their substrate molecules. The active site is the small part of the enzyme molecule that binds to its specific substrate and causes the catalysis.

Induced-fit hypothesis A model used to explain enzyme action. When an enzyme and substrate combine, the enzyme's active site changes shape to become truly complementary to the part of the substrate to which it attaches. Combination with its substrate *induces* the enzyme's active site to *fit*.

Tip

Most enzymes work within cells, in other words they are **intracellular**. Some, like the enzymes of our digestive system or of the microorganisms that cause the recycling of nutrients within an ecosystem (see Chapter 13 in *Edexcel A level Biology 2*), are secreted from the cell – they are **extracellular**.

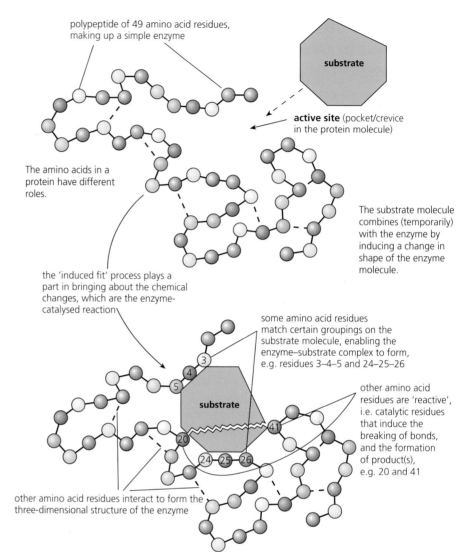

polypeptide of 49 amino acid residues, making up a simple enzyme

substrate

active site (pocket/crevice in the protein molecule)

The amino acids in a protein have different roles.

The substrate molecule combines (temporarily) with the enzyme by inducing a change in shape of the enzyme molecule.

the 'induced fit' process plays a part in bringing about the chemical changes, which are the enzyme-catalysed reaction

some amino acid residues match certain groupings on the substrate molecule, enabling the enzyme–substrate complex to form, e.g. residues 3–4–5 and 24–25–26

substrate

other amino acid residues are 'reactive', i.e. catalytic residues that induce the breaking of bonds, and the formation of product(s), e.g. 20 and 41

Specificity:
- Some amino acid residues allow a particular substrate molecule to 'fit'
- Some amino acid residues bring about particular chemical changes.

other amino acid residues interact to form the three-dimensional structure of the enzyme

Figure 2.11 The induced-fit hypothesis of enzyme action

Key term

Activation energy The energy barrier that must be overcome before reactants reach their temporary transition state. Enzymes lower the activation energy of the reactions they catalyse, making the reaction occur more readily.

An enzyme lowers the activation energy of the reaction it catalyses

As molecules react they become unstable intermediates, but only momentarily whilst in a so-called transition state. Effectively, the products are formed immediately. The amount of energy needed to raise substrate molecules to their transition state is called **activation energy**. This is the energy barrier that has to be overcome before the reaction can happen. As Figure 2.12 shows, like all catalysts, enzymes work by lowering the activation energy of the reactions they catalyse.

Figure 2.12 also shows a simplistic 'model' of the start of a chemical reaction. A boulder represents a substrate perched on a slope, prevented from rolling down by a small hump (representing the activation energy) in front of it. The boulder can be pushed over the hump, or the hump can be dug away (representing a lowering of the activation energy), allowing the boulder to roll and shatter at a lower level (representing formation of products).

'boulder on hillside' model of activation energy

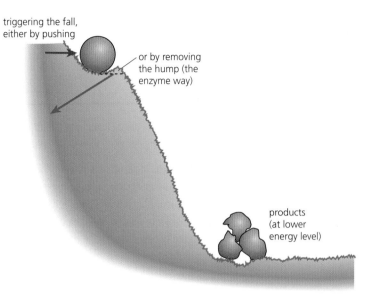

triggering the fall,
either by pushing

or by removing
the hump (the
enzyme way)

products
(at lower
energy level)

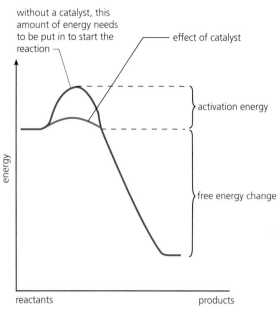

without a catalyst, this
amount of energy needs
to be put in to start the
reaction

effect of catalyst

activation energy

free energy change

energy

reactants

products

Figure 2.12 Activation energy

Test yourself

11 Distinguish between the terms *anabolism*, *catabolism* and *metabolism*.

12 Give **three** properties shown by all catalysts.

13 Do collisions between an enzyme and its substrate always result in the formation of an enzyme–substrate complex? Explain your answer.

14 Explain the terms *activation energy* and *transition state* in relation to an enzyme-catalysed reaction.

15 What is meant by the term *active site*?

The active site and enzyme specificity

Enzymes are highly specific in their action – they catalyse only one type of reaction or only a very small group of highly similar reactions. This means that an enzyme 'recognises' a very small group of substrate molecules or even only a single type of molecule. This is because the active site to which the substrate molecule binds has a precise shape and distinctive chemical properties (meaning the presence of particular chemical groups and bonds). Only particular substrate molecules, with a shape that is complementary to that of the active site, are attracted to a particular active site and can fit there. All other substrate molecules are unable to fit and so cannot bind.

Tip

The active site of an enzyme has a shape complementary to that of its substrate; it does not have the same shape.

Investigate a factor affecting the initial rate of an enzyme-controlled reaction

The background to your practical investigation

You can measure the rate of an enzyme-catalysed reaction in one of two ways:

- the amount of substrate that has disappeared from a reaction mixture in a given period of time
- the amount of product that has accumulated from the reaction mixture in a given period of time.

In your college or school laboratory, you will carry out at least one experiment to investigate the effect of an environmental variable on the rate of an enzyme-controlled reaction. It doesn't really matter which factor you investigate or which enzyme you use. Here we use the enzyme catalase as our example to show general features involved in any investigation into the effect of a variable on the rate of an enzyme-catalysed reaction.

Catalase is an enzyme found in many tissues. It catalyses the hydrolysis of hydrogen peroxide, a toxic by-product of some metabolic reactions, to water and oxygen:

$$2H_2O_2 \xrightarrow{\text{catalase}} 2H_2O + O_2$$

Catalse is particularly common in liver tissue but, if you prefer not to handle animal tissue, it is also very common in potato tissue.

Working with catalase, it is easy to measure the rate of reaction by measuring the rate at which the product (oxygen) accumulates. In the experiment illustrated in Figure 2.13, the volume of oxygen that has accumulated at 30-second intervals is recorded in a table of raw data; these data are then processed and plotted on a graph.

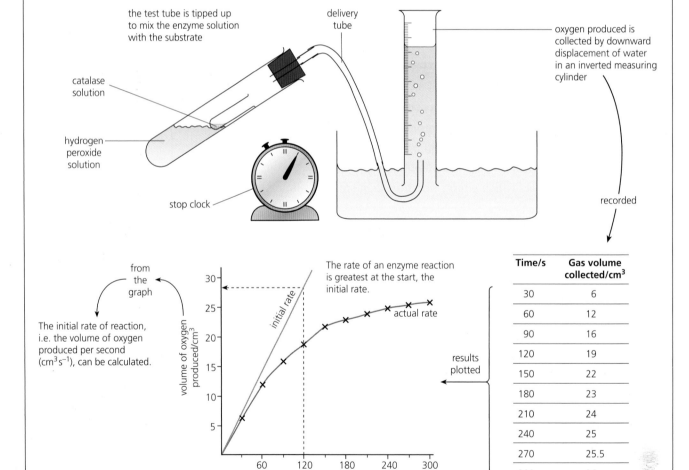

Time/s	Gas volume collected/cm^3
30	6
60	12
90	16
120	19
150	22
180	23
210	24
240	25
270	25.5
300	26

Figure 2.13 Measuring the rate of reaction, using catalase. Wear eye protection when performing this experiment

Look at Figure 2.13, showing how you could measure the initial rate of reaction of a catalase-controlled reaction.

1 The catalase solution is initially placed in a small test tube within the larger test tube containing the hydrogen peroxide solution. Explain why.

2 You measure the volume of oxygen in the measuring cylinder at 30-second intervals. How would you take your measurement to ensure it is accurate?

3 Using a measuring cylinder, such as the one in Figure 2.13, how would you estimate your error in taking your readings of volume?

4 When during your investigation would you draw your table for the raw data you collect? What 'rules' would you follow in drawing the table?

5 How could you modify this investigation to measure the effect of temperature on the initial rate of reaction?

6 Which temperatures would you use? Explain your answer.

7 How would you ensure that the temperature in your water bath remained constant?

8 Ideally, you should repeat the procedure at each temperature several times. Explain why.

9 Explain how you would process, and then plot on a graph, your data to show the effect of temperature on the rate of this enzyme-controlled reaction.

10 Why should you place 'Temperature/°C' in the left-hand column of your new table?

11 It would be difficult to determine the optimum temperature for catalase from your graph. Describe how you could obtain a more accurate value for the optimum temperature.

Key term

Accurate Measurement that is close to the true value of what is being measured.

Factors that change the rate of reaction of enzymes

Enzymes are very sensitive to environmental conditions. Here we will consider how temperature, pH, substrate concentration and enzyme concentration affect the activity of enzymes.

Temperature

Figure 2.14 shows typical results of the effect of temperature on the rate of an enzyme-catalysed reaction.

Not all enzymes have the same optimum temperature. For example, the bacteria in hot thermal springs have enzymes with optima in the region 80–100°C, whilst seaweeds of northern seas and the plants of the tundra have optima closer to 0°C. Humans have enzymes with optima at or about normal body temperature (37°C).

Other variables – such as the concentrations of the enzyme and substrate solutions – were kept constant.

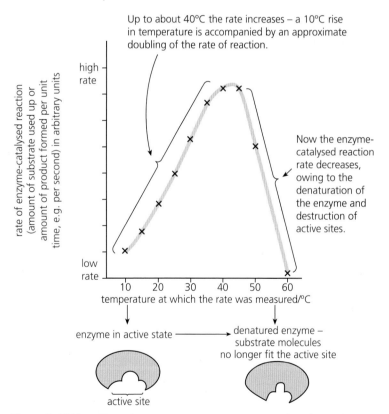

Up to about 40°C the rate increases – a 10°C rise in temperature is accompanied by an approximate doubling of the rate of reaction.

Now the enzyme-catalysed reaction rate decreases, owing to the denaturation of the enzyme and destruction of active sites.

enzyme in active state ⟶ denatured enzyme – substrate molecules no longer fit the active site

active site

Figure 2.14 The effect of temperature on an enzyme-catalysed reaction

Tip

When asked to 'explain', it might help you remember to give a reason if you develop the habit of including the word 'because' in your answer.

Key terms

Random thermal movement The movement shown by all particles, whether sub-atomic particles, atoms, or molecules, at temperatures above absolute zero. As the temperatures increases, so does the rate of random thermal movement.

Optimum temperature The temperature at which the rate of an enzyme-controlled reaction is fastest. At this point there is a balance between an increase in successful collisions between enzyme and substrate molecules and loss of active enzyme molecules as a result of their denaturation.

Optimum pH The value, or narrow range of values, over which an enzyme-catalysed reaction is fastest.

Example

How can we explain the graph in Figure 2.14?

It is tempting to describe the trend or pattern shown by a graph or by data in a table. In an examination, a question requiring this response would use the command word 'Describe'. Here, though, we are looking for an *explanation*. This means we must give *reasons* for the trend or pattern shown by the data; a simple description would not gain credit.

Two concepts are involved in explaining the effect of temperature on the rate of an enzyme-controlled reaction. Increases in temperature cause an increase in:

- the random thermal movement of particles. In other words, the higher the temperature, the more molecules, and the particles within them, move about. If they move about more, collisions between them become increasingly likely
- the rate of denaturation of protein molecules. This happens because high temperatures cause such violent movement of particles within a protein molecule that the bonds holding the protein molecule together (Figure 2.5) break, so its active site loses its critical shape.

Let's apply these two concepts to explain the shape of the curve in Figure 2.14.

Answer

As the temperature increases from 10 °C to 30 °C, the rate of reaction increases *because* the molecules of enzyme and substrate are moving more rapidly and are more likely to collide and react.

As the temperature increases from 45 °C to 60 °C, the rate of reaction slows *because* more and more enzyme molecules have been denatured and so fewer functional enzyme molecules remain available to catalyse the reaction.

At some temperature between 30 °C and 45 °C, a balance is reached between the increased rate of reaction caused by more collisions of enzyme and substrate and the decreased rate of reaction caused by denaturation of the enzyme. This temperature is known as the optimum temperature of that enzyme.

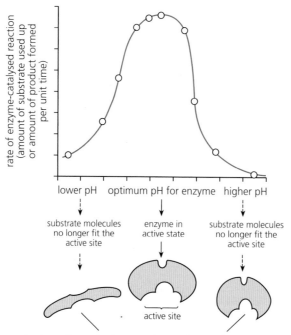

Figure 2.15 The effect of pH on enzyme shape and activity

structure of protein changes when a change of pH alters the ionic charge on $-COO^-$ (acidic) and $-NH_3^+$ (basic) groups in the peptide chain, so the shape of the active site is lost

pH

Each enzyme has a range of pH values, called its optimum pH, in which the rate of the reaction it catalyses is fastest. pH has this effect because the structure of a protein (and therefore the shape of the active site) is maintained by various bonds within the three-dimensional structure of the protein (look back to Figure 2.5 to remind yourself of these bonds). A change in pH from the optimum value alters the bonding patterns. As a result, the shape of the active site of the enzyme molecule is progressively changed.

This is shown in Figure 2.15. At the optimum pH, the active site has the appropriate shape to combine with its substrate. At pH values away from the optimum, the shape of the active site changes so that it will no longer bind with its substrate. Unlike the effect of temperature, however, the effects of pH on the active site are normally reversible. That is, provided the change in surrounding acidity or alkalinity is not too extreme, as the pH is brought back to the optimum for that enzyme, the active site may reappear.

Substrate concentration

Figure 2.16 shows the effect of increasing substrate concentration on the rate of an enzyme-catalysed reaction. The curve has two phases.

● At lower substrate concentrations, the rate increases in direct proportion to the increase in substrate concentration.
● At higher substrate concentrations, the rate of reaction becomes constant, showing no further increase as the substrate concentration increases.

Figure 2.16 also shows why these two phases occur. At low substrate concentrations, there is effectively an excess of enzyme molecules present. This means there are 'free' enzyme molecules that are available to react with added substrate molecules to form more enzyme–substrate complexes per unit time. We can say that substrate concentration is the limiting factor at this stage of the reaction.

As more substrate molecules are added, however, there comes a point at which the concentration of substrate is greater than that of the enzyme. There are no longer 'free' enzyme molecules. Now, in effect, substrate molecules have to 'queue up' for access to an active site. Adding more substrate increases the number of molecules awaiting contact with an enzyme molecule. There is now no increase in the rate of reaction, explaining the plateau in the curve shown in Figure 2.16.

Key term

Limiting factor Any factor that limits the rate at which a reaction, or process, can occur.

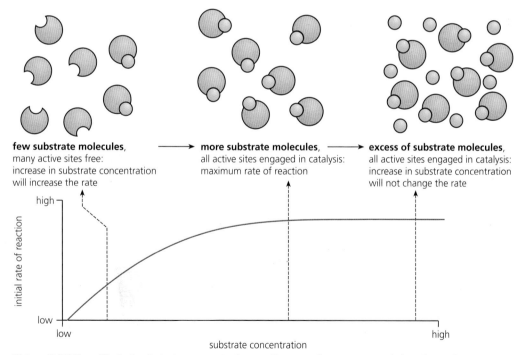

few substrate molecules, many active sites free: increase in substrate concentration will increase the rate ⟶ **more substrate molecules,** all active sites engaged in catalysis: maximum rate of reaction ⟶ **excess of substrate molecules,** all active sites engaged in catalysis: increase in substrate concentration will not change the rate

Figure 2.16 The effect of substrate concentration on the rate of an enzyme-catalysed reaction

Enzyme concentration

Figure 2.17 shows the effect of increasing enzyme concentration on the rate of an enzyme-catalysed reaction. This curve also has two phases.

● At lower enzyme concentrations, the rate increases in direct proportion to the increase in enzyme concentration.
● At higher enzyme concentrations, the rate of reaction becomes constant, showing no further increase as the enzyme concentration increases.

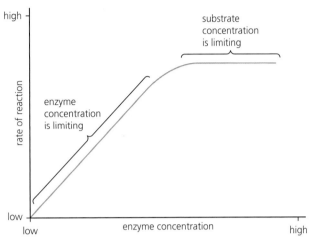

Figure 2.17 The effect of enzyme concentration on the rate of an enzyme-catalysed reaction

Figure 2.17 also shows why these two phases occur. At low enzyme concentrations, there is effectively an excess of substrate molecules present. This means there are no 'free' enzyme molecules available to react with the substrate molecules already there. This time, it is the enzyme concentration that is the limiting factor at this stage of the reaction.

As more enzyme molecules are added, more 'free' active sites become available, so more enzyme–substrate complexes can be formed and the rate of reaction increases. As even more enzyme molecules are added, however, there comes a point at which the concentration of enzyme becomes greater than that of the substrate. There are now 'free' enzyme molecules. Consequently, adding even more enzyme molecules will not increase the rate of reaction, explaining the plateau in the curve shown in Figure 2.17.

Inhibitors of enzymes

Key term

Enzyme inhibitor
A substance that slows the rate of an enzyme-controlled reaction by preventing binding of the substrate to the active site of the enzyme.

● Competitive inhibitors bind directly to the enzyme's active site, blocking access by the substrate.
● Non-competitive inhibitors bind to the enzyme at a site other than the active site and, by doing so, cause the shape of the enzyme's active site to change.

Some substances can react with an enzyme, slowing the rate of the reaction it catalyses. These substances are known as enzyme inhibitors. Studies of the effects of inhibitors have helped our understanding of:

● the chemistry of the active site of enzymes
● the natural regulation of metabolism
● the ways in which certain commercial pesticides and many drugs work (by inhibiting specific enzymes and preventing particular reactions).

There are two types of enzyme inhibitor with which we need to be familiar. Their effects are summarised in Table 2.1 and in Figure 2.18.

Table 2.1 Competitive and non-competitive inhibition of enzymes compared

Competitive inhibition	Non-competitive inhibition
Inhibitor chemically resembles the substrate molecule and binds with the active site, blocking access to substrate molecules.	Inhibitor chemically unlike the substrate molecule, but by binding to another (allosteric) site, changes the shape of the enzyme molecule, including the active site.
With a low concentration of inhibitor, increasing the concentration of substrate eventually overcomes inhibition as substrate molecules displace inhibitor and enzyme–substrate collisions become more likely than enzyme–inhibitor collisions.	With a low concentration of inhibitor, increasing concentration of substrate can neither displace inhibitor nor prevent binding of further inhibitor molecules.
For example, O_2 competes with CO_2 for the active site of rubisco.	For example, alanine non-competitively inhibits pyruvate kinase.

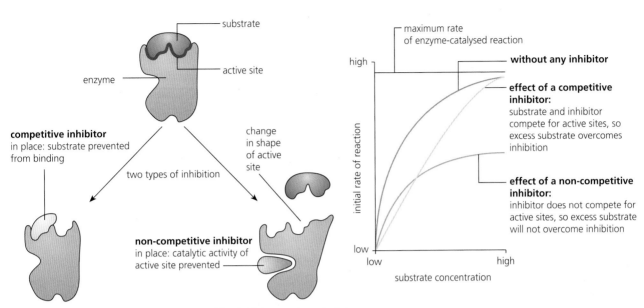

When the initial rates of reaction of an enzyme are plotted against substrate concentration, the effects of competitive and non-competitive inhibitors are seen to be different.

Figure 2.18 Competitive and non-competitive inhibitors – the principles

Competitive inhibitors

The molecules of a competitive inhibitor sufficiently resemble the shape of the true substrate so that they can bind to the active site of the enzyme, forming an enzyme–inhibitor complex. Because these inhibitors are not acted on by the enzyme and turned into 'products' as normal substrate molecules are, the enzyme–inhibitor complex tends to remain intact. However, if the concentration of the substrate molecule is increased, the inhibitor molecules are progressively displaced from the active sites – and become less likely to re-attach.

An enzyme you will meet in year 2, called ribulose bisphosphate carboxylase (or rubisco for short), is one of the most common enzymes on Earth. It catalyses the reaction between carbon dioxide and a 'CO_2-acceptor molecule' during the process of photosynthesis. Oxygen is a competitive inhibitor of this enzyme.

Non-competitive inhibitors

The molecules of a non-competitive inhibitor are quite unlike the true substrate molecule, yet can still combine with the enzyme. In this case, the attachment does not occur at the active site of the enzyme but at another (allosteric) site. As a result of this binding, the shape of the enzyme molecule changes and, with it, the shape of the active site also changes. The active site is no longer complementary to molecules of the substrate and the enzyme loses its ability to bind with substrate molecules. Unlike the case with competitive inhibition, adding more substrate does not dislodge the inhibitor, since the substrate and inhibitor are not competing for the active site of the enzyme. Consequently, non-competitive inhibition is often permanent.

One of the steps in cell respiration that you will meet in year 2 is catalysed by an enzyme called pyruvate kinase. The amino acid alanine is a non-competitive inhibitor of this enzyme. Many poisons are non-competitive inhibitors of enzymes.

Key term

End-product inhibition
A feature of a chain of enzyme-controlled reactions in which a product of a late reaction in the series inhibits the enzyme controlling an earlier reaction.

End-product inhibition in the control of metabolic pathways

Many metabolic pathways exist as a chain of reactions, each catalysed by a different, specific enzyme. Enzyme inhibition is often involved in the regulation of such pathways.

Figure 2.19 represents a chain of reactions by which a substrate (A) is converted to a useful end product (F). Each reaction in the chain is controlled by a different, specific, enzyme (a to e). The whole process can be regulated because the end product (F) is a non-competitive inhibitor of the first enzyme in the chain (a). As the concentration of end product (F) increases, it inhibits enzyme a, slowing the rate of the first reaction in the series (conversion of A to B) and, hence, slowing the entire pathway. So, in **end-product inhibition**, as the product molecules accumulate, the steps in their production are switched off. But these product molecules may now become the substrates in subsequent metabolic reactions. If so, the accumulated product molecules will be removed, and production of new product molecules will recommence.

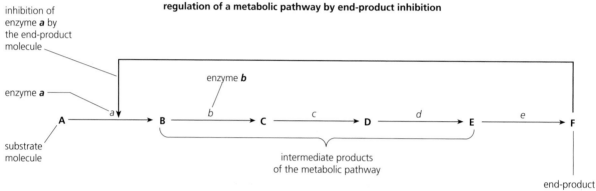

regulation of a metabolic pathway by end-product inhibition

This is an example of the regulation of a metabolic pathway by **negative feedback**.

Figure 2.19 End-product inhibition of metabolism

Test yourself

16 Outline the difference between the lock-and-key and the induced-fit models of enzyme action.

17 Explain the difference between the time of reaction and the rate of reaction.

18 Which **one** of the following statements is true? Explain your answer.

 A Enzymes only begin to denature when they are heated to boiling.

 B Enzymes only begin to denature at temperatures above their optimum temperature.

 C Enzymes begin to denature at their optimum temperature.

 D Enzymes begin to denature below their optimum temperature.

19 When investigating enzyme-catalysed reactions, scientists usually include a buffer solution in their reaction mixtures. Explain why.

20 Amylase is an enzyme that hydrolyses starch to maltose. It is secreted by the salivary glands of some humans and is also secreted by the pancreas. Amylase from the salivary glands has an optimum pH in the range 4.6 to 5.2, whereas amylase from the pancreas has an optimum pH in the range 6.7 to 7.0.

What can you conclude from this information about the nature of human amylase?

Exam practice questions

1 The graph shows the effect of temperature on the rate of an enzyme-catalysed reaction.

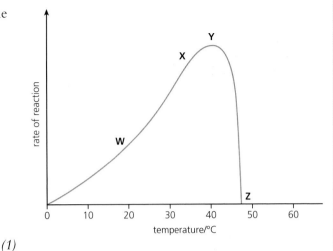

Which of the following statements is true about the graph?

A The enzyme starts to become denatured at point W.

B The enzyme starts to become denatured at point X.

C The enzyme starts to become denatured at point Y.

D The enzyme starts to become denatured at point Z. *(1)*

2 Denaturation involves breakage of which types of bond?

A Hydrogen bonds, ionic bonds and peptide bonds

B Disulfide bonds, hydrogen bonds and peptide bonds

C Disulfide bonds, ionic bonds and peptide bonds

D Disulfide bonds, hydrogen bonds and ionic bonds *(1)*

3 Which of the following statements is true about an amino acid in a weakly acidic solution?

A It will not ionise.

B It will ionise as an anion.

C It will ionise as a cation.

D It will ionise as a zwitterions. *(1)*

4 Collagen and haemoglobin are two different proteins. Explain **one** way in which the structure of each molecule is adapted for its function. *(4)*

5 The diagrams represent four different types of molecule important in biology. Identify each type of molecule. *(4)*

A

$$\begin{array}{c} H \\ | \\ O \\ | \\ H \end{array}$$

B

$$H-\overset{\overset{\displaystyle H}{|}}{\underset{\underset{\displaystyle H}{|}}{C}}-\overset{\overset{\displaystyle H}{|}}{\underset{\underset{\displaystyle H}{|}}{C}}-\overset{\overset{\displaystyle H}{|}}{\underset{\underset{\displaystyle H}{|}}{C}}-\overset{\overset{\displaystyle H}{|}}{\underset{\underset{\displaystyle H}{|}}{C}}-\overset{\displaystyle H}{C}=\overset{\displaystyle H}{C}-\overset{\overset{\displaystyle H}{|}}{\underset{\underset{\displaystyle H}{|}}{C}}-\overset{\overset{\displaystyle O}{\|}}{\underset{\displaystyle O-H}{C}}$$

C

D

> **Tip**
>
> Question 4 tests recall with understanding (AO1). Note that the command word is 'explain'; a simple description of each molecule will not gain full marks.

> **Tip**
>
> Question 5 tests recall with understanding (AO1) from Chapters 1 and 2.

6 The diagram shows a molecule of an amino acid called alanine.

$$
\begin{array}{c}
CH_3 \\
| \\
H_2N-C-COOH \\
| \\
H
\end{array}
$$

a) Draw a new diagram to show how two molecules of alanine join together to form a dipeptide. *(2)*

b) What name is given to the type of reaction by which a dipeptide is formed? *(1)*

c) How do other amino acids differ in structure from alanine? *(1)*

d) There are 20 different amino acids found in the proteins in living organisms. Theoretically, how many different dipeptides could exist? *(1)*

7 A student added $1\,cm^3$ of a dilute solution of catalase to $20\,cm^3$ of a 5% solution of hydrogen peroxide. She measured the amount of product formed at regular intervals.

a) Why did the student use a *dilute* solution of catalase? *(2)*

b) Given a 100% solution of hydrogen peroxide, describe how you would produce $20\,cm^3$ of a 5% solution of hydrogen peroxide. *(1)*

The student recorded her results in a table. She used this table to produce the sketch graph of her results, shown here.

c) Explain why the graph is described as a 'sketch graph'. *(1)*

d) Suggest why the student produced a sketch graph of her results. *(1)*

e) Explain the shape of the curve shown in the graph. *(2)*

f) Add a second curve to the graph to show the results you would expect if this student had repeated her experiment but added a non-competitive inhibitor to the starting mixture. Justify the curve you have drawn. *(3)*

8 Compare and contrast the effects of competitive inhibitors and non-competitive inhibitors. *(5)*

Stretch and challenge

9 Enzyme-catalysed reactions can be analysed quantitatively. Three commonly used measures are the temperature coefficient (Q_{10}), the maximum rate of reaction (V_{max}) and the Michaelis constant (K_m).

a) The temperature coefficient of an enzyme is found by the following equation:

$$
Q_{10} = \frac{\text{rate of reaction at temperature } (T + 10)°C}{\text{rate of reaction at temperature } T\,°C}
$$

Calculate the Q_{10} value for the enzyme-catalysed reaction shown in Figure 1.

> **Tip**
>
> You might find it helpful to refer to Figure 2.13 when answering Question 7.

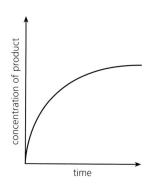
concentration of product / time

2 Proteins and enzymes

b) Figure 2 shows changes in the rate of reaction at different substrate concentrations.

Use information from Figure 2 to answer the following questions:

i) Describe how the Michaelis constant (K_m) is calculated.

ii) Explain how you could use information about the Michaelis constant to determine whether an enzyme-catalysed reaction was affected by a competitive inhibitor or a non-competitive inhibitor.

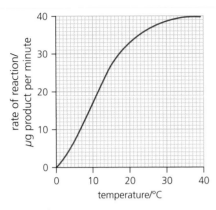

Figure 1

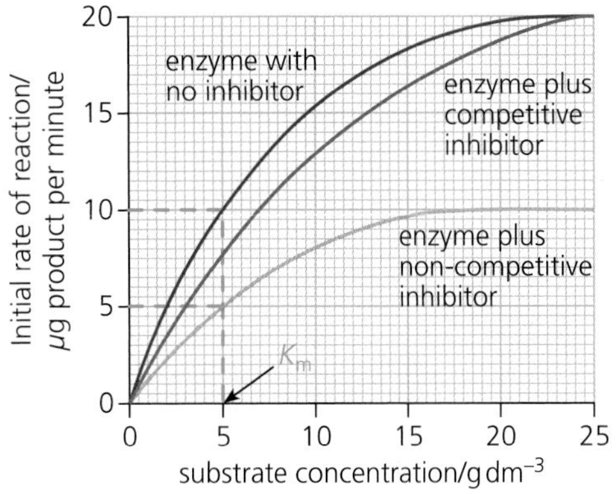

Figure 2

10 Earlier in this chapter, you were told 'The catalysts that are produced in cells are enzymes. Most are protein molecules.'

Clearly, this statement tells you that some enzymes are *not* protein molecules.

Use a search engine or an undergraduate textbook to find which group of enzymes are not proteins.

a) What type of molecule are they?

b) In what reactions are they involved and where?

c) Why do some scientists think these enzymes might have had an important role in the evolution of life on Earth?

d) Suggest why the discoverers of these non-protein enzymes had difficulty publishing their findings.

11 In Year 2 of your course, you will learn about a number of coenzymes. Find out what a coenzyme is and about the general role of coenzymes.

12 You are currently reading a paper-based book. Several enzymes were used in the production of this paper from boiled wood pulp. Carry out the research needed to produce a list of at least four types of enzyme likely to have been used in producing paper from wood pulp and give the function of each.

Nucleic acids and protein synthesis

Prior knowledge

In this chapter you will need to recall that:

→ a DNA molecule is made up of two strands forming a double helix

→ each DNA strand is a polymer made from four different nucleotides; each nucleotide consists of a common sugar and phosphate group with one of four different bases attached to the sugar

→ the sequence of bases in a cell's DNA molecule(s) determines the sequence of amino acids in the polypeptides and proteins the cell produces

→ a gene is a sequence of DNA bases that determines the sequence of amino acids of a single protein

→ a change in the sequence of bases within a gene can alter the activity of the protein for which it codes.

Test yourself on prior knowledge

1 A DNA molecule is a polymer of nucleotides.

   a) What is meant by a polymer?

   b) Name the components of a single DNA nucleotide.

2 DNA carries the genetic code for the sequence of amino acids in a protein. In what form is the genetic code carried in a DNA molecule?

3 a) Where is the DNA in a human cell?

   b) Where in a human cell is protein made?

4 What is a gene mutation?

5 Give the role of a messenger RNA molecule in a human cell.

The structure of nucleic acids

Nucleic acids are the 'information molecules' of cells. The 'information' they carry determines the sequence of amino acids in each protein a cell can produce. As you will see, the way in which 'information' about the amino acid sequence is held in nucleic acids – the genetic code – is universal. This means that it is not specific to any one organism or even to a larger group – like mammals or bacteria – alone. It makes sense in all organisms.

There are two types of nucleic acid, DNA (deoxyribonucleic acid) and RNA (ribonucleic acid). These molecules have roles in the day-to-day control of cells and organisms and in the transmission of genetic information from generation to generation. Before understanding how they do this, we need to look at the nucleotides from which nucleic acids are formed.

Structure of nucleotides

A nucleotide is the monomer from which both DNA and RNA are formed.

> **Key term**
>
> **Nucleotide** The monomer from which nucleic acids are formed. Each nucleotide comprises a pentose, a phosphate group and a purine or pyrimidine base.

Each nucleotide consists of three substances combined together (Figure 3.1):

- a pentose (ribose in RNA and deoxyribose in DNA)
- a nitrogenous base, which might be:
  - a double-ringed purine (either adenine or guanine in both DNA and RNA)
  - a single-ringed pyrimidine (either cytosine or thymine in DNA; either cytosine or uracil in RNA)
- phosphoric acid.

the components:

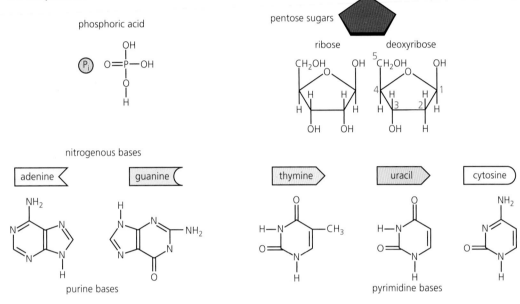

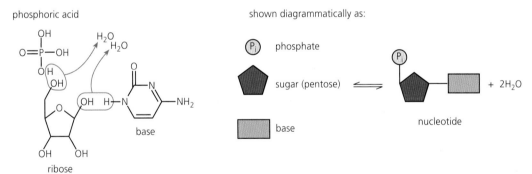

Figure 3.1 The components of nucleotides

Polynucleotide chains

Two nucleotides can be joined together by a condensation reaction, catalysed by an enzyme **DNA polymerase**. Figure 3.2 shows how this reaction results in the formation of a covalent bond, called a **phosphodiester bond**, between adjacent nucleotides. In Figure 3.3, four condensation reactions have produced a chain of five nucleotides.

Key terms

DNA polymerase The enzyme that catalyses the formation of a phosphodiester bond between two nucleotides.

Phosphodiester bond The covalent bond between two nucleotides.

condensation to form a dinucleotide...

...shown diagrammatically as:

Nucleotides become chemically combined together, phosphate to pentose sugar, by covalent bonds, with a sequence of bases attached to the sugar residues. Up to 5 million nucleotides condense together in this way, forming a polynucleotide (nucleic acid).

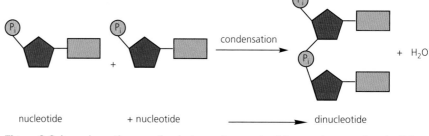

nucleotide   + nucleotide   dinucleotide

Figure 3.2 A condensation reaction between two nucleotides produces a dinucleotide

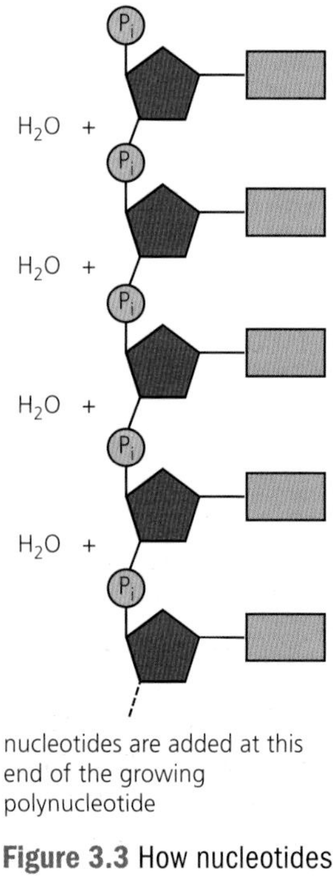

Large numbers of nucleotides become condensed together to form huge molecules – the nucleic acids, also known as polynucleotides. A nucleic acid or polynucleotide is a very long, thread-like macromolecule. You can see in Figure 3.3 how alternating sugar and phosphate molecules form the 'backbone' of the polynucleotide, with a nitrogenous base attached to each sugar molecule along the strand. Notice the label that shows where new nucleotides are added to a developing polynucleotide chain. This becomes important when we look at how DNA is copied later in this chapter.

H_2O +

H_2O +

H_2O +

H_2O +

H_2O +

nucleotides are added at this end of the growing polynucleotide

Figure 3.3 How nucleotides make up a polynucleotide chain

RNA molecules

RNA molecules are relatively short. In fact, RNA molecules tend to be between 100 and thousands of nucleotides long, depending on the particular role they have.

In every RNA nucleotide:

- the pentose is ribose
- the base is cytosine, guanine, adenine or uracil, but never thymine.

This is shown in Figure 3.4, which also shows that RNA molecules are always a single strand of nucleotides.

In the 'information business' of cells there are three functional types of RNA:

- messenger RNA (mRNA) – carries a copy of a single gene to a cell's ribosomes
- transfer RNA (tRNA) – carries individual amino acids to ribosomes during protein synthesis
- ribosomal RNA – forms part of the sub-units of ribosomes.

We will expand on the roles of these RNA molecules later in this chapter.

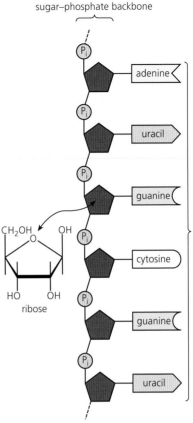

sugar–phosphate backbone

single strand of polynucleotide with ribose sugar and nitrogenous bases: adenine, uracil, guanine and cytosine

CH_2OH OH
HO   OH
ribose

Figure 3.4 RNA structure

DNA molecules

DNA molecules form extremely long strands, of the order of several million nucleotides in length.

In every DNA nucleotide:

- the pentose is deoxyribose
- the base is cytosine, guanine, adenine or thymine, but never uracil.

Unlike RNA, a DNA molecule consists of two polynucleotide strands, held together by hydrogen bonds between its bases. The two strands take the shape of a double helix (Figure 3.5). You can see that the hydrogen bonds that hold the two strands together are formed between specific bases: adenine with thymine; cytosine with guanine. This pairing, known as **complementary base pairing**, is the key to:

- the stability of the DNA double helix (although individual hydrogen bonds are weak, millions of them in a DNA molecule provide strength)
- the way in which genetic information can be transferred from DNA to RNA (mRNA)
- the way amino acids are assembled into polypeptides in the cytoplasm.

The bases of the two strands fit together only if the deoxyribose molecules to which they are attached point in opposite directions. You can see the effect of this in Figure 3.5. Because the two sugar–phosphate backbones point in opposite directions, these DNA strands are said to be **antiparallel**.

> **Key term**
>
> **Complementary base pairing** A key feature of DNA molecules in which two antiparallel polynucleotide chains are held together by hydrogen bonds between the bases adenine and thymine or the bases cytosine and guanine.

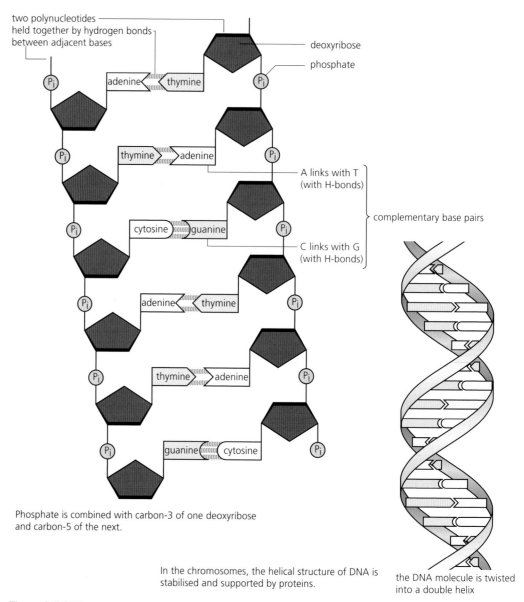

two polynucleotides held together by hydrogen bonds between adjacent bases

deoxyribose

phosphate

adenine — thymine

thymine — adenine

A links with T (with H-bonds)

complementary base pairs

cytosine — guanine

C links with G (with H-bonds)

adenine — thymine

thymine — adenine

guanine — cytosine

Phosphate is combined with carbon-3 of one deoxyribose and carbon-5 of the next.

In the chromosomes, the helical structure of DNA is stabilised and supported by proteins.

the DNA molecule is twisted into a double helix

Figure 3.5 DNA structure

Test yourself

1 Name:

   a) the enzyme that catalyses the condensation of two nucleotides

   b) the name of the product(s) of this condensation reaction

   c) the bond formed between the two nucleotides.

2 How does deoxyribose differ from ribose?

3 Other than the nature of their pentose, give **three** ways in which the structure of RNA is different from that of DNA.

4 Figure 3.3 shows that nucleotides are added only to one end of a growing polynucleotide. Suggest why they can only be added to one end.

5 How does the function of tRNA differ from that of mRNA?

DNA replication – how DNA copies itself

Every time a cell divides, a copy of its DNA passes to each 'daughter' cell formed by the division. This can happen because, prior to dividing, the cell has made accurate copies of each of its DNA molecules. We call this copying process **DNA replication**. As you will see in Chapter 5, in eukaryotic cells, DNA replication takes place in the interphase nucleus, well before the events of nuclear division.

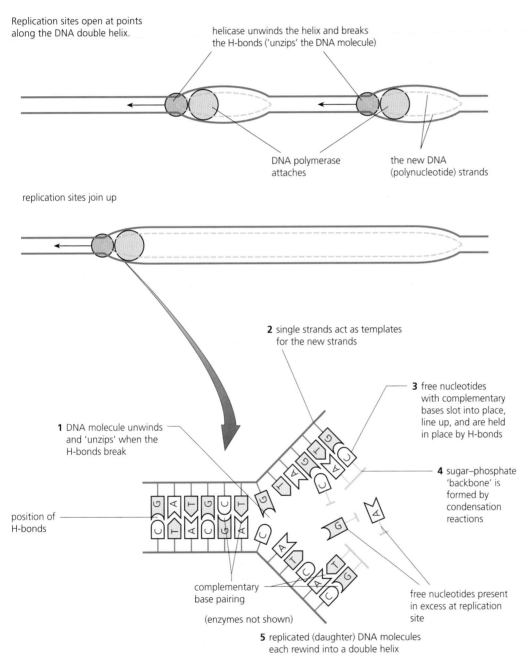

Replication sites open at points along the DNA double helix.

helicase unwinds the helix and breaks the H-bonds ('unzips' the DNA molecule)

DNA polymerase attaches

the new DNA (polynucleotide) strands

replication sites join up

2 single strands act as templates for the new strands

3 free nucleotides with complementary bases slot into place, line up, and are held in place by H-bonds

4 sugar–phosphate 'backbone' is formed by condensation reactions

1 DNA molecule unwinds and 'unzips' when the H-bonds break

position of H-bonds

complementary base pairing

(enzymes not shown)

free nucleotides present in excess at replication site

5 replicated (daughter) DNA molecules each rewind into a double helix

Figure 3.6 DNA replication

The lower diagram in Figure 3.6 provides a simple summary of the process of DNA replication. The steps are outlined in more detail on the next page.

Step 1: the double helix of a DNA molecule unwinds and the hydrogen bonds holding the strands together break. As a result, the bases on both strands become exposed.

Step 2: each of the separated strands of DNA acts as a template for the production of a new polynucleotide strand with a complementary sequence of nucleotide bases.

Step 3: free nucleotides, produced by the cell earlier, are attracted to their complementary exposed bases on each template strand and become held in place by hydrogen bonds.

Step 4: condensation reactions link the new nucleotides together to form the sugar-phosphate backbone of each new strand.

Once completed, each daughter DNA molecule rewinds into a double helix. Since one strand of each new double helix was present in the original DNA molecule and the other is a newly synthesised strand, this process is known as semi-conservative replication.

Of course, these reactions are catalysed by enzymes. The breakage of the hydrogen bonds and the unwinding of the DNA double helix (Step 1) is catalysed by the enzyme **DNA helicase**. The linking of nucleotides in the developing strands (Step 4) is catalysed by the enzyme **DNA polymerase**.

The upper diagrams in Figure 3.6 add a little more detail to this process. Notice that the points in the DNA molecule at which DNA helicase opens the double helix are called **replication sites**. You can see that more than one replication site occurs in each DNA molecule. This means that new strands of DNA are produced in segments that are then joined together. The joining of these segments is catalysed by another enzyme – **DNA ligase**.

DNA polymerase also has a role in 'proof reading' the new strands. Any 'mistakes' that start to happen (for example, the wrong bases pairing up) are corrected. As a result, each new DNA double helix is an exact copy of the original.

> **Key term**
>
> **Semi-conservative replication** The process by which two copies of a DNA molecule are made and in which both 'parent' strands remain intact and act as templates for the formation of new, complementary, strands.

Activity

The evidence for DNA replication

The structure of DNA that you have learnt about above was first proposed by James Watson and Francis Crick in 1953. At the time, they also postulated that DNA replication would be semi-conservative. It was not until 5 years later that Matthew Meselson and Franklin Stahl devised an experiment that would test this hypothesis.

Meselson and Stahl used cultures of a bacterium commonly found in human intestines, called *Escherichia coli*. They planned to allow cells in these cultures to grow and divide and then to extract DNA from these new cells. They knew that the *E. coli* cells in a culture all divide at the same time – every 60 minutes.

1 Suggest why it was important that the cells in a culture of *E. coli* divided at the same time.

The way in which Meselson and Stahl intended to analyse the DNA they collected involved the two isotopes of nitrogen. We came across isotopes in Chapter 1, where we saw that all atoms in an element have the same number of protons and electrons but might contain different numbers of neutrons. The most commonly occurring nitrogen atoms have 14 neutrons in their nuclei but some nitrogen atoms have 15. The atoms with 15 neutrons are called 'heavy' nitrogen and represented as ^{15}N, while the more common atoms with 14 neutrons are called 'light' nitrogen and re-presented as ^{14}N.

2 Which part of DNA nucleotides contain nitrogen?

Since DNA bases contain nitrogen, Meselson and Stahl intended to follow DNA replication by labelling these bases.

They cultured *E. coli* in a medium (food source) where the available nitrogen contained only the 'heavy' nitrogen isotope, ^{15}N. They continued to do this for long enough that they could be sure that all the DNA of the bacteria was entirely 'heavy'.

Now came the clever part. They transferred these bacteria to a medium containing the normal '(light)' isotope, ^{14}N, and allowed them to grow for 60 minutes.

3 What would you expect to happen during those 60 minutes?

4 What properties would you expect the new DNA strands produced by the *E. coli* to possess if replication is semi-conservative?

5 What properties would you expect the new DNA strands produced by *E. coli* to possess if the original DNA had remained intact and a completely new copy had been made from it?

Clearly, Meselson and Stahl had worked out in advance how they could detect heavy and light DNA – the differences in mass are too small to use a top-pan balance. They relied on the fact that the two DNA strands – one containing ^{14}N and the other containing ^{15}N – would have different densities. They extracted and purified DNA from bacteria that had only been grown in medium containing ^{15}N and DNA from bacteria that had been allowed to grow for 60 minutes in medium containing ^{14}N. They then placed samples of each type of DNA in separate centrifugation tubes containing a solution of a salt whose density increased from the top of the tube to the bottom of the tube. They then centrifuged the tubes.

6 What would you expect to find after centrifugation if DNA replication is semi-conservative?

Figure 3.7 shows that this is exactly what Meselson and Stahl found. It also shows the results Meselson and Stahl found when they allowed the transferred *E. coli* to divide for a second time in the medium containing ^{14}N.

1 Meselson and Stahl 'labelled' nucleic acid (i.e. DNA) of the bacterium *Escherichia coli* with 'heavy' nitrogen (^{15}**N**), by culturing in a medium where the only nitrogen available was as ^{15}**NH**$_4^+$ ions, for several generations of bacteria.

2 When DNA from labelled cells was extracted and centrifuged in a density gradient (of different salt solutions) all the DNA was found to be 'heavy'.

3 In contrast, the DNA extracted from cells of the original culture (before treatment with ^{15}**N**) was 'light'.

4 Then a labelled culture of *E.coli* was switched back to a medium providing unlabelled nitrogen only, i.e. ^{14}**NH**$_4^+$. Division in the cells was synchronised, and:
- after **one generation** all the DNA was of intermediate density (each of the daughter cells contained (i.e. *conserved*) one of the parental DNA strands containing ^{15}**N** alongside a newly synthesised strand containing DNA made from ^{14}**N**)
- after **two generations** 50% of the DNA was intermediate and 50% was 'light'. This too agreed with semi-conservative DNA replication, given that labelled DNA was present in only half the cells (one strand per cell).

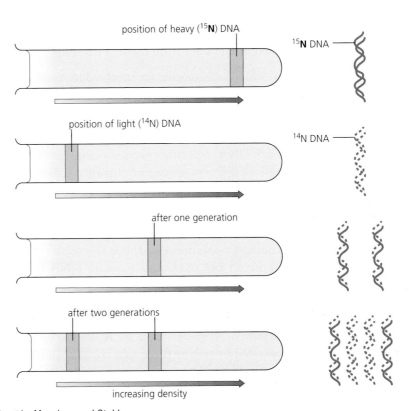

position of heavy (^{15}N) DNA

^{15}N DNA

position of light (^{14}N) DNA

^{14}N DNA

after one generation

after two generations

increasing density

Figure 3.7 A summary of the experiment carried out by Meselson and Stahl

6 Explain why DNA replication is described as semi-conservative.

7 What is the importance of base pairing during DNA replication?

8 What experimental results would you expect after one generation in the Meselson–Stahl experiment in Figure 3.7 if DNA replication had been conservative?

9 What experimental results would you expect if the Meselson–Stahl experiment in Figure 3.7 had continued for three generations?

10 What causes DNA molecules to unwind during replication?

DNA and protein synthesis – the genetic code

Key term

Gene A sequence of DNA nucleotide bases that encodes the sequence of amino acids in a functional polypeptide.

The major role of DNA is to enable a cell to make specific proteins. The huge length of a single DNA molecule codes for a very large number of proteins. Within this extremely long molecule, the relatively short length of DNA that codes for the sequence of amino acids in a single polypeptide chain is called a gene. Proteins are very variable in size and, consequently, so are genes. A very few genes are as short as 75–100 nucleotides long. Most are at least 1000 nucleotides in length, and some are more.

Most proteins contain several hundred amino acids condensed together in a linear series. There are only 20 or so amino acids that are used in protein synthesis; all cell proteins are built from them. The unique properties of each protein lie in:

● which amino acids are involved in its construction
● the sequence in which these amino acids are joined.

The genetic code

Key terms

Genetic code A combination of three nucleotide base triplets encodes an individual amino acid. Each combination of base triplets encodes the same amino acid in all organisms, i.e. this code is universal.

Degenerate code – the genetic code is said to be degenerate because some amino acids are encoded by more than one base triplet.

Each DNA molecule encodes a large number of proteins. The DNA molecules in the cells of different species of organism will have different nucleotide base sequences, encoding proteins that are unique to each species. Despite these differences, the basis of the coding is common to all organisms.

This basis of coding – the genetic code – is a sequence of three nucleotide bases coding for an amino acid. It is this triplet code that is universal – the same combination of three DNA nucleotide bases (or **DNA base triplet**) codes for the same amino acid in all organisms. With four bases (C, G, A, T) there are 64 possible different triplet combinations (4 × 4 × 4). As we have already seen, only 20 amino acids are commonly used by cells. In other words, the genetic code has many more different DNA base triplet combinations than are needed to encode 20 amino acids. Many amino acids are encoded by two or three base triplets. To reflect this, we say that the genetic code is degenerate. Also, some of the DNA base triplets represent the 'punctuations' of the code – for example, there are 'start' and 'stop' triplets.

You can see the genetic code in Figure 3.8. The table explains the abbreviations used to represent the 20 amino acids commonly present in proteins. The circle is one way of showing the nucleotide base triplets encoding each amino acid. Notice that this code uses RNA bases, rather than a DNA bases – you can see it uses uracil (U) rather than thymine (T). You might wonder why this is. As we will shortly see, the DNA code is translated into messenger RNA that is used by the ribosomes to make proteins; a ribosome 'reads' RNA bases. Figure 3.8 also introduces a new term – **codon**. We use this term to describe a nucleotide base triplet on a molecule of mRNA.

Key term

Codon A nucleotide base triplet on messenger RNA that encodes a single amino acid.

The 20 amino acids used in protein synthesis

Amino acids	Abreviations
alanine	Ala
arginine	Arg
asparagine	Asn
aspartic acid	Asp
cysteine	Cys
glutamine	Gln
glutamic acid	Glu
glycine	Gly
histidine	His
isoleucine	Ile
leucine	Leu
lysine	Lys
methionine	Met
phenylalanine	Phe
proline	Pro
serine	Ser
threonine	Thr
tryptophan	Trp
tyrosine	Tyr
valine	Val

The genetic code in circular form

The codons are messenger RNA base triplets (where uracil, U, replaces thymine, T)

Read the code from the centre of the circle outwards along a radius. For example, serine is coded by UCU, UCC, UCA or UCG, or by AGU or AGC.

In addition, some codons stand for 'stop', signalling the end of a peptide or protein chain.

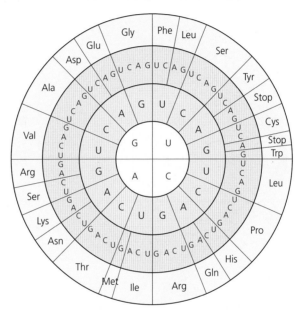

Figure 3.8 The genetic code – a universal code

Test yourself

11 Define the term *gene*.

12 Explain why different genes have different lengths.

13 Explain why the following statement is *not* true.

   'The nucleus contains the cell's genetic code.'

14 What is a codon?

15 The sequence of bases in a sample of mRNA was found to be:

   GGU, AAU, CCU, UUU, GUU, ACU, CAU, UGU

   a) Use Figure 3.8 to give the sequence of amino acids this codes for.

   b) Write out the sequence of bases in the antisense strand of DNA from which this mRNA was transcribed.

The process of protein synthesis

Key terms

Antisense strand The polynucleotide chain in a DNA molecule that is always used in protein synthesis to determine the order of amino acids in a polypeptide, i.e., it is the strand that is transcribed.

Non-overlapping The property of the genetic code in which each nucleotide base forms part of only one base triplet.

Before looking at protein synthesis in detail, look at Figure 3.9. It shows a summary of how a gene controls the production of a polypeptide. Since a gene is extremely long, the diagram represents only part of a gene. On the left-hand side, you can see the two DNA strands held together by hydrogen bonds between complementary base pairs. You can also see that the base sequence is different, depending which of the two polynucleotide chains you look at. Only one of these strands is ever used in protein production. We call it the **antisense strand**. Now look at the right-hand side of Figure 3.9; it shows the DNA triplet code for three amino acids. The first triplet is AGA and codes for the amino acid serine. Notice, though, that once AGA has been used the next triplet is CTG. This illustrates another important principle of the genetic code – it is **non-overlapping**. In other words, each base is part of only one triplet code. This means that the DNA base sequence in Figure 3.9 is read, from top to bottom AGA, CTG and TTC and not AGA, GAC, ACT, etc.

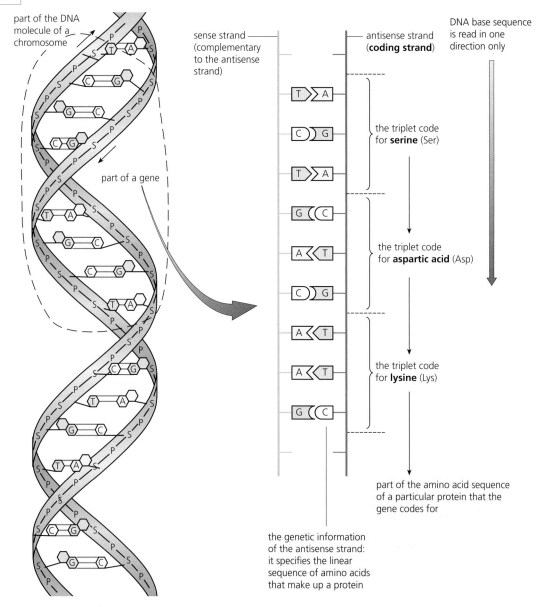

Figure 3.9 Part of a gene and how its DNA codes for amino acids

To aid understanding, we will consider protein synthesis in three stages.

Stage 1 – transcription

Figure 3.10 shows how this stage occurs in the nucleus of a eukaryotic cell and results in a gene being copied into the base sequence of **messenger RNA (mRNA)**, which then leaves the nucleus. This process is called transcription and is controlled by three groups of enzymes. The following events occur at the point where the gene is to be copied.

- The DNA double helix unwinds, and the hydrogen bonds holding the two strands together break. Just as we saw in DNA replication, DNA helicase catalyses this reaction.
- One of the separated strands of DNA, the antisense strand, acts as a template for the formation of mRNA.

Key term

Transcription The process by which the DNA nucleotide base sequence of a gene is copied into the RNA nucleotide base sequence in a molecule of messenger RNA (mRNA).

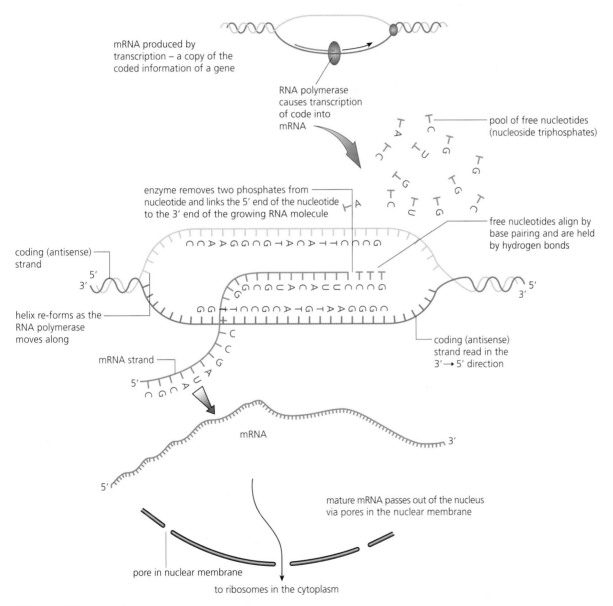

Figure 3.10 Transcription

- Free RNA nucleotides present in the nucleus pair up with the exposed nucleotides on the antisense strand. (Notice in Figure 3.10 that these free RNA nucleotides are referred to as nucleoside triphosphates. These molecules are like nucleotides but have three phosphate groups rather than just the one shown in Figure 3.1. Do not be confused by this, the presence of additional phosphates simply makes the molecules more reactive.).
- Complementary base pairing ensures that cytosine always pairs with guanine and uracil always pairs with adenine.
- RNA polymerase catalyses the formation of phosphodiester bonds between the RNA nucleotides, forming a molecule of messenger RNA.

Once the mRNA molecule is formed, it leaves the nucleus through pores in the nuclear membrane (Figure 3.10) and passes to tiny structures in the cytoplasm called **ribosomes** where the information can be 'read' and is used. Once the cell has finished transcribing this gene, this part of the DNA molecule rewinds.

Stage 2 – activation of amino acids

In this stage, the amino acids are activated for protein synthesis by combining with short lengths of a different sort of RNA, called **transfer RNA (tRNA)**. This activation occurs in the cytoplasm.

All molecules of tRNA have the shape of a clover-leaf, but there is a different tRNA for each of the 20 amino acids involved in protein synthesis. At one end of each tRNA molecule is a site where a particular amino acid can be joined (Figure 3.11). At the other end, there is a sequence of three bases called an **anticodon**. This anticodon is complementary to the codon of mRNA that codes for the specific amino acid.

The amino acid is attached to its tRNA by an enzyme. These enzymes are specific to the particular amino acids (and types of tRNA) to be used in protein synthesis. The specificity of the enzymes is a way of ensuring the correct amino acids are used in the right sequence.

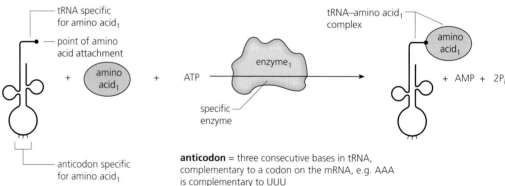

Each amino acid is linked to a specific transfer RNA (tRNA) before it can be used in protein synthesis. This is the process of amino acid activation. It takes place in the cytoplasm.

anticodon = three consecutive bases in tRNA, complementary to a codon on the mRNA, e.g. AAA is complementary to UUU

Figure 3.11 Amino acid activation

Stage 3 – translation

In this stage, a protein chain is assembled, one amino acid residue at a time (Figure 3.12). Tiny organelles called ribosomes move to the messenger RNA and move along it, 'reading' the codons from a 'start' codon. As we saw earlier in this chapter, ribosomes themselves contain RNA. In the ribosome, complementary anticodons on the amino acid–tRNAs slot into place and are temporarily held in position by hydrogen bonds. While held there, the amino acids of neighbouring amino

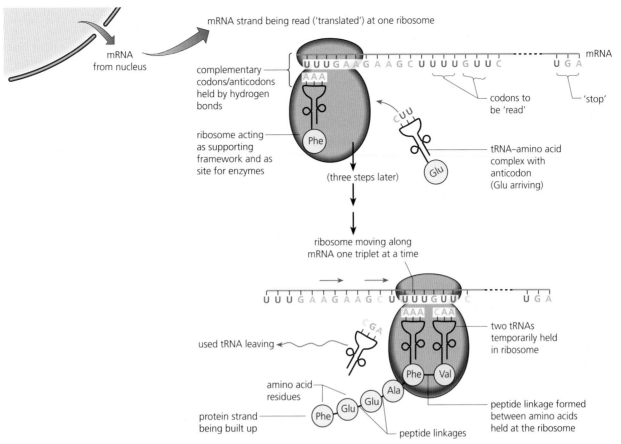

Figure 3.12 Translation

acid–tRNAs are joined by peptide bonds. This frees the first tRNA, which moves back into the cytoplasm for re-use. Once this is done, the ribosome moves on to the next mRNA codon. The process continues until a 'stop' codon occurs.

Not all DNA codes for protein: mRNA editing

The DNA of eukaryotic cells (those with a nucleus – see Chapter 4) contains many non-coding sections of DNA, called **introns**. As Figure 3.12 shows, these introns lie between coding sections of DNA, called **exons**.

The mRNA first produced during transcription includes RNA copies of the introns. Before leaving the nucleus, this **pre-mRNA** is edited to remove these introns. Figure 3.13 shows how this is done.

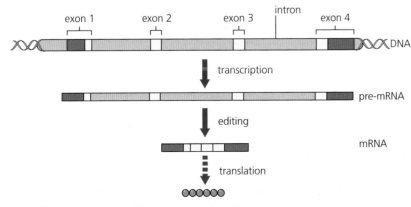

Figure 3.13 Editing of pre-mRNA in eukaryotic cells

Key terms

Introns DNA base sequences within a gene that do not code for the amino acid sequence of a polypeptide. Although copied to RNA during DNA transcription, these introns are edited out of the mRNA before it leaves the nucleus.

Exons DNA base sequences within a gene that code for the amino acid sequence of a polypeptide.

During splicing the mRNA copies of the exons can be assembled in different orders. In this way, it is possible for a single gene to give rise to mRNA molecules with different nucleotide base sequences. This enables such a gene to code for more than one polypeptide and explains, for example, how we are able to manufacture a vast number of antibody molecules from a small number of genes.

Since the DNA of prokaryotic cells (cells without a nucleus – see Chapter 4) does not contain introns, this mRNA-editing process does not occur in these cells.

DNA can change: gene mutations

We have seen that a gene is a sequence of nucleotide bases that codes for the sequence of amino acids in a polypeptide. Normally, the sequence of nucleotides in DNA is maintained without changing but, very occasionally, it does change.

A gene mutation involves a change in the number, or sequence, of bases in a particular gene. We have already noted that the enzyme machinery that brings about the building of a complementary DNA strand also 'proof reads' and corrects most errors. However, gene mutations can and do occur spontaneously during this step. Certain chemicals can also cause change to the DNA sequence of bases. So do some forms of radiation, such as X-rays. Factors that increase the chances of a mutation are called **mutagens**.

More than one type of gene mutation is possible. They include:

- **base deletion** – one or more bases lost from the sequence
- **base insertion** – one or more bases added to the sequence
- **base substitution** – one or more bases changed for a different base.

If a gene mutation involves only one base, it is described as a point mutation. Table 3.1 shows the effect of point mutations using English words, rather than DNA base triplets. Notice how in some cases a nonsense message is produce but in other cases a message with a new meaning is produced. With gene mutations, this could result in a non-functional polypeptide or a polypeptide with a different function.

Key terms

Gene mutation
A random and unpredictable change in the number, or sequence, of bases in a single gene.

Point mutation A gene mutation involving deletion, insertion or substitution of a single base.

Table 3.1 Changes in the sense of sentences using three-letter English words to represent the effect of point mutations on the code carried by DNA base triplets

Type of mutation	Effect on 'triplet code'
Normal code (no mutation)	Did you get the car
Deletion (base lost)	Did yog ett hec ar
Insertion (base added)	Did you age tth eca r
Substitution (base changed)	Did you wet the car

Sickle cell anaemia: an example of a point mutation

Sickle cell anaemia is a condition that is common among people originating from areas where malaria is endemic. It results from a point mutation in the gene that codes for the amino acid sequence of a part of the respiratory pigment haemoglobin, found in our red cells (its structure was shown in Figure 2.7). In this case, the point mutation is a base substitution – adenine replaces thymine in one base triplet – causing valine, instead of glutamic acid, to be incorporated into the polypeptide chain (Figure 3.14). The resulting, abnormal haemoglobin tends to clump together and form long fibres that distort the red cells into sickle shapes. In this condition they cannot transport oxygen efficiently and the cells may block smaller capillaries.

Anaemia is a disease typically due to a deficiency in healthy red cells in the blood.

Haemoglobin occurs in red cells – each contains about 280 million molecules of haemoglobin. A molecule consists of two α-haemoglobin and two β-haemoglobin subunits, interlocked to form a compact molecule.

The **mutation** that produces sickle cell haemoglobin (**Hg^S**) is in the gene for β-haemoglobin. It results from the substitution of a single base in the sequence of bases that make up all the codons for β-haemoglobin.

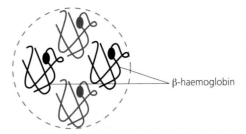
β-haemoglobin

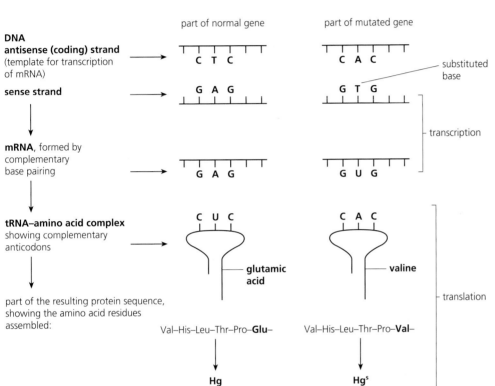

**DNA
antisense (coding) strand**
(template for transcription of mRNA)

sense strand

mRNA, formed by complementary base pairing

tRNA–amino acid complex
showing complementary anticodons

part of the resulting protein sequence, showing the amino acid residues assembled:

part of normal gene

part of mutated gene

C T C

G A G

G A G

C U C

glutamic acid

Val–His–Leu–Thr–Pro–**Glu**–

Hg

C A C

G T G

G U G

C A C

valine

Val–His–Leu–Thr–Pro–**Val**–

Hg^s

substituted base

transcription

translation

Test yourself

16 The genetic code is described as degenerate and non-overlapping. Explain what this means.

17 Distinguish between the terms *transcription* and *translation*.

18 During protein production, tRNA molecules carry amino acids to a ribosome. What ensures that the tRNA molecules are used in the correct order?

19 Explain why the RNA produced during transcription is modified before leaving a cell's nucleus.

20 What is a point mutation?

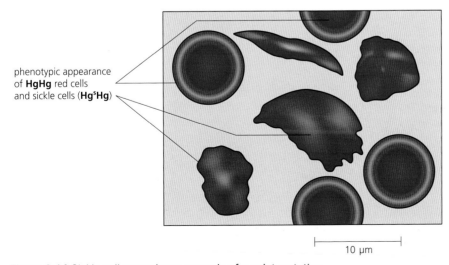

drawing based on a photomicrograph of a blood smear, showing blood of a patient with sickle cells present among healthy red cells

phenotypic appearance of **HgHg** red cells and sickle cells (**Hg^sHg**)

10 μm

Figure 3.14 Sickle cell anaemia: an example of a point mutation

Exam practice questions

1 Which of the following statements is true of base pairing in a molecule of DNA?

   A Adenine always pairs with cytosine

   B Adenine always pairs with guanine

   C Adenine always pairs with thymine

   D Adenine always pairs with uracil *(1)*

2 Which of the following statements is true of human cells?

   A The number of bases in a gene is the same as the number of amino acids in the polypeptide it codes for.

   B The number of bases in a gene bears no direct relation to the number of amino acids in the polypeptide it codes for.

   C The number of bases in a gene is three times the number of amino acids in the polypeptide it codes for.

   D The number of bases in a gene is 64 times the number of amino acids in the polypeptide it codes for. *(1)*

3 DNA is described as a stable, information-carrying molecule.

   a) What makes a DNA molecule stable? *(2)*

   b) How does a DNA molecule carry 'information'? *(2)*

4 The diagram represents part of a molecule of DNA during replication. The letter C on the diagram represents the organic base cytosine and the letter T represents the organic base thymine.

   a) The letters L, M and N represent unknown organic bases. Use information in the diagram to name each. *(3)*

   b) Explain what caused the parent molecule to split. *(3)*

5 The DNA in a bacterial cell is held in a single, circular molecule of DNA. A scientist analysed the bases in the DNA of a bacterial cell. The table shows her results.

DNA strand	Percentage of each base in each DNA strand			
	A	C	G	T
1	22			
2	15		33	

   a) Give two ways in which the DNA of a bacterial cell is different from that of a human cell. *(2)*

   b) Explain how you could find the missing values to complete the table. *(3)*

6 The diagram shows a DNA trinucleotide.

    a) Explain how you can tell that this trinucleotide
is from a DNA molecule. *(2)*

    b) Draw a circle around one nucleotide that contains a purine. *(2)*

    c) To which end of the molecule would a fourth nucleotide be
added if this were a growing
DNA strand? Explain your answer. *(3)*

7 Describe the roles of different types of RNA molecule during the
process of translation. *(5)*

Stretch and challenge

8 The diagram represents part of a DNA molecule during DNA
replication.

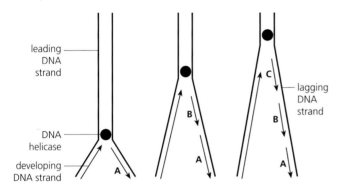

Use your knowledge of enzyme action to explain:

    a) the different patterns of replication shown in the two strands of the
DNA molecule

    b) how the fragments A, B and C become a single strand again.

9 The process of replication you have learnt about in this chapter is
catalysed by an enzyme often referred to as DNA-dependent DNA
polymerase. Carry out research to find why this name is used and how
the action of this enzyme differs from that of RNA-dependent DNA
polymerase.

Cell structure and viruses

Prior knowledge

In this chapter you will need to recall that:

→ there are two types of living cell – eukaryotic and prokaryotic
→ a eukaryotic cell has a cell surface membrane surrounding cytoplasm in which there is a nucleus containing the genetic material. The nucleus is separated from the cytoplasm by a membrane, called the nuclear envelope
→ some eukaryotic cells have a cell wall outside their cell surface membrane
→ the cytoplasm of a eukaryotic cell contains organelles. Some of these are surrounded by membrane, including mitochondria and chloroplasts. The aerobic stages of respiration take place in mitochondria; photosynthesis occurs in chloroplasts
→ a prokaryotic cell has a cell wall and a cell surface membrane surrounding its cytoplasm. It does not have a nucleus; instead its genetic material is a single, circular molecule of DNA
→ some prokaryotic cells also have smaller, circular DNA molecules, called plasmids
→ a virus is not a living organism. It is a particle that contains genetic material surrounded by a protein coat. It infects other cells and depends on these cells to produce more virus particles
→ electron microscopy has increased our understanding of sub-cellular structures.

Test yourself on prior knowledge

1 Name **two** groups of organisms that have a cell wall.
2 Name the structures that contain the genetic material in a eukaryotic cell.
3 A eukaryotic cell is surrounded by a cell surface membrane and many of its organelles are surrounded by membranes. Do these membranes have the same structure?
4 Plants have eukaryotic cells; bacteria have prokaryotic cells. Give **two** ways in which the structure of a plant cell and a bacterial cell are:
  a) similar
  b) different.
5 What is the function of a mitochondrion?
6 Some cells have a cell wall. Give **one** advantage of possessing a cell wall.

Introducing cells

Key term

Cell theory Cells are the fundamental unit of structure, function and organisation in all living organisms.

In the last chapter, we came across a unifying theory in biology, which stated that the genetic code – the base triplets encoding each amino acid – is the same in all organisms. Here we come across a second: the **cell theory**. Put simply, the cell theory states that:

● cells are the smallest unit of living organisms
● all cells are derived from the division of other (pre-existing) cells
● within cells are the sites of all the chemical reactions of life (metabolism).

Some organisms are made of a single cell; they are called **unicellular** organisms. Figure 4.1 shows three different unicellular organisms. You can see that the three look quite unalike. Among the differences are their size, and the possession or absence of a nucleus and a cell wall. One feature they all have in common is a plasma membrane surrounding their cytoplasm. We will examine these similarities and differences later in this chapter.

Key term

Unicellular Composed of a single cell.

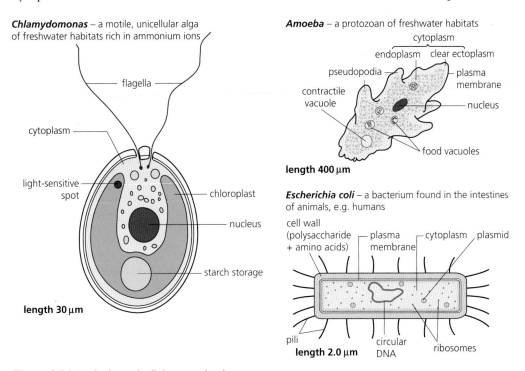

Figure 4.1 Introducing unicellular organisation

Other organisms are made of many cells, and are known as **multicellular** organisms. Much of the biology in this book is about multicellular organisms, including humans, and the processes that go on in these organisms. But remember, unicellular organisms carry out all the essential functions of life too, all within the confines of a single cell.

A feature of multicellular organisms is that, after formation, their cells develop differently; they become specialised for the functions they carry out. We call this process **differentiation**. A common outcome of this is that many fully specialised cells are no longer able to divide. But as a consequence of specialisation, cells show great variety in shape and structure, as we will see.

The cells of a multicellular organism are not arranged at random. Instead, cells with a common origin, that have differentiated to perform a particular function, group together as a **tissue**. Blood is an example of an animal tissue; you will look at blood in more detail in Chapter 11. Xylem is an example of a plant tissue; you will look at xylem in more detail in Chapter 12.

Sometimes, many tissues work together to perform a particular function. For example the heart, which pumps the blood around the body of a mammal (Chapter 11), contains epithelial tissue, muscle tissue, connective tissue and nervous tissue. A structure that performs a particular function but is made of more than one tissue is called an **organ**. As you know, the heart, together with blood vessels and blood, forms the circulatory **system**. This general pattern in which complex, multicellular organisms are organised is shown in Figure 4.2.

Key terms

Multicellular Composed of many cells. Usually the cells differentiate and become arranged into tissues, organs and systems, carrying out different functions.

Differentiation The developmental process by which the structure of the cells of a multicellular organism specialise, becoming adapted for a specific function.

Tissue A group of cells that have a common origin and a similar structure that work together to perform a single function.

Organ A structure, made of more than one type of tissue, that has a specific function.

System A group of organs and tissues that, collectively, perform a particular function.

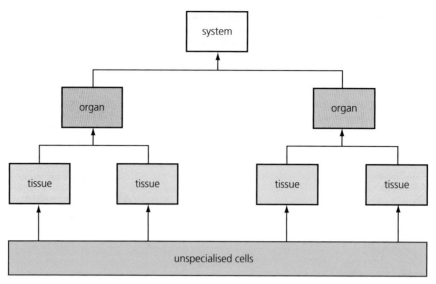

Figure 4.2 The specialised cells of complex multicellular organisms are arranged into tissues, organs and systems

Studying cells

Before going any further, we need to consider how we can study cells. Although a few are just large enough to be seen with the naked eye, cells are extremely small. To study their structure, we need to magnify them. In your college or school laboratory, you will use a compound light microscope. It is called a light microscope because it uses light to view an object. It is called a compound microscope because each of its 'lenses' contains more than one glass lens. Figure 4.3 shows an example of a compound light microscope. The ones in your college or school laboratory might be different from this, but they will have similar features.

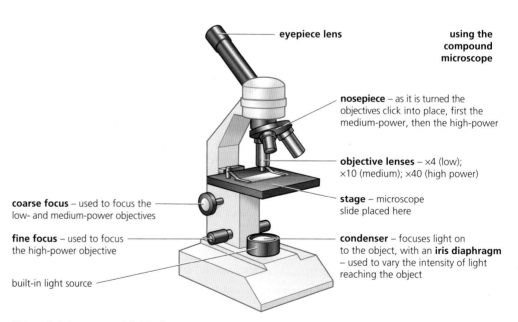

Figure 4.3 A compound light microscope

First, look at the three sets of lenses. One set focuses light before it hits the specimen to be viewed. It is called the **condenser lens** and you can see it near the base of the microscope. The microscope in Figure 4.3 has a built-in light within the condenser. Yours might be the same or you might have a concave mirror there instead. If so, you will need to use light from a window or from a bench lamp.

The other two sets of lenses focus light after it has passed through the specimen to be viewed. The one you will look down is called the **eyepiece lens**. Lower down the microscope is a second set of lenses, called the **objective lenses**. In Figure 4.3, you can see three objectives lenses housed on a nosepiece that rotates, allowing you to engage different objective lenses with different magnifications. Your microscope might have a different number of lenses but it is likely to have at least two: one that magnifies less (the low-power objective lens) and one that magnifies more (the high-power objective lens).

Now look at the side of the microscope in Figure 4.3. You can see the two knurled screws that are used to focus the lenses. Depending on the manufacturer's make of microscope, when you turn these you either move the stage or the objective lenses up and down. The upper one allows a coarse focus; if you turn this, the stage or the objective lens moves a greater distance than if you turn the lower, fine-focus screw.

Finally, there is the stage of the microscope. You will place the specimen to be viewed here and use the two clips to hold it in place. Because light passes through the specimens you view, they must be very thin. You achieve this either by cutting very thin slices of tissue or by squashing tissue. The surface you use to support these thin preparations must also allow light to pass through, so you use glass slides. (Remember to wear eye protection when using glass slides and coverslips.) There are two types of preparation you can view using a light microscope:

- A **temporary preparation**. This involves placing tissue on a glass slide, covering it with a water-based liquid to prevent it drying out, and putting an extremely thin glass **coverslip** over it.
- A **permanent preparation**. In this type of preparation, the water has been removed from the tissue and been replaced by a firmer substance. The coverslip is held in place by a resin.

Cells are usually translucent. To help you to see their structures, chemicals are added that react with cell components. They are called **stains** and, since they colour parts of the cell, staining is a key process to help you identify cell structures.

Figure 4.4 shows two cells viewed using a compound light microscope. One is a plant cell, the other a human cell. For reasons we will examine shortly, you can see very little cell detail. Cell walls and chloroplasts are visible in the plant cells. A nucleus and granules are visible in the human cell. You cannot see a membrane surrounding either cell, though we assume it must be there, and you cannot see much within the cytoplasm. Before examining why so little detail is visible, let's look at one of the core practicals you must carry out.

Test yourself

1 In biology, differentiation has a unique meaning. Explain what it means.
2 Is your stomach a tissue, an organ or a system? Explain your answer.
3 What is the function of the condenser lens in a compound microscope?
4 Why do biologists stain tissue to be viewed using a microscope?
5 How do you avoid trapping air bubbles when you are making a temporary mount of tissue?

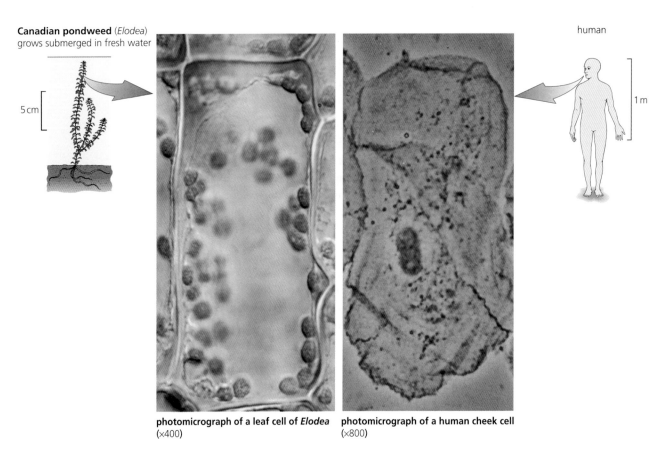

Canadian pondweed (*Elodea*) grows submerged in fresh water

5 cm

human

1 m

photomicrograph of a leaf cell of *Elodea* (×400)

photomicrograph of a human cheek cell (×800)

Figure 4.4 Animal and plant cells

Core practical 2

Use of the light microscope, including simple stage and eyepiece micrometers and drawing small numbers of cells from a specialised tissue

This core practical involves three skills: using a compound light microscope safely, measuring the actual size of cells observed using a compound light microscope and drawing a small number of these cells. Let's deal with them in that order.

Using the microscope safely

Although it looks robust, the microscope is very delicate. A slight knock might damage the alignment of the lenses within the eyepiece or objective lens arrangements. Consequently, you should treat the microscope with care.

1 If you are carrying a microscope from a storage area to your bench, use both hands – one supporting the base and the other holding the arm of the microscope.
2 Place the microscope on the bench so that its base is flat and it is far enough away from the edge of the bench to reduce the risk of it falling off. Adjust your seating so that you can comfortably adjust the focusing screws and look down the eyepiece.
3 Ensure the built-in lamp is set at its minimum setting before plugging in the power cable. Then adjust the lamp to about two-thirds of maximum setting.
4 Select a low-power objective lens by rotating the nosepiece. When the lens is correctly in place, you will hear a 'click'.
5 Look at a prepared slide to locate the specimen. Then put the slide onto the stage so that the coverslip is facing upwards and the specimen is located centrally below the objective lens.

Tip

The objective lenses of a compound microscope are very expensive. It is critical you do not push one into the specimen to be viewed when you are turning a focus screw. To avoid this, if you are looking down the eyepiece lens, only ever turn the screw to move an objective lens *away* from the specimen. If you need to move the objective lens towards the specimen, watch the bottom of the objective lens from the side of the microscope, so you can be certain it does not touch the specimen.

6 View the specimen by looking down the eyepiece lens. Focus the image of the specimen, first using the coarse focus control and then the fine focus control. To avoid eyestrain, try to keep both eyes open while looking down the eyepiece lens.

7 Notice that the image of the specimen is upside down and back-to-front compared with looking at the specimen directly. The same will be true when you move the slide around. If you push it to your left, the image will move to your right and vice versa.

8 Focus the condenser by placing a sharp object, such as a mounted needle, on the centre of the light source. Adjust the condenser lens until the specimen and the sharp object are in focus together. You are now ready to use the microscope.

Measuring the actual size of cells

Look at Figure 4.5, which summarises the method you used for measuring the size of cells. In this case, the specimen is a blood smear, but the principles are the same whatever specimen you used.

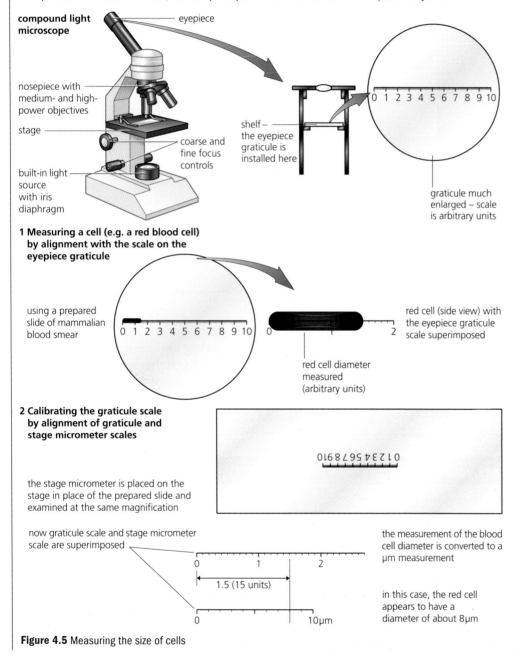

Figure 4.5 Measuring the size of cells

The first step involves inserting a graticule into the eyepiece lens.

1 Describe how you did this. Include in your description any precautions you took.
2 The graticule has a graduated line across it. How did you ensure this appeared horizontal when you placed the eyepiece back into the microscope and viewed cells?
3 Figure 4.5 shows a single red blood cell against the graticule. What can you deduce about the size of this cell?
4 How did you calibrate your graticule?
5 What is the actual diameter of the red blood cell in Figure 4.5?

Drawing small numbers of cells

view (phase contrast) of the layer of the cells (epithelium) lining the stomach wall

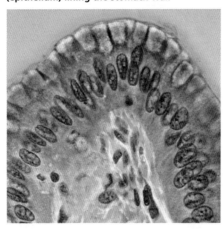

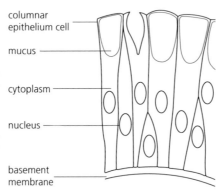

The lining of the stomach consists of columnar epithelium. All cells secrete mucus copiously.

columnar epithelium cell

mucus

cytoplasm

nucleus

basement membrane

Figure 4.6 Recording cell structure by drawing

6 How did your drawings to show the distribution of tissues differ from your drawings to show cell structure?
7 Describe how you ensured that your drawings of a small group of cells accurately represented what you saw using a light microscope.

Digital microscopy

These days, scientists rarely draw the cells and tissues they observe under a microscope. Instead they use a digital microscope or, alternatively, connect an appropriate camera using a microscope coupler or eyepiece adaptor that replaces the standard microscope eyepiece. Images can be displayed directly on a VDU monitor or saved to a computer hard drive, from which they can be retrieved and printed. This technique of **digital microscopy** is shown in Figure 4.7.

Test yourself

6 What is the function of a cell's nucleus?
7 Biologists use a basic stain when staining nuclei. Suggest why.
8 Within a multicellular organism cells show great variety in shape and structure. Explain why.
9 A cell is reported to be 0.000 38 m long. Express this measurement using a more suitable unit of length.
10 The cell theory is a unifying theory in biology. In your own words, give **three** of its component statements.

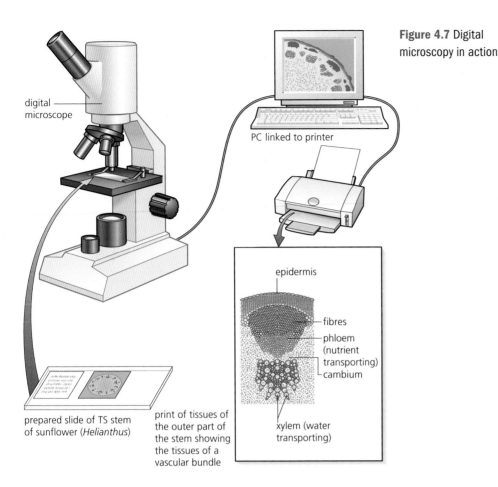

digital microscope

PC linked to printer

Figure 4.7 Digital microscopy in action

prepared slide of TS stem of sunflower (*Helianthus*)

print of tissues of the outer part of the stem showing the tissues of a vascular bundle

epidermis

fibres

phloem (nutrient transporting)

cambium

xylem (water transporting)

Magnification and resolution of an image

We can now return to our earlier observation regarding Figure 4.4, namely, that we could not see much detail of the cell structure. You might think we could overcome this by increasing the magnification, in other words the number of times larger an image is than the specimen. The magnification obtained with a compound microscope depends on which of the lenses you use. For example, using a ×10 eyepiece and a ×10 objective lens, the image is magnified 100 times (10 × 10). When you switch to the ×40 objective lens with the same eyepiece lens, the magnification becomes 400 times (10 × 40). These are the most likely orders of magnification you will use in your laboratory work.

Theoretically, there is no limit to magnification. For example, if a magnified image is photographed, further enlargement can be made photographically. This is what usually happens with photomicrographs shown in books and articles. We can find the magnification using the formula:

$$\text{magnification} = \frac{\text{size of image}}{\text{size of specimen}}$$

For example, suppose a plant cell with a diameter of $150\,\mu m$ is photographed with a microscope and its image enlarged photographically so that its diameter on the print is $150\,mm$ diameter ($150\,000\,\mu m$). The magnification is:

$$\frac{150\,000}{150} = 1000 \text{ times}$$

Key term

Magnification The extent to which an object has been enlarged by a microscope, in a drawing or in a photograph.

Tip

You need to be confident in converting units. It might help you remember how to convert units of length if you work in steps of 1000 (10^3). Thus, $1\,m = 10^3\,mm = 10^6\,\mu m = 10^9\,nm$. This series (milli, micro and nano) is also true for other units, such as units of mass and volume. Notice that if you follow this advice, you will not measure length in centimetres (cm).

If a further enlargement is made, to show the same cell at 300 mm diameter (300 000 μm), the magnification would be:

$$\frac{300\,000}{150} = 2000 \text{ times}$$

In this case, the image size has been doubled but the detail will be no greater. You will not be able to see, for example, details of cell membrane structure, however much the image is enlarged. This is because the layers making up a cell's membrane are too thin to be seen as separate structures using the light microscope.

Example

Using scale bars to determine actual size and magnification

**photomicrograph of Amoeba proteus (living specimen) –
phase contrast microscopy**

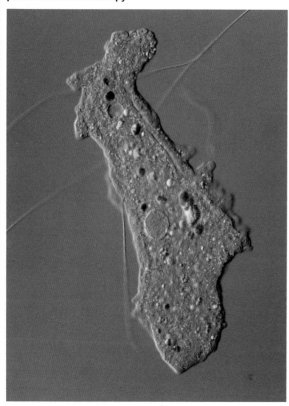

interpretive drawing

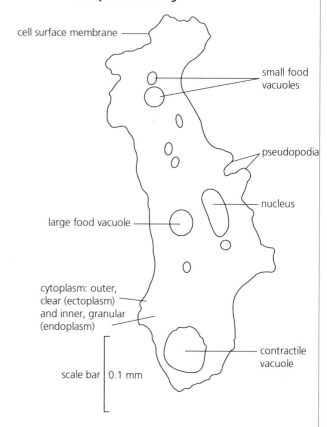

Figure 4.8 Using a scale bar to record size

Once the size of a cell has been measured, a scale bar line may be added to a micrograph or drawing to record the actual size of the structure. This has been done in the photomicrograph and drawing of a single-celled organism called *Amoeba proteus* in Figure 4.8.

1 What is the length of the scale bar in Figure 4.8?

2 What is the length of the drawing of the cell shown in Figure 4.8?

3 Use both your measurements to calculate the actual length of the cell in Figure 4.8.

4 Now calculate the magnification of the drawing in Figure 4.8, using the formula:

$$\text{magnification} = \frac{\text{size of image}}{\text{size of specimen}}$$

Answers

1 Using a rule with millimetre divisions, you should have measured the length of the scale bar as 20 mm.

2 You should have found that the cell is 100 mm long.

3 The actual length is $\frac{100}{20} \times 0.1 = 0.5\,\text{mm}$.

4 The magnification is $\frac{100}{0.5} = \times 200$.

Although we could, theoretically, increase magnification indefinitely, doing so would not show us more detail of cell structure. The problem is not magnification but the nature of light itself. Look around the room. How can you distinguish between objects within it? Apart from different colours or textures, you can see space between them. This is fine with large objects but not with tiny ones that are extremely close together. The wavelength of light is such that it cannot pass between these tiny objects. The ability to distinguish tiny objects that are extremely close together is termed resolving power, or resolution. If two separate objects cannot be resolved they will be seen as one object. Merely enlarging them will not separate them. Using a light microscope, the limit of resolution is about 0.2 µm. This means that two objects less than 0.2 µm apart will always be seen as one object however much we magnify them using a light microscope.

So, how can we see greater detail of cell structure? The answer is to use radiation with a shorter wavelength than light. Most commonly, we use electrons in an electron microscope. Because an electron beam has a much shorter wavelength than light rays, the resolving power of an electron microscope is much greater than the best light microscopes. Used with biological materials, the limit of resolution in transmission electron microscopy is about 5 nm. Look at Figure 4.9, which illustrates this point well. The detailed structure of the chloroplast can be seen using a transmission electron microscope a) but cannot be seen with a light microscope b).

Key term

Resolution The ability to distinguish between points that are very close together. The limit of resolution for a light microscope is about 2 µm (2000 nm) whereas that of a transmission electron microscope is about 5 nm.

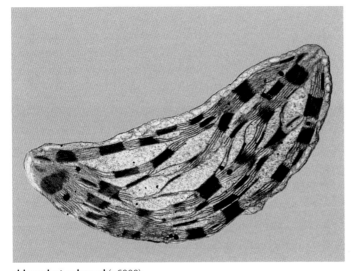

chloroplast enlarged (×6000)
a) from a transmission electron micrograph

b) from a photomicrograph obtained by light microscopy

Figure 4.9 Magnification without resolution

Electron microscopy – the discovery of cell ultrastructure

In an electron microscope, a beam of electrons is used to produce a magnified image in much the same way as the optical microscope uses light. The electron beam is generated by an electron gun and is focused using electromagnets, rather than glass lenses. Since we cannot see electrons, the electron beam is focused onto a fluorescent screen for viewing, or onto a photographic plate for permanent recording. You can see these features in Figure 4.10. Notice the outlet to a vacuum pump in the diagram. Electrons would be deflected by molecules in the air, so the large red column you can see in Figure 4.10 holds the specimen inside a vacuum.

With a **transmission electron microscope**, the electron beam is passed through an extremely thin section of material. Membranes and other structures present are stained with heavy metal ions, making them electron-opaque, so they stand out as dark areas in the image. You can see a technician using a transmission electron microscope in Figure 4.10.

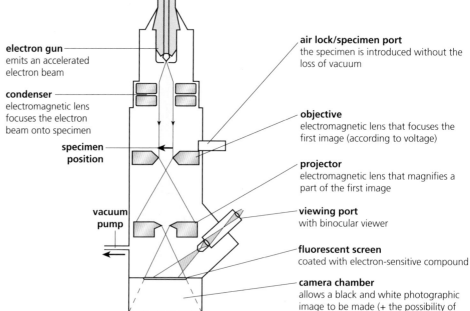

electron gun
emits an accelerated electron beam

condenser
electromagnetic lens focuses the electron beam onto specimen

specimen position

vacuum pump

air lock/specimen port
the specimen is introduced without the loss of vacuum

objective
electromagnetic lens that focuses the first image (according to voltage)

projector
electromagnetic lens that magnifies a part of the first image

viewing port
with binocular viewer

fluorescent screen
coated with electron-sensitive compound

camera chamber
allows a black and white photographic image to be made (+ the possibility of further magnification)

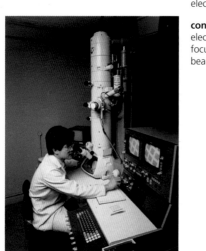

Figure 4.10 A transmission electron microscope

Key term

Cell ultrastructure The structure of cells seen using a microscope with a high degree of resolution, usually a transmission electron microscope.

Only transmission electron microscopes can resolve the fine detail of the contents of cells, the organelles and cell membranes, known as cell ultrastructure and shown in Figure 4.11. Note that the photograph of a specimen viewed using a transmission electron microscope is called a transmission electron micrograph (TEM).

Test yourself

11 A student is using a compound microscope to study plant tissue at an institution that does not have graticules or stage micrometers. Suggest how this student could use a ruler to estimate the length of a plant cell in the tissue he is viewing.

12 An *Amoeba proteus* is shown in Figure 4.8.
   a) What is the evidence that this organism is *not* a plant?
   b) In fact, *Amoeba proteus* is a eukaryotic unicell. Explain the meaning of 'eukaryotic' and 'unicell'.

13 Explain the advantage of using a transmission electron microscope rather than a compound light microscope to study cell structure.

TEM of liver cells (×15000)

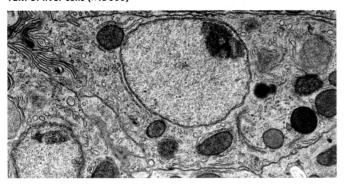

interpretive drawing

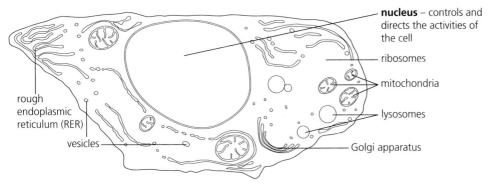

nucleus – controls and directs the activities of the cell

ribosomes

mitochondria

lysosomes

Golgi apparatus

rough endoplasmic reticulum (RER)

vesicles

Figure 4.11 Transmission electron micrograph (TEM) of mammalian liver cells with an interpretative drawing

With a scanning electron microscope, a narrow electron beam is scanned back and forth across the surface of the specimen. Electrons that are reflected or emitted from this surface are detected and converted into a three-dimensional image, such as the one in Figure 4.12.

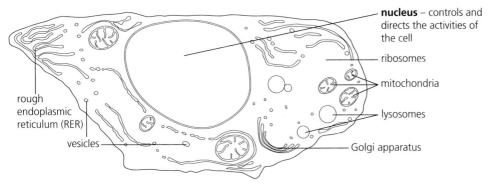

Figure 4.12 A scanning electron micrograph (SEM)

Tip

To answer Question 15 you need to convert micrometres (μm) to nanometres (nm).

Test yourself

14 Suggest **one** disadvantage of using a transmission electron microscope rather than a compound light microscope to study cells.

15 The resolving power of a light microscope is given as 2μm and that for a transmission electron microscope as 5nm. How many times greater is the resolving power of the electron microscope than the light microscope? Show your working.

The discovery of two types of cell organisation

Electron microscopy has disclosed two entirely different types of cellular organisation, based on the presence or absence of a nucleus.

The cells of animals, fungi, plants and protoctists have a large, obvious nucleus. The surrounding cytoplasm contains many different membrane-bound organelles. These cells are called eukaryotic cells (meaning cells with a 'true nucleus'). We will examine this type of cell organisation first.

In contrast, bacteria contain no nucleus and their cytoplasm does not have any membrane-bound organelles. They are called prokaryotic cells (meaning cells 'before the nucleus'). Another key difference between the cells of the prokaryotes and eukaryotes is their size. Prokaryote cells are exceedingly small – about the size of individual mitochondria or chloroplasts found in the cells of eukaryotes. We will return to prokaryotic cells later in this chapter.

Key terms

Eukaryotic cell A cell with a nucleus and membrane-bound organelles in its cytoplasm.

Prokaryotic cell A cell that does not (and never did) have a nucleus or membrane-bound organelles. Bacteria are prokaryotic.

The ultrastructure of eukaryotic cells

Key term

Organelle A structure within the cytoplasm of eukaryotic cells that performs a discrete function. With the exception of ribosomes, organelles are surrounded by at least one layer of membrane, in other words they are membrane-bound.

Today, the eukaryotic cell is seen as a 'bag' of organelles, most of which are made of membranes. The fluid around the organelles is an aqueous solution of chemicals, called the **cytosol**. The cytosol and organelles are contained within a special membrane, the **cell surface membrane**. The detailed structure of this membrane, and the processes by which it is crossed by all the metabolites that move between the cytosol and the environment of the cell, will be discussed in Chapter 9.

Our picture of the arrangement of organelles enclosed by the cell surface membrane of a cell has been built up by the examination of numerous transmission electron micrographs (TEMs). This detailed picture, referred to as the ultrastructure of cells, is represented diagrammatically in Figure 4.13. TEMs of an animal and plant cell, together with an interpretive drawing, are shown in Figure 4.14.

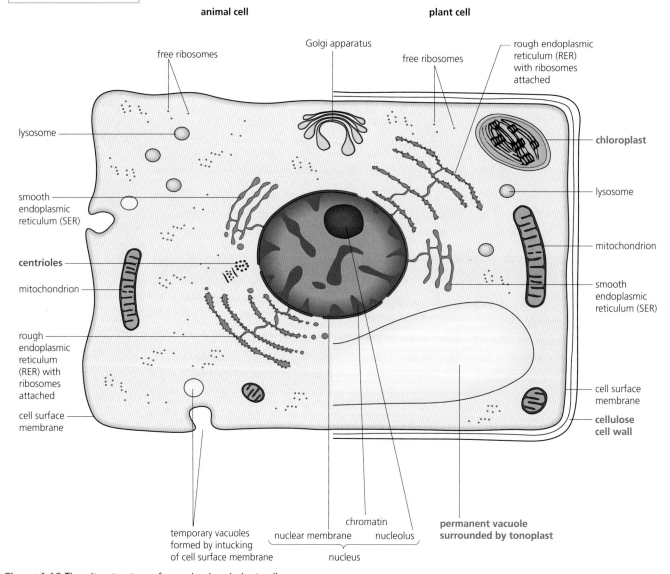

Figure 4.13 The ultrastructure of an animal and plant cell

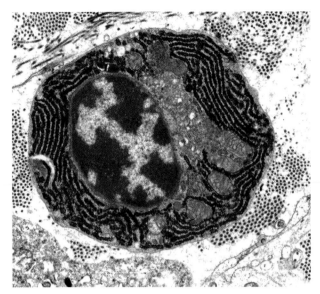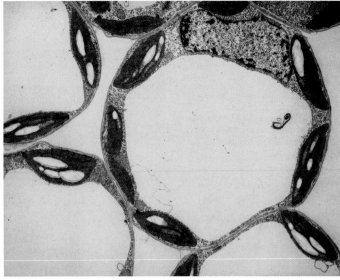

Figure 4.14 TEMs of a mammalian plasma cell and a plant palisade mesophyll cell

Introducing the organelles

Nucleus

The everyday role of the nucleus in protein synthesis has already been described in Chapter 3. It is the largest organelle in the eukaryotic cell, typically 10–20 μm in diameter. It is surrounded by a double **nuclear membrane** that contains many pores, each only about 100 nm in diameter. These nuclear pores allow movement of molecules between the cytoplasm and the nucleus, for example the movement of mRNA that you saw in Chapter 3.

The nucleus contains **chromosomes**. Figure 4.15 shows how each chromosome contains a long strand of DNA wound around beads of histone, a type of protein. The chromosomes are visible using a light microscope only at the time the nucleus divides (Chapter 5). At other times, the chromosomes appear dispersed as a diffuse network, called **chromatin**. One or more **nucleoli** (singular nucleolus) may be present in the nucleus. These rounded, dark-staining bodies are the site of ribosome synthesis. Chromatin, chromosomes and the nucleolus are visible only if stained with certain dyes.

Most eukaryotic cells contain one nucleus but there are interesting exceptions. For example, mature red blood cells of mammals (Chapter 11) and mature sieve tube elements in the phloem of flowering plants (Chapter 12) are both without a nucleus; they lose it as they mature. Voluntary muscle cells and the thin, thread-like mycelia of fungi contain cytoplasm with many nuclei (they are multinucleate).

Mitochondria

Mitochondria are rod-shaped organelles, typically 0.5–1.5 μm diameter, and 3.0–10.0 μm long. Like nuclei, each mitochondrion has a double membrane. You can see in Figure 4.16 (page 81) that the outer membrane forms a smooth boundary whereas the inner membrane is infolded to form cristae (singular *crista*). The interior of the mitochondrion contains an aqueous solution of metabolites and enzymes, called the **matrix**. Small circular molecules of DNA are also located in the matrix.

> ### Key term
>
> **Crista** One of the many folds of the inner membrane of a mitochondrion, where the chemicals involved in ATP synthesis during aerobic respiration are located.

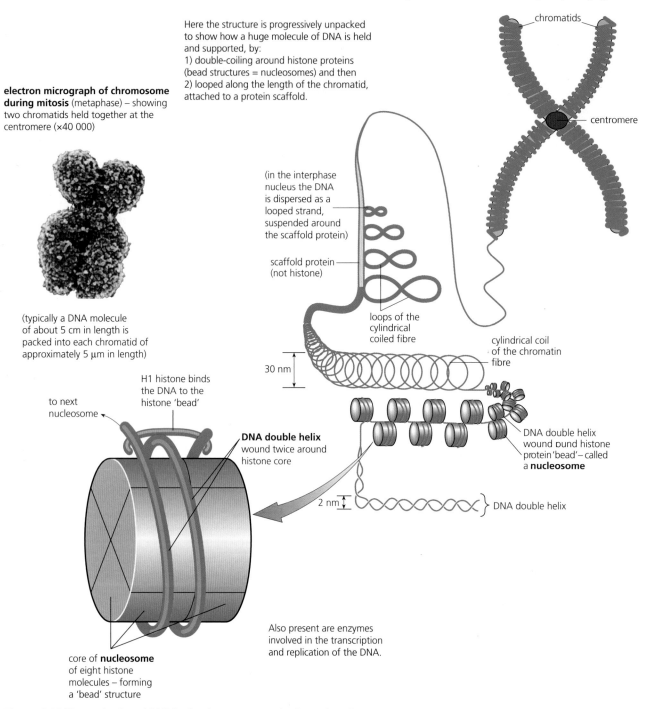

electron micrograph of chromosome during mitosis (metaphase) – showing two chromatids held together at the centromere (×40 000)

chromatids

centromere

Here the structure is progressively unpacked to show how a huge molecule of DNA is held and supported, by:
1) double-coiling around histone proteins (bead structures = nucleosomes) and then
2) looped along the length of the chromatid, attached to a protein scaffold.

(in the interphase nucleus the DNA is dispersed as a looped strand, suspended around the scaffold protein)

scaffold protein (not histone)

loops of the cylindrical coiled fibre

cylindrical coil of the chromatin fibre

(typically a DNA molecule of about 5 cm in length is packed into each chromatid of approximately 5 μm in length)

30 nm

H1 histone binds the DNA to the histone 'bead'

to next nucleosome

DNA double helix wound twice around histone core

DNA double helix wound round histone protein 'bead' – called a **nucleosome**

2 nm

DNA double helix

core of **nucleosome** of eight histone molecules – forming a 'bead' structure

Also present are enzymes involved in the transcription and replication of the DNA.

Figure 4.15 The packaging of DNA in the chromosomes of eukaryotic cells

Mitochondria are the site of the aerobic stages of respiration (*Edexcel A level Biology 2*, Chapter 1) and where most ATP is produced in cells. Not surprisingly cells that are metabolically very active, such as muscle fibres, contain very large numbers of mitochondria in their cytoplasm.

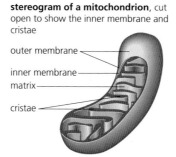
stereogram of a mitochondrion, cut open to show the inner membrane and cristae

outer membrane
inner membrane
matrix
cristae

In the mitochondrion, many of the enzymes of respiration are housed, and the 'energy currency' molecules (adenosine triphosphate, ATP) are formed.

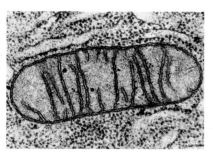

Figure 4.16 The structure of a mitochondrion

Chloroplasts

Chloroplasts are biconvex in shape, typically about 4–10 μm long and 2–3 μm wide. They are found in the cells of green plants and photosynthetic protoctists. In plants, most chloroplasts occur in the mesophyll cells of leaves (page 215), where one cell may be packed with 50 or more chloroplasts. Chloroplasts are the site of photosynthesis, the process in which light is used to synthesise sugars from carbon dioxide and water.

Look at the chloroplasts in the TEM in Figure 4.17. Hopefully, you will be able to make out the double membrane around the chloroplast on the right. The outer layer of the membrane is a smooth continuous boundary, but the inner layer becomes in-tucked to form a system of branching membranes called lamellae or **thylakoids**. In the interior of the chloroplast, the thylakoids are arranged in flattened circular piles called **grana** (singular granum). In Figure 4.17 these look a little like a stack of coins. It is here that the chlorophylls and other pigments involved in light capture are located. There are a large number of grana present. Between them the branching membranes are very loosely arranged in an aqueous environment, containing enzymes and often containing small starch grains. This part of the chloroplast is called the **stroma**. Small circular molecules of DNA are also located in the stroma.

Chloroplasts are one of a larger group of organelles called **plastids**. Plastids are found in many plant cells but never in animals. The other members of the plastid family are **amyloplasts** (colourless plastids) in which starch is stored, and **chromoplasts** (coloured plastids), containing non-photosynthetic pigments such as carotene, and occurring in flower petals and the root tissue of carrots.

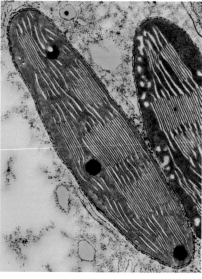

Figure 4.17 The structure of a chloroplast

Key term

Thylakoids The folded inner membranes of a chloroplast.

Ribosomes

We saw in Chapter 3 that ribosomes are the site of protein synthesis. The size of minute objects such as ribosomes is often recorded in **Svedberg** units (symbol, S). This is a measure of their rate of **sedimentation** during **centrifugation** under standardised conditions, rather than their actual size. Ribosomes in the cytoplasm of eukaryotic cells have a sedimentation rate of 80 S (an actual size of about 25 nm diameter). As we will see later, those of prokaryotic cells (and of those found within both mitochondria and chloroplasts) are slightly smaller, with a sedimentation rate of 70 S.

Figure 4.18 shows that ribosomes are built of two sub-units, and do not have membranes as part of their structures. Chemically, they consist of protein and the nucleic acid RNA. Ribosomes are found free in the cytoplasm and also bound to endoplasmic reticulum to form rough endoplasmic reticulum.

front view   side view

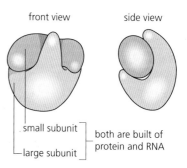

small subunit
large subunit
both are built of protein and RNA

Figure 4.18 The structure of ribosomes

TEM of RER

TEM of SER

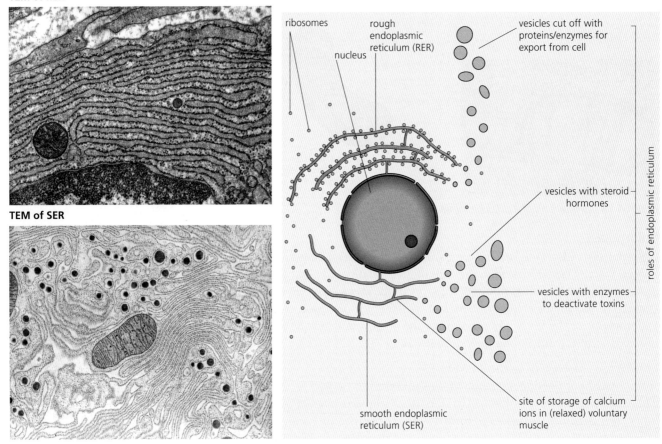

Figure 4.19 Rough endoplasmic reticulum and smooth endoplasmic reticulum

Labels (top diagram):
- ribosomes
- rough endoplasmic reticulum (RER)
- nucleus
- vesicles cut off with proteins/enzymes for export from cell
- vesicles with steroid hormones
- vesicles with enzymes to deactivate toxins
- site of storage of calcium ions in (relaxed) voluntary muscle
- smooth endoplasmic reticulum (SER)
- roles of endoplasmic reticulum

Endoplasmic reticulum

Endoplasmic reticulum consists of networks of folded single membranes forming interconnected sheets, tubes or sacs. The cytoplasm of metabolically active cells is commonly packed with endoplasmic reticulum. Figure 4.19 shows the two distinct, and separate, types of endoplasmic reticulum found in cells.

- **Rough endoplasmic reticulum** is continuous with the outer membrane surrounding the nucleus. It is called 'rough' because it has ribosomes attached to its outer surface (the surface in contact with the cytosol). You know from Chapter 3 that ribosomes link amino acids together to form polypeptide chains. In addition, the rough endoplasmic reticulum develops the tertiary and quaternary shapes of proteins that you learnt about in Chapter 2. It is in the lumen of the rough endoplasmic reticulum that, for example, the critical shape of enzyme molecules is formed. Often proteins are transferred from the rough endoplasmic reticulum to the Golgi apparatus. Sometimes this occurs by direct contact; sometimes it occurs when vesicles are formed from swellings at the margins of the rough endoplasmic reticulum that become pinched off. A **vesicle** is a small, spherical organelle bounded by a single membrane, which is used to store and transport substances around the cell.

- **Smooth endoplasmic reticulum (SER)** is separate from the rough endoplasmic reticulum and is not usually found near a cell's nucleus. It is called 'smooth' because it has no ribosomes attached to it. Smooth endoplasmic reticulum synthesises lipids, phospholipids and steroids. In the cytoplasm of voluntary muscle fibres, a special form of smooth endoplasmic reticulum is the site of storage of calcium ions, which have an important role in the contraction of muscle fibres.

Key term

Vesicle A small sac of cytoplasm enclosed by membrane. Although they are much smaller than vacuoles, there is no difference between the two structures.

Test yourself

16 What is the difference between a *double membrane* and a *lipid bilayer*?

17 Distinguish between the terms *nucleus* and *nucleolus*.

18 A mitochondrion has a smooth outer membrane and an inner membrane thrown into a large number of foldings.

a) Name the folds of the inner membrane.

b) Suggest **one** advantage of this folding.

19 Suggest why amyloplasts are so called.

20 Ribosomes are made within the nucleus of a cell. Suggest how they get to the cytoplasm.

Golgi apparatus

The Golgi apparatus consists of a stack-like collection of flattened membranous sacs, called **cisternae**. One side of the stack of membranes is formed by the fusion of membranes of vesicles from the rough endoplasmic reticulum. At the opposite side of the stack, vesicles are formed from swellings at the margins that become pinched off. These vesicles might remain within the cell or, in secretory cells, fuse with the cell surface membrane, releasing their contents. You can see vesicles being pinched off from Golgi apparatus in Figure 4.20.

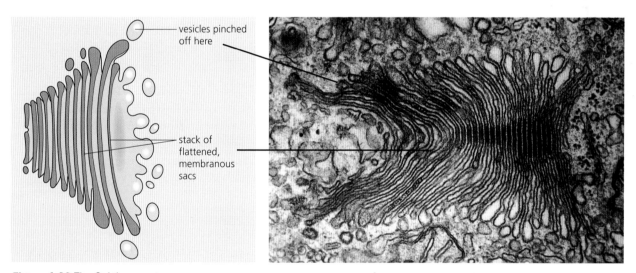

Figure 4.20 The Golgi apparatus

The Golgi apparatus occurs in all cells, but it is especially prominent in metabolically active cells, such as secretory cells. These cells produce a large number of different polymers. The Golgi apparatus is responsible for sorting, modifying and packaging these polymers for secretion or for use within the cell.

Lysosomes

Lysosomes are small spherical vesicles bound by a single membrane. They contain a concentrated mixture of about 50 hydrolytic enzymes, which are produced by the rough endoplasmic reticulum and modified in the Golgi apparatus.

Lysosomes are involved in the breakdown of imported food vacuoles, old organelles and harmful bacteria that have invaded the body and been engulfed by one of the body's defence cells. As Figure 4.21 shows, once engulfed into a larger vacuole, lysosomes fuse with the vacuole and release their hydrolytic enzymes into it. As a result, the food, organelle or bacterium is digested and the products of digestion escape into the cytosol. When an organism dies, the hydrolytic enzymes in the lysosomes of its cells escape into the cytoplasm and cause self-digestion (autolysis).

Figure 4.21 The role of lysosomes

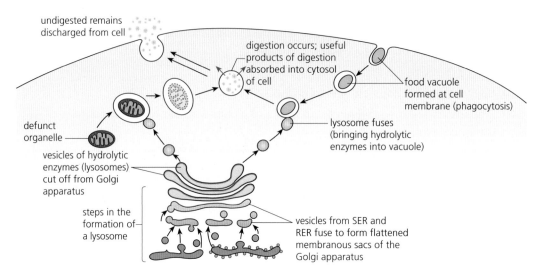

Centrioles

Although the cytosol has been described above as an aqueous solution, it is not structureless. **Microtubules** of a globular protein, called tubulin, are often present, forming a network of unbranched, hollow cylinders. These microtubules are involved in moving organelles around in the cytoplasm. They also form the **centrioles**, found in animal cells.

The centrioles occur in pairs – you can see one pair of centrioles in Figure 4.22, which also shows how they normally lie at right angles to each other, just outside the cell's nucleus. Each centriole is composed of nine bundles of microtubules. During cell division the centrioles move apart, creating the spindle (Chapter 5).

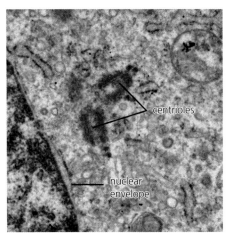

Figure 4.22 A pair of centrioles lying at right angles to each other just outside the nucleus of an animal cell

Plant cell wall

The cells of plants are surrounded by a cell wall. Strictly, the plant cell wall is not an organelle, but it is produced by the actions of organelles. The wall is composed of long, straight fibres of cellulose held together by hydrogen bonds (look back to Figure 1.13 to remind yourself of the structure of cellulose). Because these bundles are laid down at different angles, the cell wall is able to resist stretching in any direction. This prevents plant cells bursting when placed in dilute solutions (see Chapter 9). The cell wall does, however, contain spaces between the bundles of cellulose. These spaces allow the movement of water from cell wall to cell wall – the so-called **apoplast** pathway that you will learn more about in Chapter 12.

Figure 4.23 shows that when a plant cell divides, the first boundary between new cells is a gel-like layer of calcium pectate, called the **middle lamella**. Some of the endoplasmic reticulum of the parent cell becomes trapped in the gaps in this middle lamella. This trapped reticulum persists when the cellulose wall is laid down, forming plasmodesmata.

Key term

Plasmodesmata
Cytoplasmic connections between plant cells through gaps in their cell walls. They are part of the symplast pathway through which, for example, inorganic ions are able to pass from cell to cell without having to pass through cell walls or cell surface membranes.

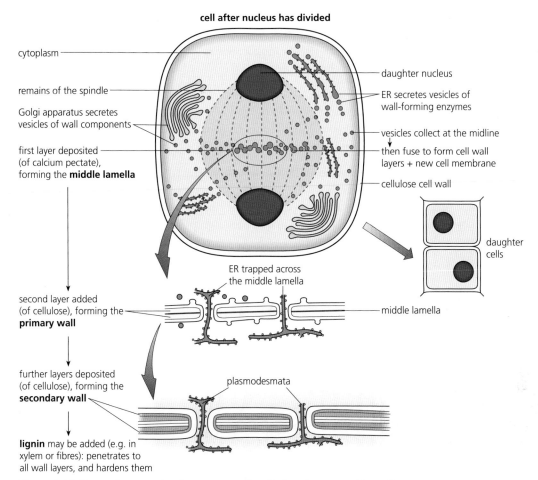

cell after nucleus has divided

cytoplasm

remains of the spindle

Golgi apparatus secretes vesicles of wall components

first layer deposited (of calcium pectate), forming the **middle lamella**

daughter nucleus

ER secretes vesicles of wall-forming enzymes

vesicles collect at the midline

then fuse to form cell wall layers + new cell membrane

cellulose cell wall

daughter cells

second layer added (of cellulose), forming the **primary wall**

ER trapped across the middle lamella

middle lamella

further layers deposited (of cellulose), forming the **secondary wall**

plasmodesmata

lignin may be added (e.g. in xylem or fibres): penetrates to all wall layers, and hardens them

Figure 4.23 Plasmodesmata develop as new cell walls form between plant cells

Permanent vacuole of plant cells and the tonoplast

The cytoplasm and cell surface membrane of a plant cell are pressed firmly against its cell wall by a large, permanent, fluid-filled vacuole, which takes up the bulk of the cell. You can see in Figure 4.24 that this vacuole is surrounded by a specialised membrane, the tonoplast. This is the barrier between the fluid contents of the vacuole (sometimes called 'cell sap') and the cytoplasm.

Key term

Tonoplast The membrane surrounding the large, fluid-filled vacuole found in plant cells.

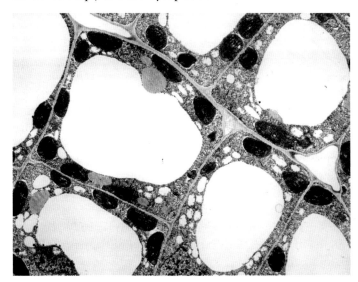

Figure 4.24 TEM of plant cells showing vacuoles and tonoplasts

The ultrastructure of prokaryotic cells

The **prokaryotes** are the bacteria and cyanobacteria (photosynthetic bacteria). These microorganisms, typically unicellular, have a fundamentally different cell structure from the cells of eukaryotes.

Key term

Nucleoid The circular DNA molecule found in prokaryotic cells. We do not refer to this as a chromosome because the DNA is neither linear nor associated with histones.

Figure 4.25 shows a scanning electron micrograph of a prokaryotic cell, in this case the intestinal bacterium *Escherichia coli*. The drawing summarises the generalised features of prokaryotic cells. Note the following distinctive features.

- Size: prokaryotic cells are exceedingly small – about the size of mitochondria and chloroplasts of eukaryotic cells.
- Absence of a nucleus: a prokaryotic cell lacks a membrane-bound nucleus. Prokaryotic cells have a single, circular DNA molecule in their cytoplasm, referred to as a nucleoid. Unlike a eukaryotic chromosome, the DNA of the nucleoid is not associated with protein.

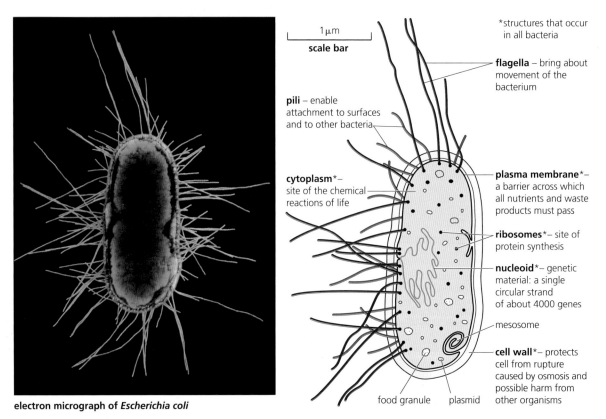

electron micrograph of *Escherichia coli*

Figure 4.25 A scanning electron micrograph of a prokaryotic cell – the bacterium *Escherichia coli* – together with an interpretive drawing of its structure

- Plasmids: in addition to the nucleoid, some prokaryotic cells have small, circular DNA molecules in their cytoplasm. These plasmids usually contain only a few genes; often they include genes conferring resistance to specific antibiotics.
- 70 S ribosomes: although these ribosomes are involved in protein synthesis, they are smaller than the 80 S ribosomes found in eukaryotic cells.
- Absence of membrane-bound organelles: the cytoplasm of prokaryotic cells lacks the range of organelles found in eukaryotes, for example mitochondria, chloroplasts, Golgi apparatus and endoplasmic reticulum.

- Cell wall: all prokaryotic cells have a cell wall. Like those of plant cells, the cell walls of prokaryotic cells prevent cells bursting when in dilute solutions. Unlike those of plant cell walls, they are not made of cellulose. Instead, they are made of **peptidoglycan**.
- Pili and flagella: where they occur, pili help prokaryotic cells to attach to surfaces or to each other and flagella help the cells to move about.

Notice that a structure called a mesosome is also shown in Figure 4.25. Many functions were proposed for these modest in-tuckings of the cell surface membrane; they have since been shown to be artefacts – caused by the chemical fixation techniques used to prepare prokaryotic cells for electron microscopy.

Key term

Artefact Something seen in a specimen being viewed using a microscope that was not present in the living cell. Artefacts result from damage caused by the processing of tissue for examination in light microscopes and electron microscopes.

Wall structure in bacteria

The rigid wall of a bacterium gives a permanent shape to the cell. It also protects the cell contents against rupture due to osmosis, for example, and it helps to protect some bacteria against harm from other organisms.

Bacterial cell walls contain polymers of amino acids and sugars, called **peptidoglycan**. All bacteria have walls of this substance, but some have additional layers on the outer surface of their wall, and these additional layers change the staining property of the wall. Figure 4.26 summarises the steps in the staining procedure that distinguishes between bacteria and explains the differences in the cell wall chemistry of the two categories of cell.

- **Gram positive** bacteria have thick walls made almost entirely of peptidoglycan. This wall becomes purple when stained by crystal violet.
- **Gram negative** bacteria have thin walls of peptidoglycan with an additional outer membrane. The high lipid content of this outer membrane prevents the crystal violet stain getting to the cell wall, so these bacteria do not become purple.

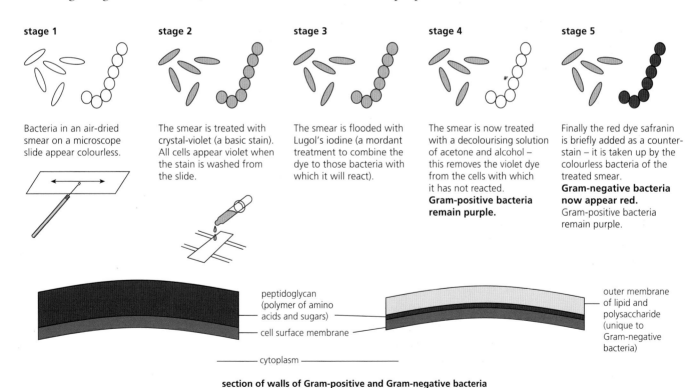

stage 1

Bacteria in an air-dried smear on a microscope slide appear colourless.

stage 2

The smear is treated with crystal-violet (a basic stain). All cells appear violet when the stain is washed from the slide.

stage 3

The smear is flooded with Lugol's iodine (a mordant treatment to combine the dye to those bacteria with which it will react).

stage 4

The smear is now treated with a decolourising solution of acetone and alcohol – this removes the violet dye from the cells with which it has not reacted. **Gram-positive bacteria remain purple.**

stage 5

Finally the red dye safranin is briefly added as a counter-stain – it is taken up by the colourless bacteria of the treated smear. **Gram-negative bacteria now appear red.** Gram-positive bacteria remain purple.

peptidoglycan (polymer of amino acids and sugars)

cell surface membrane

cytoplasm

outer membrane of lipid and polysaccharide (unique to Gram-negative bacteria)

section of walls of Gram-positive and Gram-negative bacteria

Figure 4.26 Gram staining and the difference between Gram positive and Gram negative bacteria

The stain is called Gram stain after the Dane, Hans Gram who, in 1884, devised the staining test to distinguish the two types of bacteria. The difference in the properties of these cell walls extends beyond their ability to take up the crystal violet stain. The most important is the effect of antibiotics on these two types of bacteria. The outer, lipid-rich, membrane of Gram negative bacteria is relatively impermeable to antibiotics. As a result, Gram negative bacteria are resistant to many types of antibiotic, including penicillin. Gram positive bacteria are susceptible to penicillin.

Test yourself

21 Identify **two** structures present in prokaryotic cells that are also present in all eukaryotic cells.

22 Biologists believe that mitochondria and chloroplasts evolved from free living prokaryotic organisms that formed mutually beneficial relationships with eukaryotic cells millions of years ago.

   Give **two** ways in which mitochondria and chloroplasts are similar to prokaryotes.

23 A mature red blood cell lacks a nucleus and membrane-bound organelles but is not considered to be a prokaryotic cell. Explain why.

24 Other than the evolution of antibiotic resistance, give one reason why antibiotics are not effective against all pathogenic bacteria.

▌Viruses

Although they are disease-causing agents, viruses are not regarded as living organisms. The reason for this is that they lack a metabolism of their own. Their replication depends entirely on the metabolism of cells they infect. All viruses have:

- a core of nucleic acid, around which is
- a protein coat, called a **capsid**.

Some viruses have an additional external envelope of membrane made of lipids and proteins (for example the human immunodeficiency virus, HIV, that causes acquired immunodeficiency syndrome, AIDS, in humans).

Because they lack any metabolism of their own, viral infections are difficult to treat.

Where effective antivirals have been developed, they must work by inhibiting viral replication by the host cells.

Where antivirals have not been developed, disease control must rely on preventing the spread of the virus. In the absence of an effective antiviral, control of the 2013-15 outbreak of Ebola in the West African countries of Guinea, Sierra Leone, Liberia, Senegal and Nigeria relied entirely on attempts to prevent spread of the virus.

Their inert status also makes the classification of viruses complex. Most systems of classifying them rely on features such as the nature and method of copying their nucleic acid core, the nature of their capsid, their shape and the organisms they infect. Table 4.1 shows some of these features of four viruses.

Table 4.1 Four types of virus

Name of virus	Feature				
	Host	Structure	Size/nm	Nature of nucleic acid core	Copying of nucleic acid core
λ (lambda) bacteriophage	Bacterium *Escherichia coli*	head — tail tube — tail fibre	Head diameter ~ 50–60 Tail length ~ 150	Double-stranded DNA	Double-stranded DNA transcribed to mRNA
Tobacco mosaic virus	Plants, especially those of the tobacco family	position of RNA — protein coat (capsid) of polypeptide building blocks arranged in a spiral around the canal containing RNA	Diameter ~ 18 Length ~ 300	Single-stranded RNA	RNA copied directly to form mRNA
Ebola virus	Humans (especially endothelial cells, liver cells, immune cells and dendritic cells)	RNA — outer protein coat	Diameter ~ 80 Length ~ 130 000	Single-stranded RNA	RNA copied directly to form mRNA
Human immunodeficiency virus	Humans (T helper lymphocytes)	enzymes — protein coats — single-stranded RNA — capsule — glycoprotein	Diameter ~ 120	Single-stranded RNA	RNA 'reverse transcribed' into double-stranded DNA, which is incorporated into the host cell's DNA and later transcribed to form mRNA

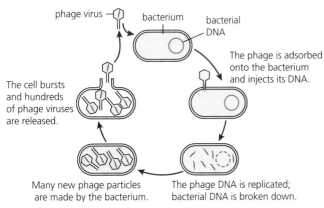

phage virus

bacterium

bacterial DNA

The phage is adsorbed onto the bacterium and injects its DNA.

The cell bursts and hundreds of phage viruses are released.

Many new phage particles are made by the bacterium.

The phage DNA is replicated; bacterial DNA is broken down.

Figure 4.27 The lytic cycle of a lambda (λ) bacteriophage

The lytic cycle

We have seen above that viruses lack any metabolism of their own; instead they rely on the metabolism of the cells they infect to produce more virus particles. Most studies of viruses have been carried out using bacteriophages – viruses that infect bacteria. Figure 4.27 shows the 'life cycle' of a bacteriophage, often referred to simply as a 'phage'. Notice that the phage infects the bacterial cell by injecting its own nucleic acid; the capsid remains outside the bacterial cell. The infected bacterium then produces more phage particles by replicating the phage nucleic acid and producing new capsids.

Eventually, the bacterial cell releases new phages that are free to infect other cells. Because the release of new phages follows lysis of the bacterial cell, this 'life cycle' is called the **lytic cycle**.

Similar events occur in other organisms when they become infected by viruses. Some pathogenic viruses, however, can undergo a period of latency. During this period, the virus does not take over the metabolism of the infected cell but remains dormant (latent) within an infected host cell. There are two types of viral latency.

- episomal latency – the viral nucleic acid remains inactive but free in the cytoplasm or nucleus of the infected cell
- proviral latency – the viral nucleic acid becomes incorporated into the DNA of the infected host cell. It is now termed a provirus but, as with episomal latency, the viral nucleic acid can be reactivated at any time.

You might have suffered a disease called chickenpox when you were a child. It is a contagious, but relatively harmless disease caused by a virus, *Varicella zoster*. If you did suffer chickenpox, you will have recovered but the nucleic acid of the virus will have remained in some of your nerve cells. This is an example of episomal latency. Although the viral DNA is inactive, any sudden stress can reactivate it, resulting in a disease called shingles. Normal aging increases the risk of shingles: in the UK, a new policy has been introduced to vaccinate elderly people against shingles, starting with those who are in their 80th year.

Key term

Viral latency A period in which, under the control of specific latency genes, a pathogenic virus remains dormant. During this time the virus, or its nucleic acid, is present inside an infected cell but does not control the cell's activities.

Test yourself

25 What is meant by the following statement: 'Viruses are inert'?

26 The human immunodeficiency virus (HIV) is classed as a retrovirus. Use information in Table 4.1 to suggest why.

27 What is the generic name given to viruses that infect bacteria?

28 Doctors are advised not to prescribe antibiotics to people suffering viral infections. Explain why.

29 Cold sores are caused by the herpes simplex virus (HSV-1). Some people suffer recurrent outbreaks of cold sores. Suggest what this shows about the way that HSV-1 infects sufferers.

Exam practice questions

1 Which one, or more, of the following statements is true?

   A Prokaryotic cells have only 70 S ribosomes.

   B Prokaryotic cells have 70 S and 80 S ribosomes.

   C Eukaryotic cells have only 80 S ribosomes.

   D Eukaryotic cells have 70 S and 80 S ribosomes.   *(1)*

2 In a photomicrograph, a cell has a diameter of 5 cm. If the magnification is 400 times, the actual size of diameter of the cell is:

   A 12.5 μm

   B 20.0 μm

   C 125.0 μm

   D 200.0 μm   *(1)*

3 A cell specialised to secrete a steroid hormone is likely to have:

   A a large amount of smooth endoplasmic reticulum

   B a large amount of rough endoplasmic reticulum

   C a large nucleus

   D few mitochondria   *(1)*

4 The diagram shows the structure of a bacterial cell.

   a) Name the structures labelled **A**, **B** and **C**. *(3)*

   b) Give one difference between the genetic material of a bacterial cell and that of a eukaryotic cell.   *(1)*

   c) Other than a difference in their genetic material, give two features that are characteristic of prokaryotes.   *(2)*

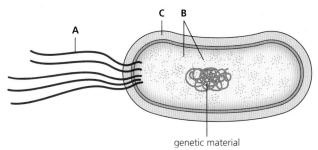

genetic material

5 Copy and complete the table showing features of different type of cells by putting a tick (✓) in an empty box if the feature is present and a cross (✗) if it is not.   *(5)*

Feature	Animal cell	Bacterial cell	Plant cell
Cell surface membrane			
Cell wall			
Tonoplast			
Ribosome			
Mitochondrion			

6 The following diagram was drawn from a photograph of a cell observed using a transmission electron microscope.

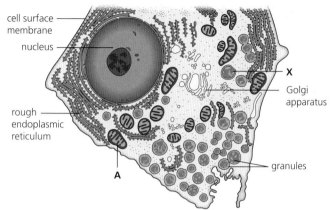

cell surface membrane
nucleus
rough endoplasmic reticulum
X
Golgi apparatus
granules
A

a) Give the evidence from the diagram that this cell was observed using a transmission electron microscope. *(3)*

b) Name the structure labelled **A**. *(1)*

c) The actual diameter of the granule labelled **X** is 1.5 μm. Calculate the magnification of this drawing. Show your working. *(2)*

7 A scientist grew cells in a tissue culture solution. After several hours, she added a small volume of a solution containing radioactively labelled amino acids. At regular intervals, she killed a sample of the cells and measured the level of radioactivity in different parts of the cells in her samples. Some of her results are shown in the table.

Time after adding radioactive amino acids/minutes	Percentage of total radioactivity found in each part of cell		
	Ribosomes	Golgi apparatus	Secretory vesicles
0	80	16	4
120	16	30	11
240	14	8	42
360	16	8	46

a) Why did the scientist add radioactively labelled amino acids several hours after starting to grow the cells in culture solution? *(2)*

b) Use your knowledge of cell structure and function to explain the data in the table. *(3)*

c) Suggest why the percentages at 360 minutes do not add up to 100 percent. *(1)*

Stretch and challenge

8 During much of the twentieth century, biologists assumed that proteins were the hereditary material; DNA was thought to be too simple. In 1952, Alfred Hershey and Martha Chase conducted a series of experiments to investigate whether protein or DNA was the hereditary material. They used bacteriophages in their experiments. Use a search engine to find how they utilised knowledge of the lytic cycle to investigate the nature of the hereditary material.

9 The cell theory is a unifying concept in biology. Use a search engine to discover how this theory was developed.

> **Tip**
>
> When quantitative data are shown in a table, it is often easier to see a trend or pattern in these data by drawing a sketch graph – unlabelled axes and roughly drawn curves.

The eukaryotic cell cycle and cell division

5

Prior knowledge

In this chapter you will need to recall that:
→ all cells arise from the division of a pre-existing cell
→ the chromosomes of a eukaryotic cell contain its genetic information
→ prior to cell division, a cell makes a copy of each of its chromosomes
→ there are two types of cell division in eukaryotic cells — mitosis and meiosis
→ mitosis is a type of cell division that occurs during growth and during the repair and replacement of cells
→ a cell that divides by mitosis produces two daughter cells that are genetically identical to each other and to itself, in other words produces clones
→ a cell that divides by meiosis divides twice and produces four daughter cells that are not genetically identical to itself or to each other
→ the four daughter cells produced by meiosis have half the number of chromosomes as the parent cell. Each also contains a combination of alleles that is different from the other three daughter cells.

Test yourself on prior knowledge

1 Name the process by which copies of chromosomes are produced.

2 Name **one** place in your body where mitosis regularly occurs.

3 Read the following statements about mitosis. One or more of them are false. Identify the true statements.

  A All cells can undergo mitosis.

  B The number of chromosomes is the same in a cell after mitosis as it was before mitosis.

  C Copies of chromosomes are made during mitosis.

  D Copies of chromosomes are separated during mitosis.

4 In Question 3 you identified one or more statements as false. Explain why they are false.

5 Where in your body does meiosis occur?

Chromosomes, cell division and the cell cycle

As you saw when considering the cell theory in Chapter 4, new cells arise by division of existing cells. Prior to cell division, a cell's DNA is copied by semi-conservative replication, as described in Chapter 3. When the cell then divides, each daughter cell receives one copy of the replicated DNA.

● In prokaryotic cells, cell division occurs by a process known as **binary fission**.
● In eukaryotic cells, cell division is part of a regulated process called the **cell cycle**. It consists of three main stages: interphase, mitosis and cytokinesis.

Key terms

Asexual reproduction
The formation of new organisms that does not involve the fusion of gametes.

Centromere A narrow region occupying a specific position on each chromosome. This is the only site on each chromosome to which the microfibres of the spindle can attach during mitosis. Following DNA replication, the centromere temporarily holds together the two copies of each chromosome (the chromatids).

Somatic cell Any cell in the body of a multicellular organism other than a germ cell (one that gives rise to gametes) or undifferentiated stem cell.

Homologous chromosomes A pair of chromosomes in a diploid cell that have the same shape and size. More importantly, they carry the same genes in the same order, although not necessarily the same alleles of each gene.

Diploid A eukaryotic cell is said to be diploid (represented as **2n**) if it contains two copies of each chromosome. In sexually reproducing organisms, one copy comes from each parent.

The cell cycle in unicellular organisms: asexual reproduction

Unicellular eukaryotic organisms, such as yeast or amoeba (Figure 4.8, page 74) grow quickly under favourable conditions. They then divide in two. Since this division results in the production of new organisms, it is a form of reproduction. As no formation or fusion of gametes is involved, it is called asexual reproduction. This cycle of growth and division is repeated rapidly, at least whilst conditions remain supportive.

The cell cycle in multicellular organisms: growth and repair

In multicellular organisms, the life cycle of individual cells is more complex. Here, life begins as a single cell that grows and divides, forming many cells. These new cells allow the organism to grow, eventually making up the adult organism. Only certain of these cells, however, retain the ability to grow and divide throughout life. Even when growth has stopped, these cells are able to replace old or damaged cells. Most of the cells of multicellular organisms, however, become specialised and are then unable to divide further.

Chromosomes – a reminder

As you saw in Chapter 3, the genetic information the nucleus holds in its chromosomes exists as a sequence of nucleotide bases in deoxyribonucleic acid (DNA). We considered the structure of chromosomes in Chapter 4. Look back to Figure 4.15 (page 80) to remind yourself of this structure. Most of the time, chromosomes are long, thin structures that cannot be resolved by light microscopes. Instead, they appear as a diffuse network called chromatin. At the time a nucleus divides, however, the chromosomes become highly coiled. Only in this **condensed** state can they be resolved by light microscopes.

Whilst we will consider other processes occurring during the cell cycle, the behaviour of the chromosomes is so important that we will first concentrate on them. There are five features of chromosomes that it is helpful to note at the outset.

1 **The shape of a chromosome is characteristic**

   Each chromosome has a particular, fixed length. Somewhere along the length of the chromosome there is a characteristically narrow region called the centromere. A centromere can occur anywhere along the chromosome but it is always in the same position on any given chromosome. The position of the centromere, as well as the length of a chromosome, enables us to identify individual chromosomes in photomicrographs. You can see that the centromere in Figure 5.1 is at the mid-point of its chromosome.

2 **Chromosomes occur in homologous pairs**

   The chromosomes of a somatic cell occur in pairs of homologous chromosomes. They are called *homologous* because the two chromosomes have the same shape and, more importantly, have the same genes in the same order. You can see a pair of homologous chromosomes represented in Figure 5.1. During sexual reproduction, one of each pair came originally from the gamete of one parent and the second from the gamete of the other parent. So, for example, your somatic cells have 23 pairs of homologous chromosomes. At the moment of fertilisation, you inherited one copy of each from your mother's egg cell and the other copy of each from your father's sperm cell. Cells in which the chromosomes are in homologous pairs are described as diploid. We represent this as **2n,** where the symbol '***n***' represents one set of chromosomes.

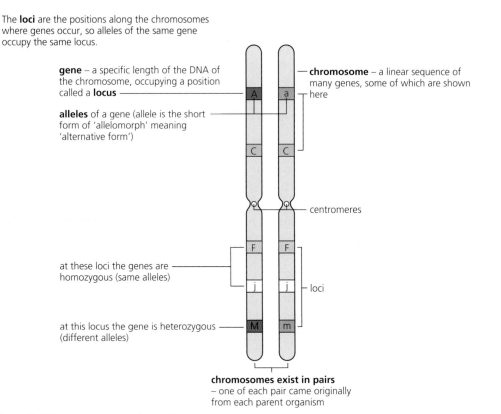

The **loci** are the positions along the chromosomes where genes occur, so alleles of the same gene occupy the same locus.

gene – a specific length of the DNA of the chromosome, occupying a position called a **locus**

alleles of a gene (allele is the short form of 'allelomorph' meaning 'alternative form')

chromosome – a linear sequence of many genes, some of which are shown here

centromeres

at these loci the genes are homozygous (same alleles)

loci

at this locus the gene is heterozygous (different alleles)

chromosomes exist in pairs – one of each pair came originally from each parent organism

Figure 5.1 The loci of a pair of homologous chromosomes

3 For each species, the number of chromosomes is fixed

For any one species, the number of chromosomes in each somatic cell is normally constant. For example: in the somatic cells of a human there are 46 chromosomes; in a mouse, 40; in an onion, 16; and in a sunflower, 34. Note that these characteristic chromosome numbers are all even numbers; this must be the case because the chromosomes are present in homologous pairs (feature 2).

Combining these first three features, we refer to the characteristic shape and number of chromosomes in a cell as its **karyotype**.

4 Homologous chromosomes might not carry the same alleles of genes

In Chapter 3, a gene was defined as a sequence of DNA nucleotide bases that encodes the sequence of amino acids in a functional polypeptide. We noted above that a diploid cell has homologous pairs of chromosomes. Thus, in each diploid cell, there are two copies of each gene. These copies lie in the same positions, or loci (singular *locus*), on the two homologous chromosomes. This is shown in Figure 5.1. Two or more different copies of a gene are called **alleles**. They are different because the order of nucleotide bases differs by one or more bases.

The two genes at any one locus in a diploid cell might have the same nucleotide base sequence. If so, this cell would be described as **homozygous** for this gene. Alternatively, the two genes might have different base sequences, in other words they would be two different alleles of the same gene. If so, the cell would be described as **heterozygous** for this gene. We can refer to diploid organisms as homozygous or heterozygous for a particular gene as well.

Key terms

Karyotype The number and shape of chromosomes in the somatic cell of an organism.

Locus (plural **loci**) The position that a particular gene occupies on a specific chromosome.

Allele One of two or more different forms of the same gene. Different alleles of the same gene have slightly different nucleotide base sequences.

Homozygous Diploid cells have two copies of each gene. If the two copies are the same allele of the gene, the cell is said to be homozygous for this gene.

Heterozygous In a diploid cell, if the two copies of a gene are different alleles, the cell is said to be heterozygous for this gene.

Let's use Figure 5.1 to put features 1, 2 and 4 into context. If you look at Figure 5.1, you can see an imaginary pair of homologous chromosomes (feature 2). As you would expect, they have the same shape – their centromeres are in the same position so that the lengths of their 'arms' is the same (feature 1). The diagram also represents five loci, in other words the positions of five genes. The chemical nature of the gene at each locus is represented by a single letter. At the loci C, F and J, the genes are the same: these loci are homozygous. In contrast, loci A and M are heterozygous (feature 4).

Key term

Chromatid Following DNA replication, a cell has two copies of each of its chromosomes. When these become visible during cell division and can be seen to be held together by their centromeres, they are briefly called chromatids.

5 Chromosomes are copied prior to division, so appear double

Between nuclear divisions, while the chromosomes are still uncoiled and cannot be resolved by light microscopes, a cell makes a copy of each chromosome. Chapter 3 described the semi-conservative replication of DNA – the process by which this copying occurs. Until separated during nuclear division, the two identical copies of each chromosome are held together by their centromeres. While held together, they are referred to as chromatids; after their separation during nuclear division, they are referred to as chromosomes again.

Look at the left-hand side of Figure 5.2, which puts all the above features into context. The photomicrograph shows the condensed chromosomes of a somatic cell from a human male.

human chromosomes of a male (karyotype)
(seen at the equator of the spindle during nuclear division)

chromosomes arranged as homologous pairs in descending order of size

homologous chromosomes

each chromosome has been replicated (copied) and exists as two chromatids held together at their centromeres

images of chromosomes cut from a copy of this photomicrograph can be arranged and pasted to produce a **karyogram**

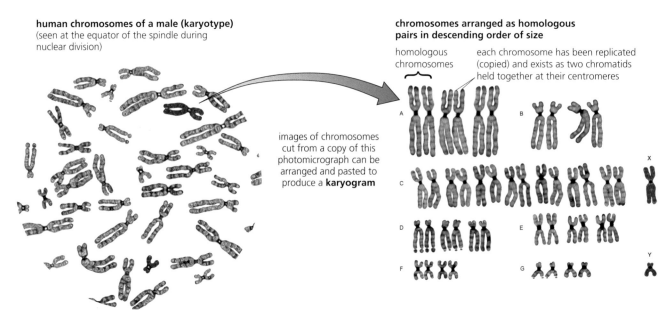

Figure 5.2 Karyogram of a human male

You saw in point 5 above that, prior to cell division, DNA replication results in the formation of two chromatids held together by a centromere. Take any of the chromosomes in Figure 5.2 and you will see this to be true.

You saw in point 3 above that each cell has a fixed number of chromosomes. As this is a human cell, can you count 46 chromosomes?

You saw in point 2 above that the chromosomes in a somatic cell occur as homologous pairs. *Can you find any homologous pairs in the left-hand image Figure 5.2?*

If not, in the right-hand image of Figure 5.2, the pairs have been sorted for you by cutting each chromosome from a second copy of the photograph, pairing them together, arranging them in descending order of size and numbering them. A photograph (or drawing) like this is called a **karyogram**. Any karyogram of a normal human male you will ever see has these chromosomes arranged and numbered in the same way. It could be used by genetic counsellors (in conjunction with other techniques) to detect the presence of abnormalities in a patient's chromosomes, for example Down's syndrome.

You saw in point 1 above that each chromosome has a characteristic shape. Look at the chromosomes in group A in Figure 5.2. The two chromatids held together in each chromosome, as well as the two chromosomes, are the same size: their centromeres are in the same position, so that the length of their 'arms' is the same. This is not just true of chromosome pairs in *this* human male; it is true of the chromosomes of every human male and, indeed, of every human female. Exactly the same is true of the chromosomes in groups B to G.

Now look at the final pair. You can see that it is not numbered. Rather, one is labelled **X** and the other **Y**. These are known as the sex chromosomes, since the **Y** chromosome determines the development of male characteristics. We will return to this issue later. All the other chromosomes (pairs in groups A to G) are called autosomes.

Key terms

Sex chromosome A chromosome that carries genes that determine the sex of an organism. In mammals, the **Y** chromosome determines maleness.

Autosome Any chromosome other than a sex chromosome.

Test yourself

1 Sister chromatids and homologous chromosomes both carry the same genes in the same order. Explain the difference between them.

2 What is meant by the term 'allele'?

3 Give **two** functions of a centromere.

4 Explain why the chromosomes within a cell cannot normally be seen using a compound light microscope.

5 What does the karyotype tell you about a cell?

The cell cycle

Although the process of cell division is continuous, it is often described as a series of discrete stages. Figure 5.3 shows how one cell cycle of a eukaryotic cell can be described as:

- **interphase** – the time between divisions
- **mitosis** – separation of the chromatids of each chromosome to form two new nuclei
- **cytokinesis** – division of the cytoplasm.

The length of this cycle depends partly upon conditions external to the cell, such as temperature, supply of nutrients and of oxygen. Its length also depends upon the type of cell. In cells at the growing point of a young stem or of a developing human embryo, the cycle is completed in less than 24 hours. The epithelial cells that line your intestine typically divide every 10 hours, whereas your liver cells divide every year or so. Nerve cells do not normally divide after they have differentiated. In specialised cells such as these, the genes needed to initiate and control cell division are 'switched off', so they cannot divide.

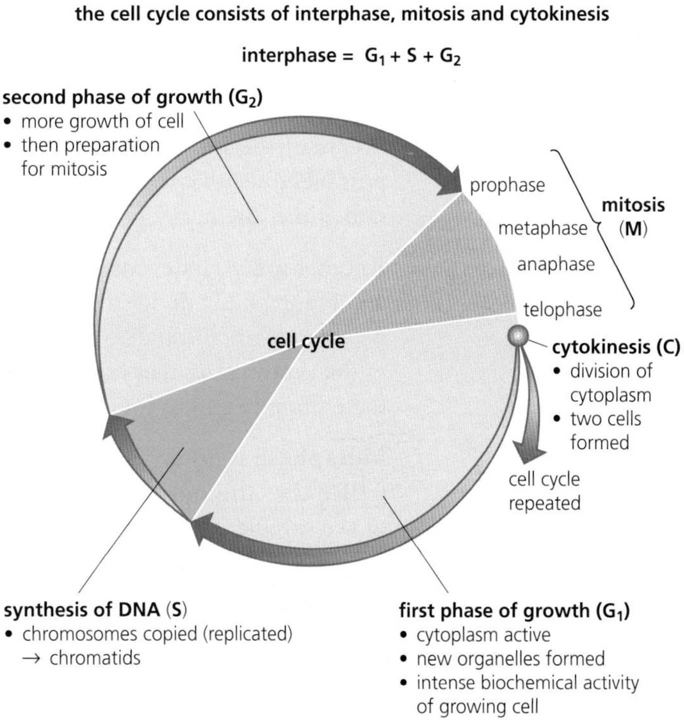

the cell cycle consists of interphase, mitosis and cytokinesis

interphase = G_1 + S + G_2

second phase of growth (G_2)
- more growth of cell
- then preparation for mitosis

prophase

mitosis (**M**)
metaphase
anaphase
telophase

cytokinesis (**C**)
- division of cytoplasm
- two cells formed

cell cycle

cell cycle repeated

synthesis of DNA (**S**)
- chromosomes copied (replicated) → chromatids

first phase of growth (G_1)
- cytoplasm active
- new organelles formed
- intense biochemical activity of growing cell

Figure 5.3 The stages of the cell cycle

Interphase

Interphase is always the longest part of the cell cycle, but is of extremely variable length. It is recognisable in light micrographs because no clear chromosomes can be seen in the nucleus. The chromatin results in a diffuse staining of the nucleus. At first glance, the nucleus appears to be 'resting', but this is not the case at all. Table 5.1 summarises the intense activity occurring in a cell during interphase.

Table 5.1 The 'stages' of interphase

G_1 – the 'first growth' phase	In the nucleus, some genes are 'switched on' and their base sequence is transcribed to pre-mRNA molecules.
	Editing of pre-mRNA to mature mRNA is also occurring.
	The cytoplasm increases in volume (grows) by producing new proteins and cell organelles, including mitochondria and endoplasmic reticulum.
S – synthesis	In the nucleus, the semi-conservative replication of DNA occurs.
	New histones are synthesised and attach to the replicated DNA in the nucleus.
	Each chromosome becomes two chromatids attached at the centromere
	Growth of the cell continues.
G_2 – the 'second growth' phase	In the nucleus, replicated DNA is 'double checked' for errors and corrected if any errors are found.
	If correction is not possible, the cell cycle is normally halted at this G2 phase.
	Cell growth continues by further synthesis of proteins and cell organelles.

Nuclear division – mitosis

When cell division occurs, the nucleus divides first. In mitosis, the chromosomes, present as the chromatids formed during interphase, are separated and distributed to two daughter nuclei.

Figure 5.4 presents mitosis in an animal cell as a process in four phases, but this is for convenience of description only. Mitosis is always one continuous process with no breaks between the phases.

In **prophase**, the chromosomes increasingly shorten and thicken by a process of super-coiling (look back to Figure 4.15 to remind yourself of this coiling). As a result, they eventually become visible as long, thin threads. Only towards the end of prophase is it possible to see that they consist of two chromatids held together at the centromere. At the same time, the nucleolus gradually disappears and the nuclear membrane breaks down.

Another important event occurs during prophase. The centrioles, described in Chapter 4 (see Figure 4.22), divide and move to opposite ends of the cell. As they do so, they radiate a network of microtubules, called the **spindle**. Some of the spindle fibres attach to each side of the centromeres that hold together the chromatids in each chromosome. Contraction of these fibres begins to move the chromosomes apart.

Metaphase is instantly recognisable in light micrographs as the spindle fibres have now pulled the chromosomes into the centre of the cell, where they line up on the equator of the spindle.

In **anaphase**, the centromere of each chromosome divides; the spindle fibres attached to them shorten, resulting in the two chromatids being pulled by their centromeres to opposite poles of the spindle. Once separated, the chromatids are referred to as chromosomes again. In Figure 5.4, you can see an early part of anaphase; the chromatids have only just left the equator of the spindle.

> **Tip**
>
> When examining light micrographs of cells undergoing mitosis, first look for those cells in metaphase. These are easiest to spot because the chromosomes are neatly lined up on the equator of the spindle.

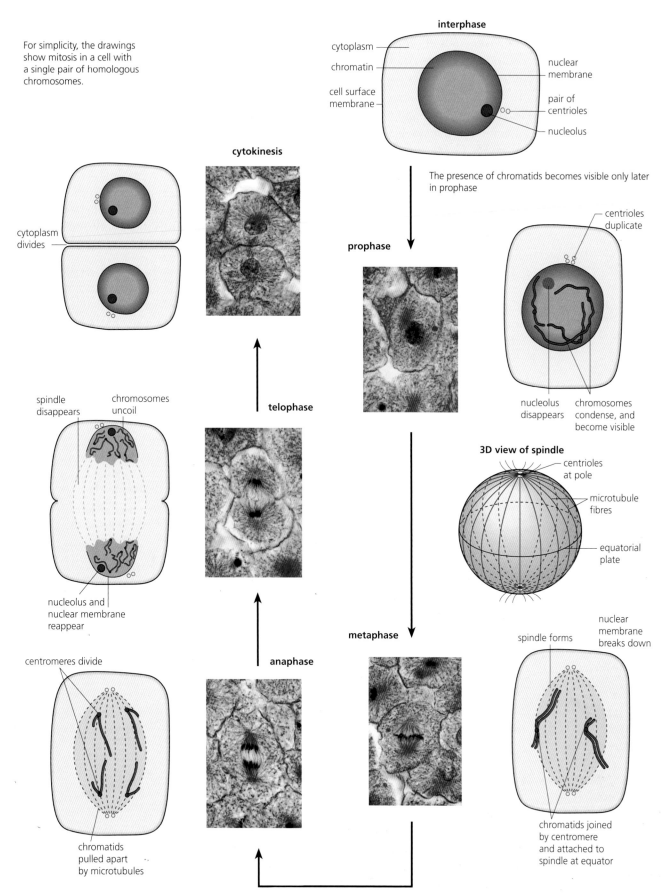

For simplicity, the drawings show mitosis in a cell with a single pair of homologous chromosomes.

interphase

cytoplasm

chromatin

cell surface membrane

nuclear membrane

pair of centrioles

nucleolus

The presence of chromatids becomes visible only later in prophase

prophase

centrioles duplicate

nucleolus disappears

chromosomes condense, and become visible

cytokinesis

cytoplasm divides

3D view of spindle

centrioles at pole

microtubule fibres

equatorial plate

telophase

spindle disappears

chromosomes uncoil

nucleolus and nuclear membrane reappear

anaphase

centromeres divide

chromatids pulled apart by microtubules

metaphase

spindle forms

nuclear membrane breaks down

chromatids joined by centromere and attached to spindle at equator

Figure 5.4 Mitosis in an animal cell

Chromosomes, cell division and the cell cycle

In **telophase**, a nuclear membrane reforms around both groups of chromosomes at opposite ends of the cell. The chromosomes 'de-condense' by uncoiling, becoming chromatin again. You can see this uncoiling has already begun in telophase in Figure 5.4. One or more nucleoli reform in each nucleus.

Cytokinesis

Division of the cytoplasm, known as **cytokinesis**, usually follows telophase. Vesicles from the Golgi apparatus are involved in cytokinesis in both animal and plant cells, but in different ways.

During cytokinesis in animal cells, a **cleavage furrow** (pinch) develops in the middle of the cell. You can see this in Figure 5.4. Contraction of this cleavage 'pinches' the cytoplasm in half. As this happens, cell organelles become distributed between the two developing cells.

This 'pinching' does not happen during cytokinesis in plant cells. Instead, as you can see in Figure 5.5, vesicles from the Golgi apparatus collect along the line of the equator of the spindle, known as the cell plate. These vesicles secrete a gel-like layer of calcium pectate, called the **middle lamella** (see Figure 4.23, page 85). Onto this, they secrete transverse layers of microfibres of cellulose, forming the primary cell wall. Subsequently, more layers are added, often at right angles, forming the secondary cell wall. Many cell walls also become impregnated with lignin. You will see the importance of lignin in Chapter 12.

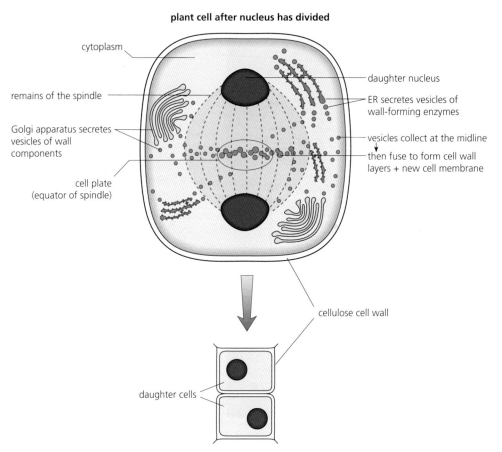

plant cell after nucleus has divided

cytoplasm

daughter nucleus

remains of the spindle

ER secretes vesicles of wall-forming enzymes

Golgi apparatus secretes vesicles of wall components

vesicles collect at the midline

then fuse to form cell wall layers + new cell membrane

cell plate (equator of spindle)

cellulose cell wall

daughter cells

Figure 5.5 Cytokinesis in a plant cell

The significance of mitosis

The 'daughter' cells produced by mitosis are clones. This means that they are genetically identical to each other and to the parent cell from which they were formed. This occurs because:

- an exact copy of each chromosome is made by accurate DNA replication during interphase, when two chromatids are formed
- the two chromatids remain attached by their centromeres during prophase of mitosis, when each becomes attached to a spindle fibre
- when the centromeres divide during anaphase, the chromatids of each pair are pulled apart to opposite poles of the spindle. Thus, one copy of each chromosome moves to each pole of the spindle
- the chromosomes at the poles form the new nuclei – two to a cell at this point
- two cells are then formed by division of the cytoplasm at the mid-point of the cell, each with an exact copy of the original nucleus.

> **Key term**
>
> **Clones** Two or more cells, or organisms, that are genetically identical to each other. In eukaryotic organisms, clones are produced by mitosis.

Test yourself

6   Look back to the drawing and photograph of a eukaryotic cell in Chapter 4 (page 77). In which part of the cell cycle is it all depicted? Explain your answer.

7   Precisely in which part of the cell cycle does DNA replication occur?

8   In which part of the cell cycle are the products of DNA replication separated?

9   Early in mitosis, a chromosome is seen as two chromatids. At the end of mitosis the same chromosome is single. Is there any difference between these two structures? Explain your answer.

10 Plant cells lack centrioles, yet they produce a spindle during mitosis. Use your knowledge of cell structure to suggest how.

Where mitosis is commonly observed

You have seen above that asexual reproduction in some eukaryotic unicells involves mitosis. Consequently, you could observe mitosis by examining cultures of these unicells growing in favourable conditions.

Alternatively, you could examine tissues of multicellular organisms that are actively growing. In mammals, these tissues include epithelial tissues, such as the skin and the lining of the intestine. In flowering plants, these tissues are found at mersitems – the actively growing tips of shoots and roots – and the **cambium** within the vascular bundles (Chapter 12). For this reason, our core practical examines mitosis in actively growing root tips of a flowering plant.

> **Key term**
>
> **Meristem** A group of plant cells that are able to divide by mitosis. Primary meristems are found at the tips of growing shoots and roots. Secondary meristems develop in woody plants, leading to an increase in diameter of roots and shoots.

Making a temporary squash preparation of a root tip to investigate the stages of mitosis under a microscope

Mitosis can be observed in any tissue that is actively dividing. A tissue in which it is easy to observe mitosis is the growing tip of a plant root. Any plant root will do; onion and garlic bulbs are commonly used in the laboratories of colleges and schools.

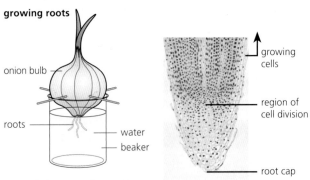

growing roots

onion bulb

roots

water

beaker

growing cells

region of cell division

root cap

Figure 5.6 The tip of a growing root

Before reading on, follow the steps in the protocol below which describes how to prepare plant tissue for examination under a light microscope, starting off with an onion bulb.

Risk assessment

The two stains most commonly used for this experiment are toluidine blue and orcein, both of which are low hazard. Toluidine blue is used in a simple and low hazard aqueous solution, whereas corrosive carboxylic acids (acetic or proprionic) are required for the orcein stain to adhere to the chromosomes.

Even though its use is more hazardous, orcein is often used in preference to toluidine blue, because the stain produces greater definition of chromosome structure.

The hazards of using the acetic orcein stain can be very much reduced by dispensing a small volume of the stain into a stoppered bottle. Containment of the stain prevents spillage.

The bottle should be firmly stoppered until it needs to be opened. The bottle should always be placed on an impervious surface when it is being handled.

Procedure

1. Onion or garlic cloves are suspended over water, until roots can be seen growing from the "blunt" surface. The roots must be no longer than 1 cm, and must still be growing when the roots are harvested. The roots will, therefore, have lots of dividing cells in a meristem region.
2. Fill a small bottle with 1 M hydrochloric acid, and place it in a thermostatically controlled water bath set at 55 °C. Leave the bottle for 15 minutes, so that the acid warms to the temperature of the water bath.

Tip

Heating the root tips in hydrochloric acid increases the permeability of the cell, so allowing the stain to penetrate more easily.

Place a garlic clove in the bottle, so that the roots are submerged in the 55 °C hydrochloric acid. Leave the roots in the acid for 5 minutes.

When the garlic clove has been in the 55 °C hydrochloric acid for 5 minutes, take it out and rinse the roots thoroughly in tap water. This can be done by either by rinsing under the tap, or by immersing the roots in a beaker of water.

3. Label a small bottle or vial, including a hazard warning. Place a pinprick hole in the bottle's cap, to prevent the lid popping off when the bottle/vial is heated. Place a small volume of the stain in the bottle. There should be enough for a layer of fluid; this is approximately 2 mm in depth.

4. Use a pair of sharp scissors on the garlic clove to cut off the root tips (5–10 mm at the pointed end) so they fall, or can be placed, in the acetic orcein. Use the scissors to make sure that the root tips are immersed in the stain. Then place the lid on the vial. Place the acetic orcein vial in the 55 °C water bath. Leave it there for 5 minutes.

Tip

The time and temperature of staining are important. The stain requires 5 minutes at 55 °C for the stain to diffuse into the cell and bind with the chromosomes. In an open watch glass, the acetic acid will often evaporate before the stain has penetrated the cells.

The stopper in the container used in this method means that the acetic acid does not evaporate.

5. After the root tips have been in the stain for 5 minutes, use forceps to take them out of the vial, and place them on a microscope slide.

Tip

Squashing the root spreads the cells, so that there is just a single layer to be seen under the microscope. This makes the individual cells easier to see.

Add a drop of water to the root tip on the slide. Tease the root tip with needles, to spread out the cells. Place a cover slip over the root tip.

Wrap the slide in several layers of a thick paper towel. Use your thumbs to gently press on the slide and cover slip through the paper towel. You should be just able to feel the glass through the towel. Remove the paper towel.

6 Place the slide onto the stage of the microscope, and view the slide under the high power objective lens.

Questions

1 What safety precautions did you take when using the scalpel?

2 In step 2, the root tip is heated in hydrochloric acid. What was achieved by doing this?

3 When examining a temporary preparation, how did you recognise an air bubble trapped under the cover slip?

4 Explain why lateral movements should be avoided in Step 5.

5 Why is paper towel placed between the thumb and the cover slip during Step 5.

6 Assuming that the condenser lens of the microscope (if present) is already focused, list the steps you followed to view the plant tissue using the high-power objective lens.

7 If you pushed the slide to your left, which way did cells in the tissue appear to move?

8 Explain why the preparation is called a 'temporary squash' preparation.

9 Suppose you wish to adapt the method described to find the number of cells in each of the stages of mitosis. How would you adapt the method to ensure that the values you find are accurate?

10 How could you determine an appropriate number of fields of view to observe?

Meiosis – a different type of nuclear division

Meiosis is a type of nuclear division with quite different outcomes from mitosis. Figure 5.7 illustrates three differences.

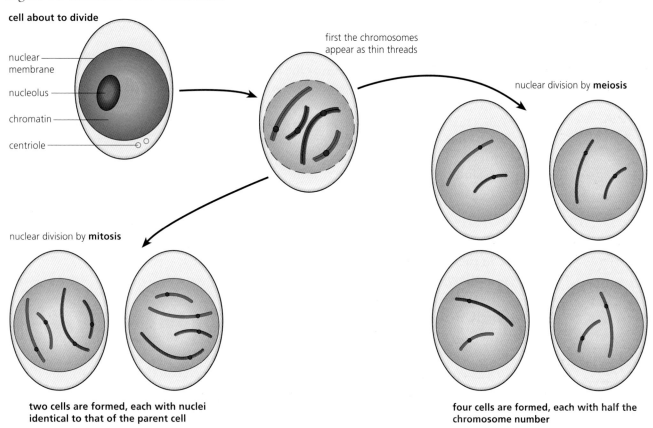

Figure 5.7 Mitosis and meiosis – the significant differences in outcome

- Meiosis normally produces four daughter cells whereas mitosis produces two.
- Meiosis produces daughter cells that have half the number of chromosomes as the parent cell, whereas mitosis produces daughter cells that have the same chromosome complement as the parent cell.
- Meiosis produces daughter cells that are genetically different, whereas mitosis produces daughter cells that are genetically identical to each other and to the parent cell, in other words are clones.

Meiosis occurs at some point in the life cycle of all organisms that reproduce sexually. In humans, for example, meiosis occurs in the gonads – testes and ovaries – and results in the formation of haploid **gametes** that fuse at fertilisation to form a diploid **zygote** (Chapter 6). This is not universally true of all organisms, however. In most fungi, for example, the parent cell is haploid and gametes are formed by mitosis.

> ### Key term
>
> **Haploid** A eukaryotic cell is haploid (represented by '***n***') if it contains only one chromosome from each of its homologous pairs.

An overview of meiosis

Meiosis involves two divisions of the nucleus, known as **meiosis I** and **meiosis II**.

Look at Figure 5.8, which provides an overview of what happens during both these divisions. To make the diagram simple to understand, Figure 5.8 shows only a single pair of homologous chromosomes.

Start with the first two diagrams at the top of Figure 5.8. You can see that, like mitosis, during the interphase that precedes meiosis, the chromosomes have been replicated to form two chromatids held together by their centromeres. The third diagram, however, shows that in meiosis the homologous chromosomes then pair up. By the end of meiosis I, shown in the fifth diagram in Figure 5.8, homologous chromosomes have been separated into two new cells. Notice, though, that each chromosome still consists of two chromatids, held together at their centromere. It is during meiosis II that these chromatids are separated.

To summarise, meiosis consists of two nuclear divisions but only one replication of the chromosomes. The important points to remember are that:

- in **meiosis I**, homologous chromosomes are separated
- in **meiosis II**, the chromatids of each chromosome are separated.

Meiosis – the detail

As with mitosis, once meiosis starts it usually proceeds as a continuous process. Figure 5.9 provides a more detailed version of meiosis, showing the stages you need to know. Again, to make things simpler to follow, the diagrams show only one pair of homologous chromosomes. The separate steps in Figure 5.9 relate to stages seen in tissue that has been killed, stained and examined under a light microscope. The stages have the same names and sequence as those in mitosis – prophase, metaphase, anaphase and telophase.

during interphase

cell with a single pair of
homologous chromosomes

centromere

chromosome number = 2 (diploid cell)

replication (copying) of
chromosomes occurs

$2n$

during meiosis I

homologous chromosomes
pair up

$2n$

homologous chromosomes
separate and enter different
cells – chromosome number
is halved

breakage and reunion of parts of
chromatids have occurred and the
result is visible now, as chromosomes
separate (**crossing over**)

now
haploid cells

n

n

during meiosis II

chromosomes separate
and enter daughter cells

cytokinesis

division of cytoplasm

product of meiosis is
four haploid cells

n n n n

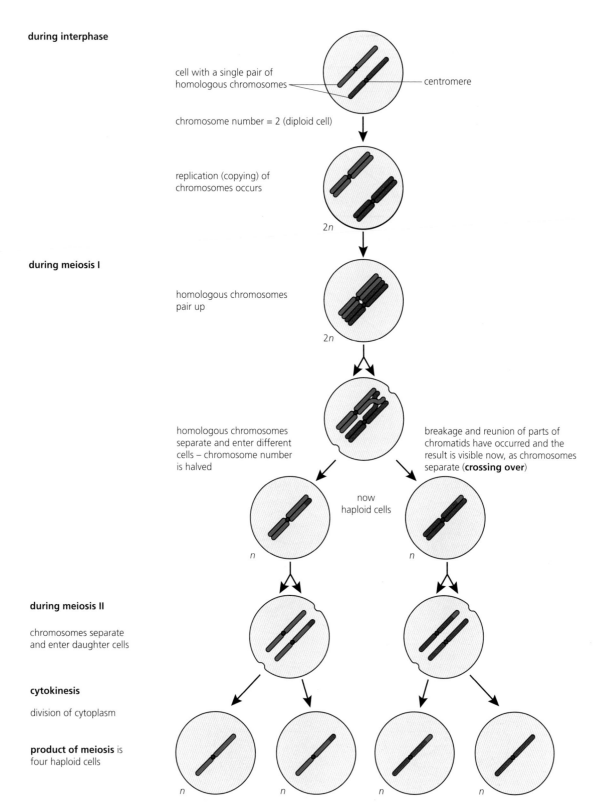

Figure 5.8 An overview of meiosis

prophase I (early)
During interphase the chromosomes replicate into chromatids held together by a centromere (the chromatids are not visible). Now the chromosomes condense (shorten and thicken) and become visible.

prophase I (late)
Homologous chromosomes repel each other. Chromosomes can now be seen to consist of chromatids. Sites where chromatids have broken and rejoined, causing crossing over, are visible as chiasmata.

anaphase I
Homologous chromosomes separate. Whole chromosomes are pulled towards opposite poles of the spindle, centromere first (dragging along the chromatids).

prophase II
The chromosomes condense and the centrioles duplicate.

anaphase II
The chromatids separate at their centromeres and are pulled to opposite poles of the spindle.

prophase I (mid)
Homologous chromosomes pair up (becoming **bivalents**) as they continue to shorten and thicken. Centrioles duplicate.

metaphase I
Nuclear membrane breaks down. Spindle forms. Bivalents line up at the equator, attached by centromeres.

telophase I
Nuclear membrane re-forms around the daughter nuclei. The chromosome number has been halved. The chromosomes start to decondense.

metaphase II
The nuclear membrane breaks down and the spindle forms. The chromosomes attach by their centromere to spindle fibres at the equator of the spindle.

telophase II
The chromatids (now called chromosomes) decondense. The nuclear membrane re-forms. The cells divide.

MEIOSIS I

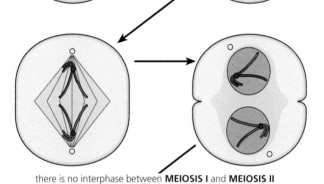

there is no interphase between **MEIOSIS I** and **MEIOSIS II**

MEIOSIS II

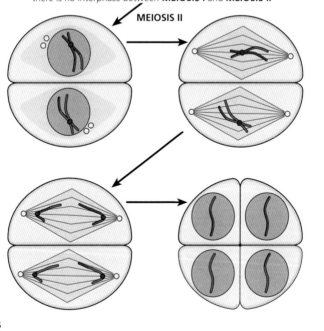

Figure 5.9 The stages of meiosis

Meiosis I

In **prophase I**, the chromosomes become visible as they condense. At the same time, the two members of each pair of homologous chromosomes pair up. Remember, members of a homologous pair of chromosomes have the same linear sequence of genes. Gene-by-gene, the pairing is very precise. When we see them like this, we refer to each pair as a bivalent.

As members of each bivalent continue to shorten, their chromatids often become entangled, causing a stress on the DNA molecules. As a result of this stress within a bivalent, individual chromatids frequently break. The broken ends rejoin more or less immediately. When non-sister chromatids from homologous chromosomes break and rejoin they do so at exactly corresponding sites, so that a cross-shaped structure called a chiasma is formed at one or more places along a bivalent. In the fourth and fifth diagrams in Figure 5.8, you can see the effect of one chiasma. As a result of breakage, one portion of a blue chromatid and the corresponding portion of a red chromosome have broken and joined the 'wrong' chromatid. This event is known as a crossing over because lengths of genes have been exchanged between chromatids. Remember that, although homologous chromosomes have the same loci in the same order, the loci might have different alleles of the same gene. Thus, new combinations of alleles can be produced as a result of crossing over.

Next the spindle forms. As in mitosis, chromatids become attached by their centromeres to the fibres of the spindle. As they contract, these fibres pull the chromosomes, still held together in their homologous pairs, until they come to lie at the equatorial plate of the spindle: this is **metaphase I**.

During **anaphase I**, further contraction of the spindle fibres pulls the homologous chromosomes apart, one to each pole of the spindle. At this stage, however, the sister chromatids remain attached by their centromeres. The pairing between them has started to break down and so, as you can see in Figure 5.9, they both become clearly visible. You can see the outcome of this in the fifth diagram in Figure 5.9.

Telophase I then follows. Two new nuclei are formed and the spindle breaks down. We now have two new cells, each with two nuclei containing a single set of condensed chromosomes, still made of two chromatids. These new cells do not go into interphase, but rather continue smoothly into meiosis II.

Meiosis II

During **prophase II**, two new spindles form at right angles to the old one. Contraction of the fibres in each spindle pulls the chromosomes to the centre of their respective spindles to reach **metaphase II**. Following division of the centromeres, further contraction of the spindle fibres pulls individual chromatids to opposite poles of their respective spindle during **anaphase II**. Now there are four groups of chromosomes, each with half the number of the original parent cell. During **telophase II**, new nuclei form around these groups of chromosomes. The chromosomes become long and thin again, the nuclear membranes reform and the cytoplasm divides to form new cells.

Key terms

Bivalent The name given to the two homologous chromosomes in a diploid cell when they are seen paired together during meiosis I. As each chromosome is currently present as two chromatids, there are four chromatids in a single bivalent.

Non-sister chromatids Chromatids on the two different members of a homologous pair of chromosomes.

Chiasma (plural chiasmata) A point seen during meiosis I at which the non-sister chromatids appear interlocked. A chiasma is the result of non-sister chromatids within a bivalent becoming entwined, breaking and re-joining to the fragment from the non-sister chromatid earlier in prophase I.

Crossing over The process by which non-sister chromatids exchange genes following formation of a chiasma

Test yourself

11 A diploid cell has 20 pairs of homologous chromosomes. How many chromosomes will be present in a cell produced from this cell by:

a) mitosis

b) meiosis?

12 In which of the two divisions of meiosis are sister chromatids separated?

13 Describe how you could distinguish between a cell in metaphase of mitosis and a cell in metaphase I of meiosis.

14 The X and Y chromosomes in Figure 5.2 look very different but are able to pair together during meiosis. What does this suggest about their nature?

Test yourself

15 Look at Figure 5.10, which depicts meiosis without crossing over.

  a) The cell depicted in Figure 5.10 has two pairs of homologous chromosomes. Without crossing over, how many genetically different daughter cells could it produce?

  b) Use your answer to part a) to devise a formula that will enable you to calculate the number of genetically different daughter cells that could be produced by a cell in which the number of pairs of homologous chromosomes is *n*.

  c) Use the formula you have derived to represent the number of genetically different egg or sperm cells that could result from meiosis in a human.

Independent assortment is illustrated in a parent cell with two pairs of homologous chromosomes (four bivalents). The more bivalents there are, the more variation is possible. In humans, for example, there are 23 pairs of chromosomes giving over 8 million combinations.

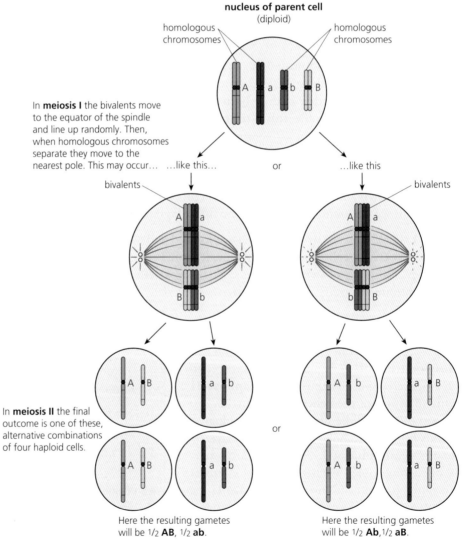

Figure 5.10 Genetic variation resulting from the independent assortment of homologous chromosomes

Meiosis and genetic variation

You have seen that mitosis produces clones – genetically identical cells. Meiosis does not and is a major source of genetic variation in organisms. The cells produced by meiosis are genetically different for two reasons.

There is **independent assortment** of homologous chromosomes. This happens because the bivalents line up at the equator of the spindle in meiosis I entirely at random. As a result, which chromosome of a given pair is pulled to which pole is unaffected by (independent of) the behaviour of the chromosomes in other homologous pairs. Look at Figure 5.10, which represents independent assortment in a parent cell with a diploid number of four chromosomes. To make the process easier to follow, the pairs of homologous chromosomes have been coloured differently. You can see that, depending on which way the two pairs of homologous chromosomes are pulled apart can result in completely different combinations of chromosomes in the daughter cells. We see that independent assortment *alone*, generates a huge amount of variation in the coded information carried by the different haploid cells produced by meiosis.

As described above, there is **crossing over** of segments of the non-sister chromatids of the members of a pair of homologous chromosomes. Figure 5.11 shows more detail of this crossing over. Using letters to represent the alleles of three genes, it also shows how two crossover events result in new combinations of alleles on the chromosomes of the haploid cells produced.

Finally in the random fusion of gametes that occurs during fertilisation, further genetic variation is generated, but that is an issue for later chapters.

The effects of genetic variation are shown in one pair of homologous chromosomes. Typically, two, three or more chiasmata form between the chromatids of each bivalent at prophase I.

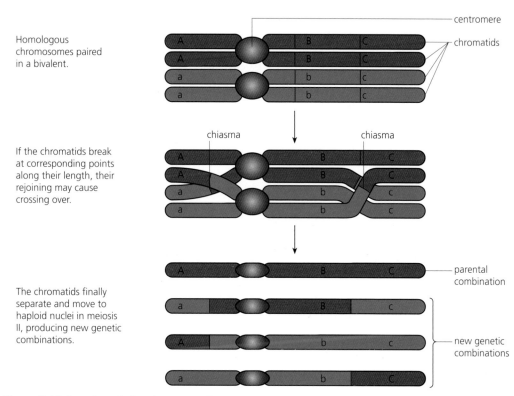

Figure 5.11 Genetic variation due to crossing over between non-sister chromatids

You looked at gene mutations in Chapter 3 (page 62). Chromosome mutations can also occur.

In a **chromosome mutation**, a change in the number or the sequence of genes can be brought about in a number of different ways. They include:

- **chromosome translocation**, in which part of a chromosome breaks and rejoins a completely different chromosome. Although these translocations are usually harmless in humans, carriers have an increased risk of producing gametes with unbalanced translocations that might, after fertilisation, result in miscarriages. About 5 per cent of Down's syndrome cases are caused by gametes in which part of the long arm of chromosome 21 has broken away and re-joined chromosome 14 (a partial chromosome translocation)

- chromosome non-disjunction, in which the members of a homologous pair fail to be separated during meiosis. Look at the lower part of Figure 5.12. It illustrates what happens if one pair of chromosomes, in this case the longer pair, fails to separate during meiosis. You can see how the two upper daughter cells in Figure 5.12 contain two copies of this longer chromosome whilst the two lower daughter cells contain no copy of that chromosome at all. If these cells are gametes that fuse with a normal gamete, fertilisation could produce:
  - either a zygote with more copies of this chromosome than usual (polysomy)
  - or a zygote with only one copy of this chromosome (monosomy).

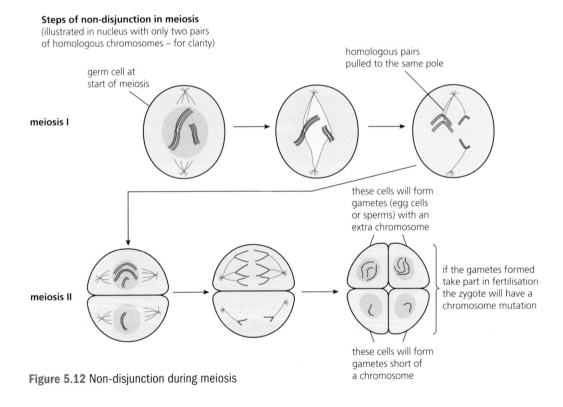

Steps of non-disjunction in meiosis
(illustrated in nucleus with only two pairs of homologous chromosomes – for clarity)

homologous pairs pulled to the same pole

germ cell at start of meiosis

meiosis I

these cells will form gametes (egg cells or sperms) with an extra chromosome

meiosis II

if the gametes formed take part in fertilisation the zygote will have a chromosome mutation

these cells will form gametes short of a chromosome

Figure 5.12 Non-disjunction during meiosis

Key terms

Chromosome non-disjunction The failure of homologous chromosomes to separate properly during the first division of meiosis. It results in daughter cells (gametes) with too many, or too few, chromosomes.

Polysomy A term used to describe a diploid cell, or diploid organism, with more than two copies of a particular chromosome.

Monosomy A term used to describe a diploid cell, or diploid organism, with only one copy of a particular chromosome.

Turner's syndrome – an example of monosomy in humans

In humans, monosomy is usually partial, in other words only part of a chromosome is lost. Full monosomy, that is, lack of one complete chromosome, is usually lethal. Only one case of full monosomy is known in humans – Turner's syndrome. Women with Turner's syndrome have only one X chromosome (often represented **XO**), rather than the normal two (represented **XX**). In the UK, about 1 in 2000 girls is affected by this condition. There is a wide range of symptoms associated with Turner's syndrome but in almost all cases, the women have undeveloped ovaries, resulting in a lack of periods and infertility.

Down's syndrome – an example of polysomy in humans

Polysomy is also rare. In the UK, about 750 babies are born each year with Down's syndrome and an estimated 60 000 people in the UK live with this condition. Most cases of Down's syndrome result from the non-disjunction of chromosome 21 during gamete production. The resulting zygote inherits three copies of chromosome 21, giving them a total of 47 chromosomes. You can see this in the karyogram of a person with Down's syndrome in Figure 5.13. As with Turner's syndrome, the symptoms of Down's syndrome are variable.

Key terms

Turner's syndrome
A condition usually caused by chromosome non-disjunction of chromosome pair 23 during gamete formation, resulting in the conception of a girl with only one X chromosome per cell. A small number of cases result from the loss of a substantial part of one X chromosome during gamete formation.

Down's syndrome A condition usually caused by the non-disjunction of chromosome pair 21, resulting in the conception of a child with three copies of this chromosome per cell. About 5 per cent of cases are caused by translocation of a large part of chromosome 21 to chromosome 14 during gamete formation.

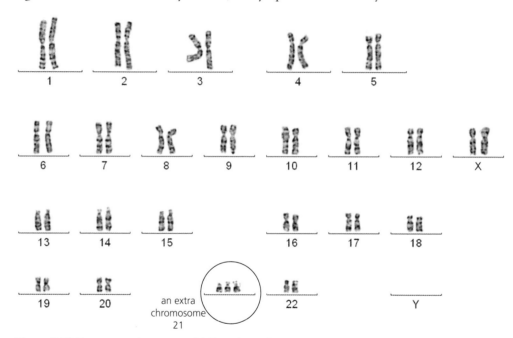

Figure 5.13 Karyogram of a person with Down's syndrome

Test yourself

16 Crossing over can occur between sister chromatids during meiosis I. Suggest what effect this would have.

17 A chromosome translocation in the gamete of a parent could, after fertilisation, result in the miscarriage of an embryo. Suggest why.

18 How many chromosomes will be present in the somatic cells of a girl with Turner's syndrome? Explain your answer.

19 Down's syndrome is often referred to as trisomy 21. Explain why.

20 Although the chances of having a baby with Down's syndrome are higher for older parents, more babies with Down's syndrome are born to younger mothers.

a) Suggest why the chances of having an affected baby increase with the age of the parents.

b) Explain why, despite the lower risk, more affected babies are born to younger mothers.

Exam practice questions

1 In mitosis, sister chromatids are separated during:

   A anaphase   **C** prophase

   B metaphase   **D** telophase   *(1)*

2 The cultivated potatoes that humans eat are tetraploid, i.e., they contain four copies of each chromosome ($4n$). A daughter cell produced by meiosis in a potato cell will be:

   A haploid (n)   **C** tetraploid ($4n$)

   B diploid ($2n$)   **D** octoploid ($8n$)   *(1)*

3 Copy and complete the table comparing and contrasting features of meiosis and mitosis.   *(6)*

Feature	Meiosis	Mitosis
Number of nuclear divisions		
Homologous chromosomes pair together		
Sister chromatids are separated		
Number of chromosomes in daughter cells is the same as in parent cell		
Is a major source of genetic variation		
Involves formation of a spindle of fibres		

4 The graph represents changes in the relative mass of DNA in a cell during a single cell cycle. The curve has been labelled at six points, A to F.

   a) Explain the shape of the curve between points **B** and **C**.   *(2)*

   b) Give the two letters, **A** to **F**, between which each of the following occurs:

      i) anaphase

      ii) S phase

      iii) cell increases in mass.   *(3)*

   c) Does the curve show that cytokinesis occurred? Explain your answer.   *(2)*

5 The following statements relate to meiosis.

   i Genetic variation is caused by random segregation of homologous chromosomes.

   ii Genetic variation is caused by crossing over.

   iii Genetic variation is caused by gene mutation.

5 The eukaryotic cell cycle and cell division

Which statements best describe the sources of genetic variation in meiosis?

A i and ii   **C** ii and iii

B i and iii   **D** i, ii and iii   *(1)*

6 The diagram shows a cell with two pairs of homologous chromosomes. Each diagram shows this cell in one stage of division.

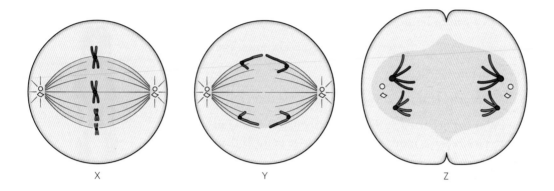

X   Y   Z

a) Is the cell in the diagram an animal cell or a plant cell? Justify your answer. *(1)*

b) Other than size and shape, how does a chromosome of one homologous pair differ from a chromosome of the other homologous pair? *(1)*

c) Identify the type and stage of cell division in cells X, Y and Z. *(3)*

d) Explain the reason for your answer for cell Z. *(2)*

7 The diagram shows the life cycle of three organisms: yeast (a single-celled fungus); human (a multicellular animal); *Ulva* (a multicellular protoctist). The diagram shows whether each stage is haploid (*n*) or diploid (2*n*).

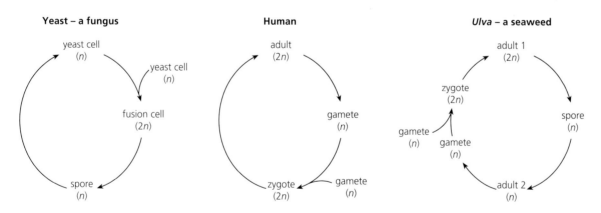

a) On a copy of the diagram write the letter **M** by each arrow on the three life cycles that represents meiosis. Write the letter **T** by each arrow on the three life cycles that represents mitosis. *(3)*

b) Is it true to say that gametes are produced by meiosis? Use evidence from the diagram to justify your answer. *(2)*

c) The two adults in the life cycle of *Ulva* look identical. Outline how you could tell which is which? Details of any procedure(s) are **not** required. *(2)*

8 The photo shows a pair of homologous chromosomes during meiosis.

a) Explain what is meant by the term 'homologous chromosomes'. *(1)*

b) Name the structure labelled A. *(1)*

c) Explain the appearance of the chromosomes in the diagram. *(4)*

d) The same event has occurred at the points labelled B and C in the diagram. This event is more likely to occur at point B than at point C. Suggest why. *(2)*

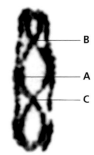

Stretch and challenge

9 A group of scientists investigated mitosis in one species of flowering plant.

- They planted seeds in pots of sawdust. After 2 weeks, they collected healthy root tips from the germinated seeds at 2-hourly intervals throughout a 24-hour period.

- They left each sample of root tips for 24 hours in $2\,cm^3$ of a 3:1 mixture of ethanol:ethanoic acid.

- After this time, they transferred the root tips into warm, dilute hydrochloric acid for 5 minutes.

- They then removed about 1 mm from the end of each root tip, placed each on a glass slide containing orcein and squashed it under a cover slip.

- Finally, they examined the slides under a compound light microscope and counted the cells in each stage of mitosis.

a) Suggest one reason why a group of scientists, rather than a single scientist, carried out this work.

b) In each case, explain why the scientists:

   i) placed the root tips in a mixture of ethanol and ethanoic acid

   ii) transferred the root tips into warm, dilute hydrochloric acid

   iii) removed about 1 mm from the end of each root tip

   iv) placed orcein on the glass slide

   v) squashed the root tips.

The table shows the scientists' results.

Time of day/ hours	Percentage of cells in each stage of mitosis				
	Prophase	Metaphase	Anaphase	Telophase	Interphase
06:00	41.2	8.8	7.4	6.6	36.0
08:00	35.0	11.9	11.3	13.0	28.8
10:00	28.0	18.7	14.5	15.4	22.4
12:00	29.7	22.0	19.2	11.4	17.7
14:00	27.8	24.8	17.9	10.8	18.7
16:00	23.6	25.6	18.9	14.2	17.7
18:00	17.2	23.4	18.8	20.1	20.5
20:00	18.1	15.1	20.1	21.1	25.6
22:00	27.2	13.6	14.7	20.4	30.2
24:00	17.5	12.5	12.5	14.5	43.0
02:00	22.5	11.6	12.2	6.9	46.8
04:00	28.9	11.7	9.1	8.0	42.3

c) Explain why the scientists converted their raw data to the percentage values shown in the table.

d) As this investigation was carried out by professional scientists, explain how they would have ensured their data were accurate.

e) At what time of day was most DNA synthesis occurring? Explain your answer.

f) Describe how the percentage of cells in prophase changed over the period shown in the table.

g) Explain the relationship between the percentage of cells in prophase and metaphase between 06:00 and 16:00 hours.

10 At the beginning of this chapter, we saw that the cell cycle is a regulated process. A protein called p53 is one of many factors involved in this regulation. Find out what you can about p53 and its involvement in regulation of the cell cycle and in cancer.

6

Sexual reproduction in mammals and plants

Prior knowledge

In this chapter you will need to recall that:

→ sexual reproduction involves fertilisation, in which haploid gametes fuse to produce a diploid zygote

→ in some organisms, fertilisation is external and in others internal

→ in animals, female gametes are called ova (eggs) and male gametes are called spermatozoa (sperm)

→ the sperm of mammals are small and motile whereas the ova of mammals are larger and not motile

→ in mammals, the development of a zygote into an embryo and fetus occurs within the uterus of the female parent

→ during embryological development, the placenta acts as the major exchange surface of the developing offspring

→ in flowering plants, male gametes are associated with pollen grains that pass to the female parts of a flower during the process of pollination

→ the ova of a flowering plant remain within the flowers of the parent plant

→ following fertilisation in a flowering plant, an embryo is contained within a seed which, in turn, is contained with a fruit.

Test yourself on prior knowledge

1 A gamete is haploid. What is meant by 'haploid'?

2 The offspring produced by sexual reproduction between the same parents show great variation. Give **three** ways in which this variation is produced.

3 In mammals and in flowering plants, the ova remain within the parent plant. Give **two** advantages of retaining the ova within the body of the parent.

4 In flowering plants, pollination is not the same as fertilisation. Explain why.

5 Which of the following plant organs is **not** a fruit: apple; cucumber; lettuce; tomato? Explain your answer.

Key terms

Gametes Cells carrying half the normal chromosome number; two gametes fuse during sexual reproduction.

Fertilisation The fusion of the nuclei of two gametes to form the nucleus of a zygote, containing the chromosomes from both gametes.

Zygote The cell formed by the fusion of two gametes.

Reproduction is the production of new individuals by an existing member or members of the same species. It is one of the fundamental characteristics of life.

As you saw in Chapter 5, reproduction can be asexual or sexual. The distinctive feature of sexual reproduction is that a genetically novel individual is formed as a result of the mixing of chromosomes from two individuals – the parents. The parental chromosomes are carried by cells called gametes. At fertilisation, the nuclei of two gametes fuse to form a new nucleus inside a cell called a zygote. At some point in the life cycle of sexually reproducing organisms, meiosis must occur so that future gametes contain only half the chromosome number again. Mammals and flowering plants can reproduce by sexual reproduction.

Sexual reproduction in mammals

The events involved in sexual reproduction are basically the same in all mammals. We will use ourselves as examples of mammals, so the following account relates to humans.

The sexes are separate in humans. In biology, the **sex** of an organism has a precise meaning and it has a genetic basis. In Figure 5.2 (page 96) you saw a karyogram of a human male. It showed a pair of sex chromosomes – a male has one X chromosome and one Y chromosome. This would be true of all mammals and it is the possession of a Y chromosome that determines the sex of a male. We can, in fact, be a little more precise than this. It is one particular gene on the Y chromosome, called *SRY*, which determines that a zygote will develop into a male.

Key term

Sex The sex of an organism has a genetic basis. In mammals, it depends on the absence or presence of an activated gene, *SRY*, located on the Y chromosome. The karyotype XY is male and XX is female.

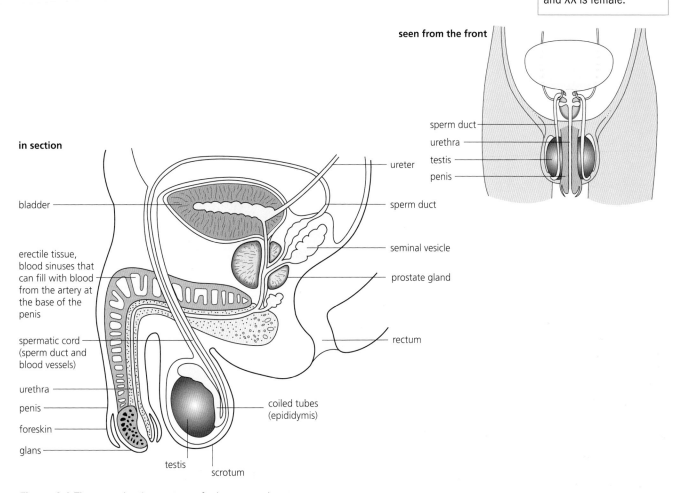

Figure 6.1 The reproductive system of a human male

A male zygote has a Y chromosome. Activation of the *SRY* gene on this chromosome causes the gonads to develop into testes. As a result, the associated reproductive structures develop into a male reproductive system, shown in Figure 6.1. In contrast, a female zygote lacks a Y chromosome. In the absence of an activated *SRY* gene, she develops ovaries and, as a consequence, the other associated structures of the female reproductive system, shown in Figure 6.2.

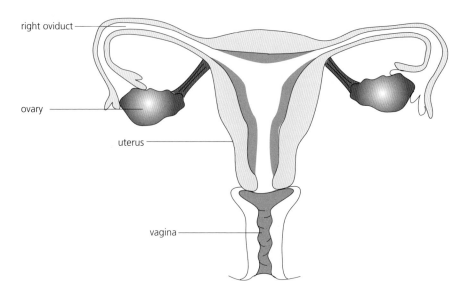

seen from the front

right oviduct

ovary

uterus

vagina

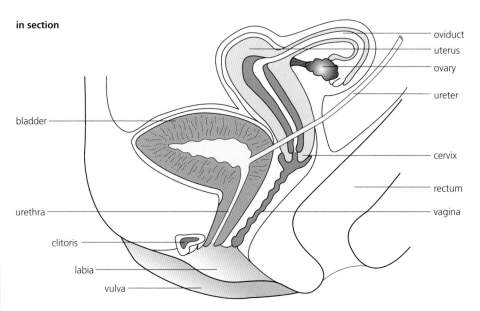

in section

oviduct
uterus
ovary
ureter

bladder

cervix

rectum

urethra

vagina

clitoris

labia

vulva

Figure 6.2 The reproductive system of a human female

Gametogenesis in humans

Gametogenesis is the process of gamete production. The female and male gametes are produced in organs called gonads. The female gonads are **ovaries** and the gametes they produce are called **ova** (singular: ovum or, simply, egg cell). The male gonads are **testes** and the gametes they produce are called **spermatozoa** (singular: spermatozoon or, simply, sperm cell).

Whilst there are several differences in outcome, the process of gametogenesis in females and males shares a common sequence of three phases:

● **multiplication** (proliferation), in which cells present in a layer called the germinal epithelium divide by mitotic cell division. This division is repeated to produce many cells capable of becoming gametes

Key terms

Gametogenesis The process by which gametes are produced. In animals, egg cells are produced by oogenesis and sperm cells are produced by spermatogenesis.

Gonad An animal organ that produces gametes.

Germinal epithelium A single layer of cells (epithelium) that undergoes regular mitosis, producing other cells.

- then each of these cells undergoes **growth**
- finally **maturation** occurs, which involves meiosis and results in the formation of haploid gametes.

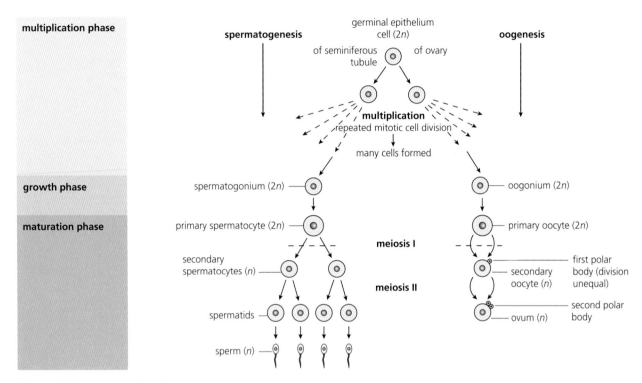

Figure 6.3 Gametogenesis in humans

You can see these phases in Figure 6.3. The early stages look similar. In the multiplication phase of both, mitosis of cells in the germinal epithelium gives rise to cells that, in turn, undergo repeated mitosis to produce large numbers of either **oogonia** or **spermatogonia**. This is followed by a growth phase in which these cells develop into either **primary oocytes** or **primary spermatocytes** that can enter the maturation phase and undergo meiosis.

If you look at what happens after the formation of primary oocytes and primary spermatocytes in Figure 6.3, you can see a difference between oogenesis and spermatogenesis.

The primary oocyte produces one haploid secondary oocyte and a smaller, haploid, polar body, which is lost. Meiosis II of the secondary oocyte has a similar outcome – a large ovum and a smaller polar body, which is lost. Only one gamete – the ovum – is produced from a primary oocyte.

In contrast, the primary spermatocyte produces haploid secondary spermatocytes that are the same size, both of which undergo meiosis II to produce four gametes.

What Figure 6.3 does not show is the timing of oogenesis and spermatogenesis, and herein lies another difference. In a human male, the entire process of spermatogenesis shown in Figure 6.3 occurs in his testes only after puberty. He then produces millions of sperm cells every day throughout his life. In a female, all the stages up to production of primary oocytes occur in her ovaries whilst she is a fetus. As a result, a girl is born with thousands of primary oocytes already formed in her ovaries. Unusually, almost all of these cells remain in the prophase of meiosis II. The stimulus needed for them to complete meiosis II is fertilisation.

Table 6.1 shows a fuller comparison of oogenesis and spermatogenesis.

Table 6.1 A comparison of oogenesis and spermatogenesis

Oogenesis	Spermatogenesis
Oogonia formed in the embryonic ovaries, long before birth.	Spermatogonia formed from the time of puberty, throughout adult life.
Oogonia become surrounded by follicle cells, forming tiny primary follicles, and remain dormant within the ovary cortex. Most fail to develop further – they degenerate.	All spermatogonia develop into sperm, nurtured by the Sertoli cells in the seminiferous tubules of the testes.
Each month from puberty until the menopause, a few primary oocytes undergo meiosis I to become secondary oocytes. Only one of these secondary oocytes forms a Graafian follicle – the others degenerate.	Millions of sperm are formed *daily*.
One ovum is formed from each oogonium (the polar bodies degenerate too).	Four sperm are formed from each spermatogonium.
The Graafian follicle releases a secondary oocyte into the oviduct at ovulation.	Sperm are released from the body by ejaculation.
Meiosis II reaches prophase and then stops until a male nucleus enters the secondary oocyte, triggering completion of meiosis II.	Meiosis I and II go to completion during sperm production.
The fertilised ovum is non-motile and becomes lodged in the endometrium of the uterus where cell divisions (cleavage) lead to embryo formation.	Sperm are small, motile gametes.

Test yourself

1 How does a gamete differ from a somatic cell from the same organism?

2 Explain why biologists refer to a person's sex rather than gender.

3 Trisomy can result in humans with the karyotype XXY. What sex will this person be? Explain your answer.

4 State the function of a germinal epithelium.

5 Give **two** ways in which the *processes* of oogenesis and spermatogenesis differ.

The ovary and secondary oocyte

Figure 6.4 puts the process of oogenesis into the context of an ovary.

You can see in Figure 6.4 that the germinal epithelium is an outer layer of the ovary. You can also see how oocytes develop inside structures called follicles. Each follicle starts as a layer of cells around the oocyte but eventually develops into a fluid-filled sac, called a **Graafian follicle**. You can also see in Figure 6.4 that ovulation involves the release from a follicle of a secondary oocyte, still surrounded by a layer of follicle cells. In a woman of reproductive age, ovulation occurs about every 28 days. The empty Graafian follicle becomes filled with hormone-secreting cells, forming a yellow body, or **corpus luteum**. If the secondary oocyte is not fertilised, the corpus luteum quickly degenerates.

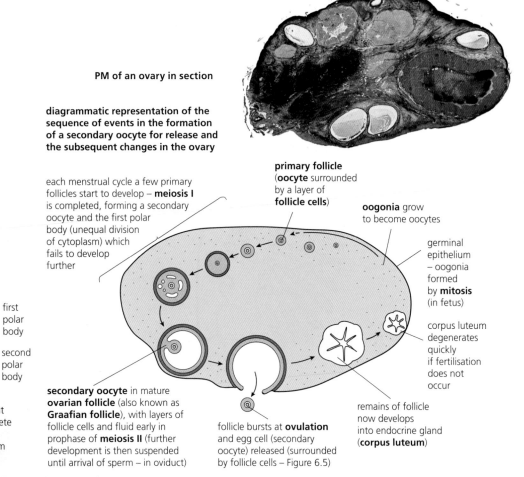

summary of changes from oogonium to ovum
– steps in the growth and maturation phases of gametogenesis in the ovary

diagrammatic representation of the sequence of events in the formation of a secondary oocyte for release and the subsequent changes in the ovary

each menstrual cycle a few primary follicles start to develop – **meiosis I** is completed, forming a secondary oocyte and the first polar body (unequal division of cytoplasm) which fails to develop further

primary follicle (oocyte surrounded by a layer of **follicle cells)**

oogonia grow to become oocytes

growth — oogonium

maturation — primary oocyte (2*n*)

meiosis I —

first polar body
— secondary oocyte (*n*)

meiosis II

— ovum (*n*)
second polar body

Secondary oocyte **begins** meiosis II but this does not complete until sperm nucleus penetrates cytoplasm of oocyte.

secondary oocyte in mature **ovarian follicle** (also known as **Graafian follicle**), with layers of follicle cells and fluid early in prophase of **meiosis II** (further development is then suspended until arrival of sperm – in oviduct)

germinal epithelium – oogonia formed by **mitosis** (in fetus)

corpus luteum degenerates quickly if fertilisation does not occur

remains of follicle now develops into endocrine gland (**corpus luteum**)

follicle bursts at **ovulation** and egg cell (secondary oocyte) released (surrounded by follicle cells – Figure 6.5)

Figure 6.4 A human ovary and stages in oogenesis

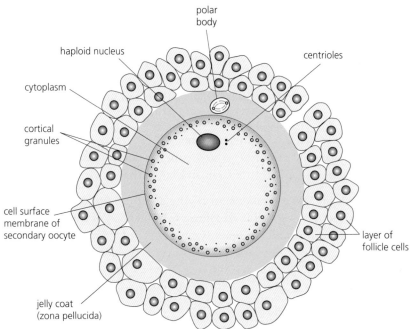

polar body

haploid nucleus

centrioles

cytoplasm

cortical granules

cell surface membrane of secondary oocyte

jelly coat (zona pellucida)

layer of follicle cells

Figure 6.5 A secondary oocyte after ovulation

Figure 6.5 is an enlarged drawing of a secondary oocyte, surrounded by follicle cells. At about 150 μm in diameter, the oocyte is much larger than the follicle cells. Notice the **cortical granules** – vesicles around the outside of the cytoplasm of the oocyte. As you will see later, these vesicles play an important role in ensuring the oocyte can be fertilised by only one sperm cell. Also notice the layer of glycoprotein, called the **zona pellucida**, between the oocyte and the follicle cells. You will see the importance of this when we consider fertilisation.

Sexual reproduction in mammals

The testis and spermatozoa

Figure 6.6 puts the process of spermatogenesis into the context of a testis. Unlike an ovary, a testis is filled with tiny tubules, called **seminiferous tubules**, in which sperm are made. You can see in the lower drawing of Figure 6.6 that the germinal epithelium is close to the outer edge of each seminiferous tubule. You can also see spermatocytes and mature spermatozoa. Notice how the heads of the spermatozoa are positioned against large **Sertoli cells**. These cells nourish sperm as the mature.

photomicrograph of TS of seminiferous tubule (×1000)

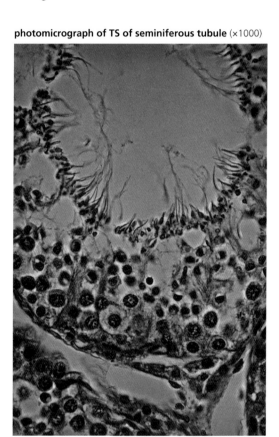

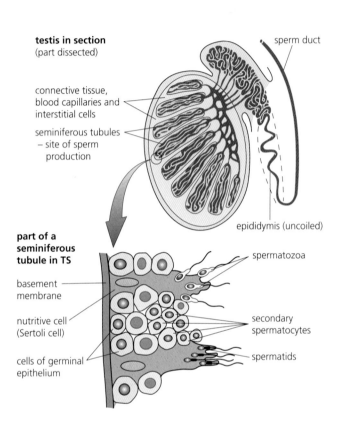

Figure 6.6 A seminiferous tubule and stages in spermatogenesis

Figure 6.7 shows the structure of an individual sperm cell. You can see it is divided into three sections:

- a **head**, containing a haploid nucleus and an acrosome
- a **middle piece**, packed with mitochondria – the organelles that produce ATP
- a **tail**, containing microtubules in an arrangement similar to that of a cilium.

The sperm is a very small cell: its 'head' is about 5 μm long and 3 μm wide and its 'tail' is about 50 μm long. Although not visible in Figure 6.7, the head of the sperm is coated with glycoproteins picked up from the epididymis. These glycoproteins must be removed to make the sperm capable of fertilisation. This occurs inside the uterus and oviduct of the female.

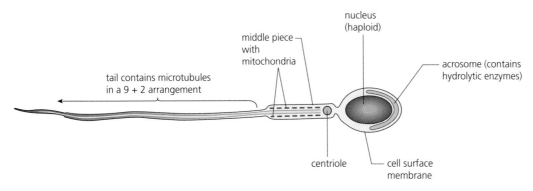

middle piece
with
mitochondria

nucleus
(haploid)

acrosome (contains
hydrolytic enzymes)

tail contains microtubules
in a 9 + 2 arrangement

centriole

cell surface
membrane

Figure 6.7 A mature spermatozoon

Test yourself

6 Use information in the text to calculate the ratio of the diameter of a human ovum to the width of the head of a human sperm cell.

7 In most animals, the sperm is much smaller than the ovum. Suggest **two** advantages of this difference in size.

8 Does ovulation involve release of the ovum from a human ovary? Explain your answer.

9 Give the role of the polar bodies produced during oogenesis.

10 What is the difference between a primary oocyte and a secondary oocyte?

Fertilisation in humans

Following ovulation, the secondary oocyte is captured by the oviduct (Figure 6.8) and is moved along it by the beating action of cilia on the cells lining the oviduct. Since the oocyte quickly becomes disorganised, successful fertilisation can only occur in the upper part of the oviduct. Of the millions of sperm deposited in the vagina, only a few hundred actually reach the upper part of the oviduct. While in the uterus and oviduct, these sperm have become capable of fertilisation by a process called capacitation. This process involves two events that change the head of the sperm.

- The cell surface membrane of the sperm head is stripped of the glycoproteins it acquired during its time in the epididymis.
- The acrosome reaction occurs. During this reaction, the acrosome swells and fuses with the cell surface membrane of the sperm, releasing its hydrolytic enzymes. Of equal importance, the exposed remains of the membrane around the acrosome acquire the potential to fuse with the cell surface membrane of the secondary oocyte.

Both events are important in enabling sperm cells to become fully capable of fertilisation.

Figure 6.8 shows the events involved in fertilisation. One or more of the few capacitated spermatozoa to reach the secondary oocyte begin to pass between the follicle cells surrounding the oocyte. Once through, these spermatozoa encounter the jelly-like layer, the zona pellucida. As a result of the acrosome reaction mentioned above, enzymes released from the acrosomes hydrolyse the glycoprotein from which the zona pellucida is made. This allows the passage of the spermatozoa to the surface membrane of the oocyte.

Key terms

Capacitation Changes to the head of a sperm cell, making it fully capable of fertilisation. This process occurs within a few hours of a sperm cell entering the uterus of a female.

Acrosome reaction One event during capacitation in which a sperm's acrosome releases its hydrolytic enzymes into the zona pellucida and its cell surface membrane becomes capable of fusing with the cell surface membrane of the secondary oocyte.

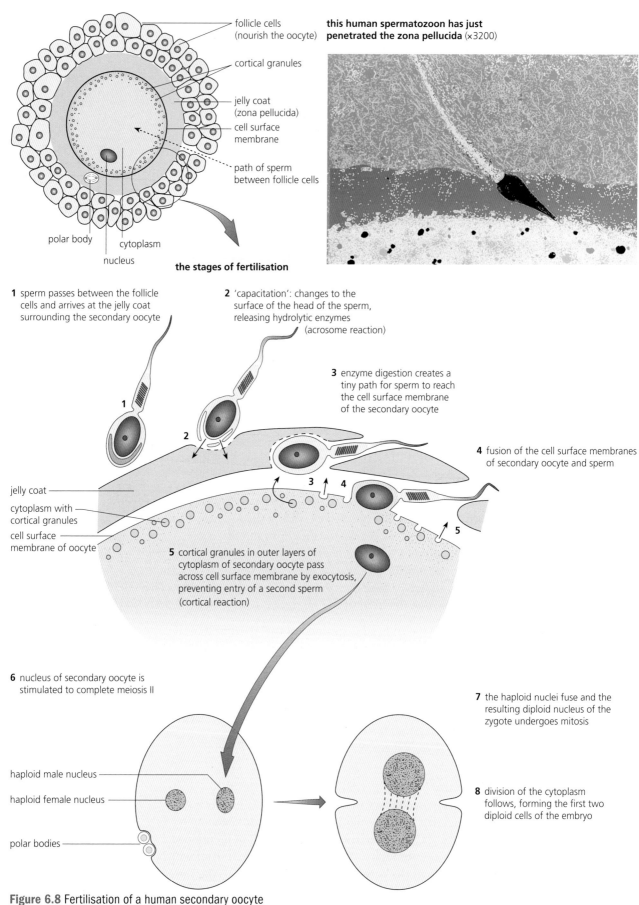

follicle cells
(nourish the oocyte)

cortical granules

jelly coat
(zona pellucida)

cell surface
membrane

path of sperm
between follicle cells

polar body

cytoplasm

nucleus

**this human spermatozoon has just
penetrated the zona pellucida** (×3200)

the stages of fertilisation

1 sperm passes between the follicle
cells and arrives at the jelly coat
surrounding the secondary oocyte

2 'capacitation': changes to the
surface of the head of the sperm,
releasing hydrolytic enzymes
(acrosome reaction)

3 enzyme digestion creates a
tiny path for sperm to reach
the cell surface membrane
of the secondary oocyte

4 fusion of the cell surface membranes
of secondary oocyte and sperm

jelly coat

cytoplasm with
cortical granules

cell surface
membrane of oocyte

5 cortical granules in outer layers of
cytoplasm of secondary oocyte pass
across cell surface membrane by exocytosis,
preventing entry of a second sperm
(cortical reaction)

6 nucleus of secondary oocyte is
stimulated to complete meiosis II

7 the haploid nuclei fuse and the
resulting diploid nucleus of the
zygote undergoes mitosis

haploid male nucleus

haploid female nucleus

polar bodies

8 division of the cytoplasm
follows, forming the first two
diploid cells of the embryo

Figure 6.8 Fertilisation of a human secondary oocyte

6 Sexual reproduction in mammals and plants

The head of a spermatozoon now comes to lie tangentially in contact with the cell surface membrane of the secondary oocyte. Microvilli on the surface of the oocyte then engulf the sperm head. As soon as this happens, there is a sudden increase in the concentration of calcium ions within the cytoplasm of the oocyte. This increase in calcium ions causes:

- the **cortical reaction**, in which the cortical granules are released from the oocyte. This causes the zona pellucida to harden, the result being that no other sperm can now cross the cell surface membrane of the oocyte. This reaction prevents polyspermy – the entry of more than one sperm into the oocyte
- the secondary oocyte to complete meiosis II, producing a haploid **ovum** and another smaller, haploid polar body.

The haploid nucleus from the sperm now fuses with that of the ovum, forming the diploid nucleus of a cell that is now called a **zygote**.

Key term

Cortical reaction
Release of cortical granules from a fertilised secondary oocyte, preventing the entry of further sperm cells.

Early development and implantation

Following fertilisation, the zygote is moved down the oviduct by ciliary action. It arrives in the uterus around 3 to 4 days later. As shown in Figure 6.9, during this

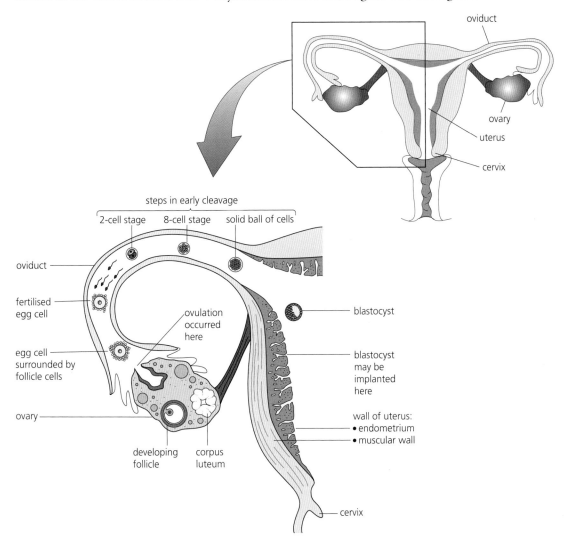

Figure 6.9 The site of fertilisation and early stages of development

Key terms

Blastocyst A fluid-filled ball of cells (called blastomeres) formed by the repeated cell division of a zygote. In humans, a blastocyst is formed 4 to 5 days after fertilisation.

Blastomeres The cells formed by cleavage divisions of the zygote as it is moved down the oviduct.

time, mitosis and cell division occur. Each of these cell divisions is referred to as a **cleavage division** and results in new cells, called blastomeres. Notice in Figure 6.9 that no growth occurs at this stage. You can see that, as more cleavage divisions occur, the blastomeres become progressively smaller. After about 4 days, the zygote has become a solid ball of tiny blastomeres. Between now and the 7th day after fertilisation, a number of important changes occur. You can see some of these in the upper part of Figure 6.10. The ball of blastomeres now contains about 128 cells and it has grown. Instead of being a solid ball, it has now formed a hollow ball of cells, called a blastocyst. The hollow ball has an outer layer of cells, now called **trophoblasts**, an inner cell mass of blastomeres and a fluid-filled **blastocoel**.

What you cannot see in Figure 6.10 is that something more fundamental has occurred.

Firstly, the cells have differentiated to form different types of cell. It is the inner cell mass that is destined to become the **embryo** that will give rise exclusively to a **fetus**. The outer trophoblasts are destined to become a membrane (the amnion) that helps to nourish the embryo and fetus.

blastocyst at about day 7

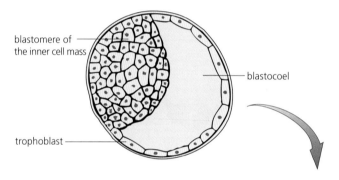

blastomere of the inner cell mass

blastocoel

trophoblast

implanted (14 days after fertilisation)

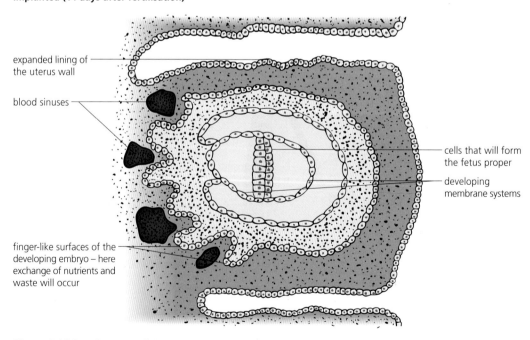

expanded lining of the uterus wall

blood sinuses

cells that will form the fetus proper

developing membrane systems

finger-like surfaces of the developing embryo – here exchange of nutrients and waste will occur

Figure 6.10 Development of the blastocyst and implantation

Secondly, control of development has switched from the mother to the blastocyst. During cleavage in the oviduct, the cytoplasm of the secondary oocyte, including all the cell organelles and mRNA it contained, was used by the blastocyst. Not until about the 8-cell stage does the blastocyst destroy its inherited maternal mRNA and begin to produce its own.

In Figure 6.9, we saw a blastocyst 'floating' in the uterus. From uterine secretions, this blastocyst is able to obtain oxygen and other metabolic substrates and to excrete metabolic wastes, such as carbon dioxide. As you will see in Chapter 9, however, there is a limit to the size this blastocyst could grow depending only on these secretions. It is important at this stage that the blastocyst embeds itself in the lining of the uterus, or endometrium. Look at the endometrium in Figure 6.9. You can see that it is relatively thick and has many invaginations. A developing blastocyst becomes embedded in one of these invaginations in a process called **implantation**. Once implanted, the blastocyst starts to receive nutrients directly from the endometrium. Clearly, as the blastocyst grows, the rate at which it uses nutrients will increase. This in turn is matched by an increase in uptake of nutrients facilitated by the development of a **placenta**.

The placenta contains both maternal and embryonic tissue. In the lower part of Figure 6.10, you can see the beginnings of the placenta. Blood vessels in the maternal tissue have begun to fuse to form the **blood sinuses** you can see in the drawing. The trophoblast layer of the implanted blastocyst increases its surface area by producing finger-like villi. Not only do these villi increase the surface area for exchange, they also help to anchor the blastocyst in the endometrium.

The successfully implanted blastocyst will remain in the uterus for about 270 to 290 days, a period called **gestation**. For the first 2 months, it is referred to as an **embryo**, after which it is referred to as a **fetus**.

Key term

Endometrium The lining of the uterus. By the time of ovulation, the endometrium has become thicker and contains more blood vessels. This adaptation enables an embryo to obtain nourishment for its growth and development.

Test yourself

11 Sponges release their gametes into the seawater in which they live; fertilisation is external. In contrast, fertilisation in mammals is internal. Suggest **two** advantages of internal fertilisation over external fertilisation.

12 Suggest **one** disadvantage of internal fertilisation.

13 Give **three** events that occur during capacitation that enable a sperm to fertilise an egg cell.

14 What is the role of the trophoblast layer in a blastocyst?

15 During the first few days, the mother controls the development of a blastocyst. Explain how.

Sexual reproduction in flowering plants

Flowering plants contain their reproductive organs in their flowers (or **inflorescences**). Many plants have flowers that are brightly coloured and conspicuous, like the buttercup shown in Figure 6.11. These features are adaptations that attract insects, on which this plant depends to transfer its pollen to another flower. Others plants, that rely on the wind to transfer their pollen, usually have dull, inconspicuous flowers, for example grasses.

Key term

Hermaphrodite

A multicellular organism that possesses both male and female reproductive structures. Most flowering plants are hermaphrodites. Fewer animals are hermaphrodites, but this condition is common in organisms such as slugs, snails and is seen in some fish.

Figure 6.11 shows a vertical section (VS) through the flower of a meadow buttercup (*Ranunculus acris*), a plant common in the UK. Like many plants, this buttercup is a hermaphrodite, in other words it carries both female and male reproductive structures.

The female structures are known as carpels. You can see in Figure 6.11 that each **carpel** consists of a **stigma** – a platform on which pollen grains may land – a **style** that supports the stigma, and an **ovary**. Inside the ovary, you can see an **ovule** that contains the female gametes. The male structures are known as **stamens**. Again, as you can see in Figure 6.11, each stamen consists of an **anther** and a **filament**. The anther produces pollen grains that contain the male gametes, and the filament supports the anther in a position that will eventually enable it to shed its pollen grains.

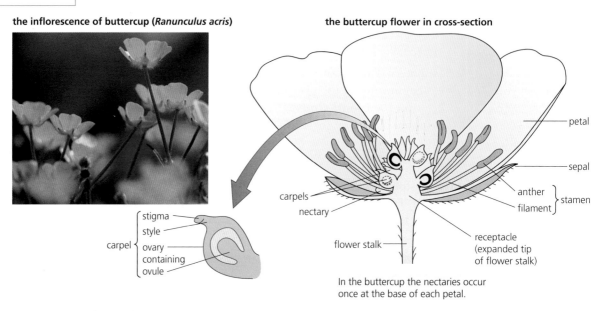

the inflorescence of buttercup (*Ranunculus acris*)

the buttercup flower in cross-section

In the buttercup the nectaries occur once at the base of each petal.

Figure 6.11 The flower of a meadow buttercup (*Ranunculus acris*)

Formation of female gametes

Figure 6.12 a) A vertical section (VS) through an ovary containing a single ovule; b) VS through an immature ovule

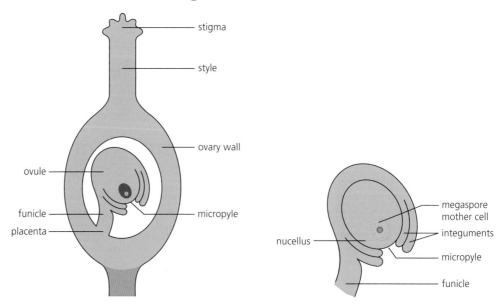

Depending on the species, one or more ovules develop inside an ovary. Figure 6.12 shows an ovary with just one ovule. Within its central mass of tissue, called the **nucellus**, is a large, diploid **megaspore mother cell** surrounded by two layers of cells, called **integuments**. As these integuments grow, they almost enclose the nucellus, leaving only a tiny hole called the **micropyle**.

Figure 6.13 shows how a megaspore mother cell divides by meiosis to produce four haploid cells, the **megaspores**. Three of them disintegrate. The one that survives grows until it almost fills the nucellus. The nucleus of this surviving megaspore divides three times by mitosis to form a cell containing eight haploid nuclei. This single cell with eight nuclei is an immature **embryo sac**. The cytoplasm of the embryo sac then divides. You might expect it to divide to form eight cells. In fact it doesn't. Instead, it divides to form seven cells. The three nearest the micropyle are the ovum and, on either side, the **synergids**. The three furthest from the micropyle are the **antipodal cells**. The two remaining nuclei remain in the centre of the embryo sac. These are the two **polar nuclei**. You can see all these in the mature embryo sac shown in Figure 6.13.

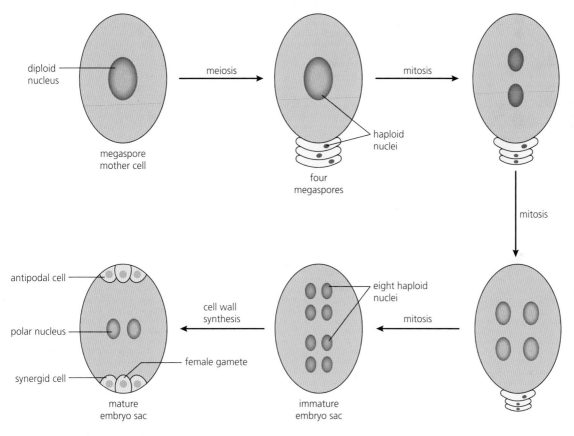

Figure 6.13 The formation of a mature embryo sac

Formation of male gametes

Depending on the species, each anther has two or four lobes. Within the tissues in each lobe are a number of diploid **microspore mother cells**. Each of these cells divides once by meiosis to produce four haploid **microspores**. Each microspore then divides once by mitosis to produce two haploid nuclei:

- the tube nucleus
- the generative nucleus, which is the male gamete.

The microspore – a single cell containing these two nuclei – is now called a **pollen grain**. You can see a large number of pollen grains in the anther of a lily flower in Figure 6.14. If you look closely, you should be able to see both nuclei in some of these pollen grains.

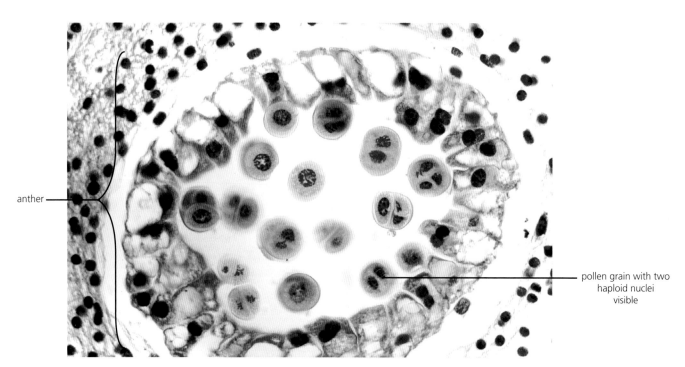

anther

pollen grain with two haploid nuclei visible

Figure 6.14 Vertical section (VS) through an anther of a lily flower

Key term

Pollination Transfer of pollen from an anther to a stigma. Depending on the species of plant, the transfer of pollen could be by wind, water, insects, birds or mammals. The structure of flowers is closely adapted to their method of pollen transfer.

Pollination

Pollination occurs when a pollen grain lands on the stigma of a plant of the same species. **Self-pollination** occurs when pollen is transferred from anther to stigma of the same plant. **Cross-pollination** occurs when pollen is transferred from the anther of one plant to the stigma of another plant. During cross-pollination, the pollen is carried from one plant to another by insects, other animals or the wind, depending on the species of plant. Pollen grains have shapes and surface protrusions that are unique to each plant species. Only if a pollen grain lands on a stigma with complementary patterns can the pollen grain germinate to form a pollen tube.

Core practical 4

Investigate the effect of sucrose concentrations on pollen tube growth or germination

Background information

When a ripe pollen grain lands on the stigma of a plant of the same species, it is dry. Under appropriate conditions, a pollen grain will absorb water and germinate, producing a pollen tube. Depending on the time you have available, you could investigate the effect of different conditions on the germination of pollen grains or on the rate of pollen tube growth. To save time, you could work in groups, each member of the group investigating a different sucrose concentration, and then pool your data.

Carrying out the investigation

Aim: To investigate the effect of sucrose concentrations on the germination of pollen grains.

Risk assessment: It is the responsibility of your centre to carry out an appropriate risk assessment. CLEAPSS *Hazcards* might be helpful. Wear eye protection when performing this experment.

1　Remove the lids from two clean Petri dishes. Place a filter paper in each of the two Petri dishes. Moisten the paper with water and replace the lids.

2　Use the pollen culture medium and $1.2\,mol\,dm^{-3}$ sucrose solution to make up a range of solutions of different sucrose concentrations. You will need no more than $10\,cm^3$ of each sucrose solution.

3　Prepare a table in which you can record your raw data.

4　Use a clean dropper pipette to add a drop of one of your sucrose concentrations to each of two clean microscope slides.

5　Take a flower that is shedding pollen grains. Gently rub its anthers using the tip of a mounted needle so that pollen grains fall onto the drop of sucrose solution on one of your slides.

6　Repeat step 5 so that you have pollen grains in the solution on your second slide.

7　Note the time and carefully place the microscope slides into the Petri dishes, one to each dish. Replace the lids.

8　At regular intervals, examine each slide using a light microscope. Make a note of the time at which the pollen grains begin to germinate.

9　After each observation, carefully return the slides to the Petri dishes and replace the lids.

10　Repeat steps 4 to 9 using another of your sucrose solutions.

Questions

1　One student decided to make her sucrose solutions in the following way. She labelled six test tubes 1 to 6 and added $5\,cm^3$ of pollen culture medium to tubes 2 to 6. After this, she placed $10\,cm^3$ of $1.2\,mol\,dm^{-3}$ sucrose solution into tube 1. She then transferred $5\,cm^3$ of solution from tube 1 to tube 2. She thoroughly mixed the contents of tube 2 and then transferred $5\,cm^3$ of this solution to tube 3. She continued this procedure until she had added $5\,cm^3$ of solution from tube 5 to tube 6. What was the concentration of her solution in tube 4?

2　How did you prepare your table for raw data in step 3?

3　Explain why you should record the time in seconds.

4　Did pollen grains stick to your mounted needle in step 5? If so, how did you remove them?

5　Why were you told to keep the slides in the Petri dishes between observations?

6　How could you use this method to find the sucrose solution at which germination is fastest?

7　What additional equipment would you need to investigate the effect of sucrose concentration on the rate of growth of pollen tubes?

8　How would you calculate the rate of growth of pollen tubes from your raw data?

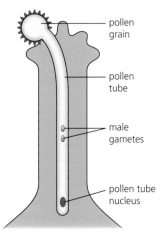

pollen grain

pollen tube

male gametes

pollen tube nucleus

Figure 6.15 Growth of a pollen tube

Fertilisation

To achieve fertilisation, the male gamete within the pollen grain must fuse with the female gamete in the embryo sac. When a mature pollen grain lands on the stigma of an appropriate plant, it absorbs water, swells and splits open. Once open, it forms a pollen tube that grows through the tissues of the stigma, style and ovary towards the embryo sac. In Figure 6.15, you can see the pollen tube nucleus at the tip of the pollen tube, controlling growth of the tube. Growth of the pollen tube is made possible because it digests the recipient plant's tissues as it moves through the style. The absorbed products of digestion provide the raw materials for the growth of the pollen tube. As the generative nucleus follows the pollen tube nucleus down the pollen tube, it divides by mitosis to produce two haploid nuclei. These nuclei are the **male gametes**. You can see them in Figure 6.15.

When it reaches the ovule, the pollen tube grows through the micropyle and into the embryo sac. The tip of the pollen tube breaks down and the pollen tube nucleus disintegrates. Figure 6.16 shows this stage. The two male gametes, present only as nuclei, move into the embryo sac and an event unique to flowering plants then occurs – a **double fertilisation**. One male nucleus fuses with the female gamete, forming a diploid zygote. The other male nucleus fuses with the two polar nuclei to form a triploid **primary endosperm cell**. The other five nuclei within the embryo sac disintegrate. During subsequent development:

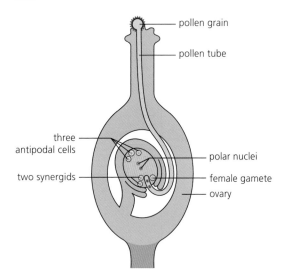

pollen grain

pollen tube

three antipodal cells

two synergids

polar nuclei

female gamete

ovary

Figure 6.16 Double fertilisation in a flowering plant

- the zygote will develop into an embryo plant, with an embryonic root (radical) and embryonic shoot (plumule)
- the primary endosperm cell forms a mass of food tissue called the endosperm
- the embryo sac develops into a seed, with its integuments forming the seed coat
- the wall of the ovary develops into a fruit.

Test yourself

16 What is meant by the term 'hermaphrodite'?

17 Name the cell that gives rise to the embryo sac in a flowering plant.

18 Give **one** way in which a male gamete of a flowering plant is different from a male gamete of a human.

19 Explain the difference between pollination and fertilisation in a flowering plant.

20 Double fertilisation is unique to flowering plants. What is meant by double fertilisation?

Exam practice questions

1 The following mammalian cells are stages in oogenesis. Give the letter, or letters, representing any cell that is diploid.

   A cell in germinal epithelium

   B oogonium

   C primary oocyte

   D secondary oocyte   (1)

2 In human reproduction, the cortical reaction prevents:

   A capacitation of sperm cells

   B disintegration of the polar bodies

   C meiosis II occurring in the secondary oocyte

   D polyspermy   (1)

3 The cell from which the female gamete is produced in a flowering plant is the:

   A antipodal cell

   B micropyle

   C microspore

   D synergid   (1)

4 The photograph shows a false-colour transmission electronmicrograph of a human sperm cell.

   a) Copy and complete the table to shows the name and function of each of the labelled structures A, B and C.   (6)

Structure	Name	Function
A		
B		
C		

b) The head of the sperm cell shown in the micrograph is 3 μm wide. Calculate the magnification of this electronmicrograph. *(2)*

5 A cucumber plant has a haploid chromosome number of 22.

a) Copy and complete the table to show the number of chromosomes in each of the cells identified. One row has been completed for you. *(3)*

Cell	Number of chromosomes
Petal of flower	
Male gamete	22
Zygote	
Endosperm	

b) Explain the answer you gave in the table for the cell in the endosperm. *(2)*

6 During *in vitro* fertilisation (IVF), sperm cells from a donor are mixed in a Petri dish with secondary oocytes.

a) Suggest why early attempts to achieve IVF were unsuccessful. *(4)*

b) Describe what would happen to the secondary oocyte during the 4 days following successful IVF. *(4)*

7 A group of students germinated seeds on moist filter paper. They investigated changes in the dry mass of two components of seeds during germination. Their results are shown in the table.

Time since sowing/days	Mean dry mass/mg	
	Embryo	Endosperm
0	5.1	44.6
2	5.2	43.9
4	6.9	35.8
6	12.7	17.4
8	19.6	11.3

a) Explain why the students recorded mass as dry mass. *(2)*

b) Suggest how the students could have determined the dry mass of the embryos. *(2)*

c) Describe how you would use the data in the table to calculate the rate of growth of the embryo during the first 4 days. *(2)*

d) Explain the relationship between the dry mass of the embryo and the dry mass of the endosperm. *(3)*

8 Primroses are flowering plants that are common in gardens in the UK. Within a population of primroses there are two types of plant – those with only pin-eyed flowers and those with only thrum-eyed flowers. The diagram shows the difference between these two types of flower.

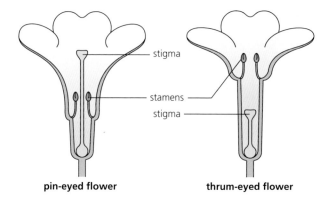

pin-eyed flower   thrum-eyed flower

a) Describe the differences between a pin-eyed flower and a thrum-eyed flower. *(2)*

b) Primroses secrete a sugary fluid, called nectar, at the base of their flowers. Bees collect this nectar by inserting their long mouthparts into both types of flower. As they do so, pollen from the flower sticks to their mouthparts.

    i) Use information from the diagram to suggest how the structure of pin-eyed and thrum-eyed primrose flowers ensures that cross-pollination occurs. *(4)*

    ii) Suggest the biological advantage of cross pollination. *(3)*

Stretch and challenge

9 Release of sperm is called ejaculation and the fluid released is called the ejaculate.

a) In humans, the mean volume of a single ejaculate is $3.4\,cm^3$ and the mean concentration of sperm is $100\,000$ sperm mm^{-3}. Use these figures to calculate the mean number of sperm in a single ejaculate. Give your answer in standard form.

b) If a man has fewer than $20\,000$ sperm mm^{-3} of ejaculate, he is likely to be infertile. Suggest why this is the case.

A clinical technician can assess the fertility of a man using the nomogram shown in the diagram.

To use the nomogram, the technician:

- draws a straight line between the observed number of sperm (scale A) and the percentage of sperm that are motile after 2 hours (scale C)

- she draws a straight line from the intersection of this line with scale B, to scale E

- reads the fertility index where this second line crosses scale D.

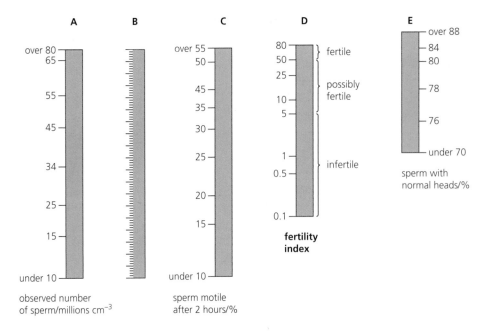

A
over 80
65
55
45
34
25
15
under 10

observed number
of sperm/millions cm^{-3}

B

C
over 55
50
45
35
30
25
20
15
under 10

sperm motile
after 2 hours/%

D
80 } fertile
50 }
25
} possibly
10 } fertile
5
1
} infertile
0.5
0.1

**fertility
index**

E
over 88
84
80
78
76
under 70

sperm with
normal heads/%

c) Use the nomogram to assess the fertility of a man whose ejaculate
contained 25 million sperm per cm^3, 35% of which were motile
after 2 hours and 82% of which had normal heads.

10 A technician investigated the growth of pollen tubes. She added pollen
grains to a suitable growth medium. At regular intervals, she removed
a sample from the culture medium and placed drops of this sample
onto individual glass slides. She then examined the slides under a light
microscope and measured the lengths of the pollen tubes. Her results are
shown in the first graph. The bars represent one standard deviation of the
mean (1 × SD).

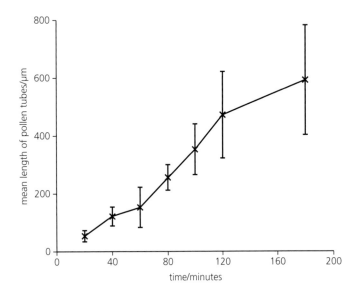

a) How would the technician modify the light microscope in order to measure the length of pollen tubes?

b) What can you conclude from the data in the graph?

The technician repeated her experiment to investigate the effect of certain inhibitors on the growth of pollen tubes. She used three growth media:

• Medium **A** – normal growth medium

• Medium **B** – normal growth medium plus actinomycin D, an inhibitor of DNA transcription

• Medium **C** – normal growth medium plus cycloheximide, an inhibitor of mRNA translation.

Her results are shown in the second graph.

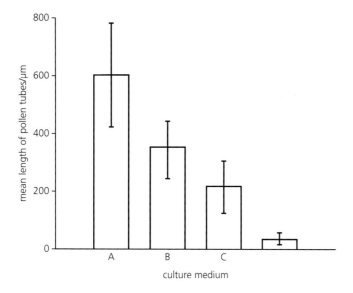

c) Use information in this question to suggest explanations for the effects of actinomycin D and cycloheximide on the growth of pollen tubes.

Classification

7

Prior knowledge

In this chapter you will need to recall that:

→ living things are normally given two names: a genus name and a species name; this is known as binomial classification

→ accurate classification of living organisms is extremely important to all scientists

→ identifying separate species is often very difficult

→ there are many undiscovered species on Earth

→ the basic structure of DNA is a double helix held together by hydrogen bonding

→ the genetic code is made up of four nucleotides containing the bases adenine, guanine, cytosine and thymine (for details, see Chapter 3).

Test yourself on prior knowledge

1 State the binomial names of the following organisms:
   a) lion
   b) tiger
   c) daisy
   d) dandelion
   e) common edible mushroom.

2 Lions and tigers are closely related and could interbreed. Give **two** reasons why they do not normally interbreed.

3 List the full classification of the lion from kingdom to species.

4 Name the **three** components of a DNA nucleotide.

5 Which of the bases in DNA are purines?

6 How are the two strands of DNA held together?

7 What are the main differences between DNA and RNA?

8 What are the names of the five kingdoms that have been used in one method of classification?

9 What are the differences between prokaryotes and eukaryotes?

The range of living things and their classification

There are vast numbers of living things in the world – almost unlimited diversity in fact. No other aspect of life is more characteristic than this great variety of different organisms. Up to now, about 2 million species have been described and named. Meanwhile, previously unknown species are being discovered all the time. We will return to this issue of the diversity of living things – referred to as 'biodiversity' – in the next chapter (Chapter 8).

What we mean by 'species'

The term **species** refers scientifically to a particular type of living thing. We are now confident that living things change with time, and that species have evolved, one from another (Chapter 8).

Later in this chapter we will return to the issue of defining the term 'species' because there are limitations to our definition. First of all we shall locate species in the context of an agreed scheme of classification.

Key term

Species A group of living organisms with similar characteristics that interbreed to produce fertile offspring.

Taxonomy – the classification of diversity

Classification is essential to biology because there are too many different living things to study and compare unless they are organised into manageable categories.

Biological classification schemes are the invention of biologists, based upon the best available evidence at the time. With an effective classification system in use, it is easier to organise our ideas about organisms and make generalisations.

The science of classification is called **taxonomy**. The word comes from 'taxa' (sing. = taxon), which is the general name for groups or categories within a classification system. The scheme of classification has to be flexible, allowing newly discovered living organisms to be added into the scheme where they fit best. It should also include fossils, since we believe living and extinct species are related.

The process of classification involves:

- giving every organism an agreed name
- imposing a scheme upon the diversity of living things.

Key term

Taxonomy The science of classification of living things.

The binomial system of naming

Many organisms have local names, but these often differ from locality to locality around the world, so they do not allow observers to be confident they are talking about the same thing. For example, in America the name 'robin' refers to a bird the size of the European blackbird – altogether a different bird from the European robin. Instead, scientists use an international approach called the binomial system (meaning 'a two-part name'). By this system everyone, anywhere in the world, knows exactly which organism is being referred to.

Figure 7.1 'Magpie' species of the world

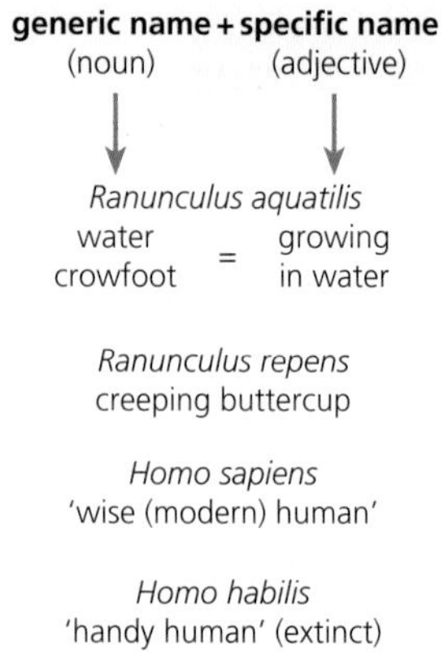

generic name + specific name
(noun)   (adjective)

Ranunculus aquatilis
water   =   growing
crowfoot   in water

Ranunculus repens
creeping buttercup

Homo sapiens
'wise (modern) human'

Homo habilis
'handy human' (extinct)

Figure 7.2 Naming organisms by the binomial system

So each organism is given a scientific name consisting of two words in Latin (Figure 7.1). The first (a noun) designates the **genus**, the second (an adjective) the species. The generic name begins with a capital letter, followed by the specific name. Conventionally, this name is written in *italics* (or is <u>underlined</u>).

As shown in Figure 7.2, closely related organisms have the same generic name; only their species names differ. You will see that when organisms are frequently referred to the full name is given initially, but thereafter the generic name is shortened to the first (capital) letter. Thus, in continuing references to humans in an article or scientific paper, *Homo sapiens* would become *H. sapiens*.

The scheme of classification

In classification, the aim is to use as many characteristics as possible in placing similar organisms together and dissimilar ones apart. Just as similar species are grouped together into the same genus (plural = genera), so, too, are similar genera grouped together into families. This approach is extended from families to orders, then classes, phyla and kingdoms. This is the hierarchical scheme of classification; each successive group containing more and more different kinds of organism. The taxa used in taxonomy are given in Figure 7.3.

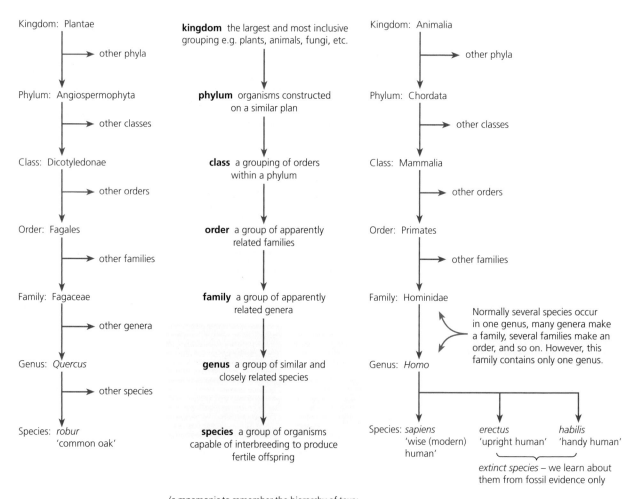

(a mnemonic to remember the hierarchy of taxa:
King **P**eter **C**alled **O**ut **F**or **G**enuine **S**cientists)

Figure 7.3 The taxa used in taxonomy, applied to the genera from two different kingdoms

The features of organisms selected in classification

The *quickest* way to classify living things is on their immediate and obvious similarities and differences. For example, we might classify together animals that fly, simply because the essential organs – wings – are so easily seen. This would include almost all birds and many insects (as well as the bats and certain fossil dinosaurs). However, resemblances between the wings of the bird and the insect are superficial. Both are aerofoils (structures that generate 'lift' when moved though the air); they are built from different tissues and have different origins in the body. We say that the wings of birds and insects are **analogous structures**.

Consequently, they illustrate only superficial resemblances. A classification based on analogous structures is an artificial classification.

Alternatively, a natural classification is based on similarities and differences due to close relationships between organisms because they share common ancestors. The bone structure of the limbs of all vertebrates suggests they are modifications of a common plan we call the pentadactyl limb. So, there are many comparable bones in the human arm, the leg of a horse and the limb of a mole – in other words they are **homologous structures** (Figure 7.4).

Key terms

Genus A division of biological classification that is ranked above species but below family.

Analogous structures These resemble each other in function but differ in their fundamental structure.

Homologous structures These are structures built to a common plan but adapted for different purposes.

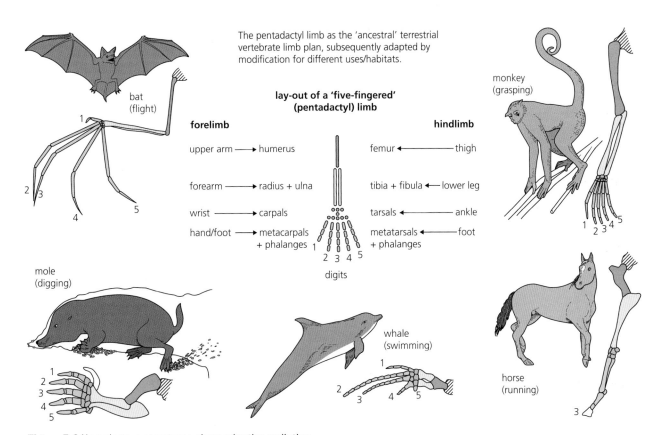

Figure 7.4 Homologous structures show adaptive radiation

The ideal solution – a classification system based on evolutionary relationships

A natural classification based on homologous structures is believed to reflect evolutionary relationships. It is what taxonomists work towards.

Today, similarities and differences in the biochemistry of organisms have become important in taxonomy in addition to structural features. For example, the composition of nucleic acids (and cell proteins) often shows relationships more accurately than structural features. Large molecules like nucleic acids are subjected to changes with time – we call these changes in nucleic acids mutations. Biochemical changes like mutations in DNA occur at a more or less constant rate, and can be used as a 'molecular clock'. It is possible to estimate the relatedness of different groups of organisms by the amount of variation in their DNA – which is a function of time since particular organisms share a common ancestor. Since the rate of change can be reliably estimated, the extent of change is a function of the time that has passed between the separations of evolutionary lines.

DNA sequencing

DNA sequencing used to be a laborious laboratory analysis, where pieces of DNA some 50 000–100 000 bases long would take a year to analyse. Today the whole process has been refined and is now carried out automatically by very expensive machines capable of doing the same analysis in a few hours. Despite this automation the principles of the analysis are very similar:

1 The DNA molecule is cut into pieces at very specific points by enzymes.

2 The pieces of DNA are chemically modified and tagged with fluorescent dyes, which give a different colour for each base.

3 These pieces of DNA are then separated by electrophoresis and the bases recognised by the colour of their fluorescence.

4 When the sequence of these pieces of DNA are recorded they are then linked together to make up the base sequence of the whole DNA molecule.

Extracting and cutting up DNA

DNA can be extracted from tissue samples by mechanically breaking up the cells, filtering off the debris and breaking down cell membranes by treatment with detergents. The protein framework of the chromosomes is then removed by incubation with a protein-digesting enzyme (protease). The DNA, now existing as long threads, is isolated from this mixture of chemicals by precipitation with ethanol and is thus 'cleaned'. The DNA strands are then re-suspended in aqueous, pH-buffered medium. They are now ready for 'slicing' into fragments.

The DNA is sliced or chopped into fragments by addition of restriction endonucleases (restriction enzymes). These enzymes occur naturally in bacteria, where they protect against viruses that enter the bacterium by cutting the viral DNA into small pieces, thereby inactivating it. (Viral DNA might otherwise take over the host cell.) Viruses that specifically parasitise bacteria are called bacteriophages or phages. Restriction enzymes were so named because they restrict the multiplication of phage viruses.

> ## Key term
>
> **DNA sequencing** The process of determining the exact order of the nucleotides in a DNA molecule.

Many different restriction enzymes have been discovered and purified, and today they are used widely in genetic engineering experiments. A distinctive and important feature of restriction enzymes is that they cut at particular base sequences (Figure 7.5) and are of two types, forming either 'blunt ends' or 'sticky ends' to the cut fragments. Sticky ends are single-stranded extensions formed in the double-stranded DNA after 'digestion' with a restriction enzyme that cuts in a staggered fashion. In DNA profiling, a selected restriction enzyme is used to cut at specific base-sequence sites.

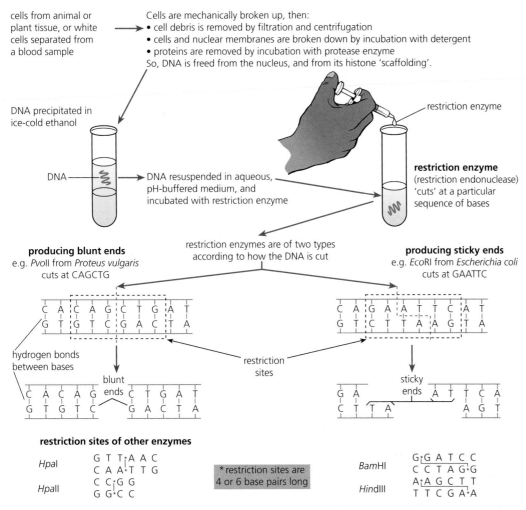

Figure 7.5 Isolating and cutting of DNA

Isolating DNA fragments – electrophoresis

Electrophoresis is a process used to separate particles, including biologically important molecules such as DNA, RNA, proteins and amino acids. It is typically carried out on an agarose gel (a very pure form of agar) or on polyacrylamide gel (PAG). Both these substances contain tiny pores, which allow them to act like a molecular sieve. Small particles can move through these gels quite quickly, whereas larger molecules move much more slowly.

Biological molecules separated by electrophoresis also carry an electrical charge. In the case of DNA, phosphate groups in DNA fragments give them a net negative charge. Consequently, when DNA molecules are placed in an electric field they migrate towards the positive pole.

Key term

Electrophoresis
A technique of separating charged ions in a fluid by applying a potential difference.

So, in electrophoresis, separation occurs according to the size and the charge carried. This is the double principle of electrophoretic separations. Separation of DNA fragments produced by the actions of restriction enzymes is shown in Figure 7.6. Note that the bands of DNA fragments are not visible on the gel until, in this case, a DNA-binding fluorescent dye as been added.

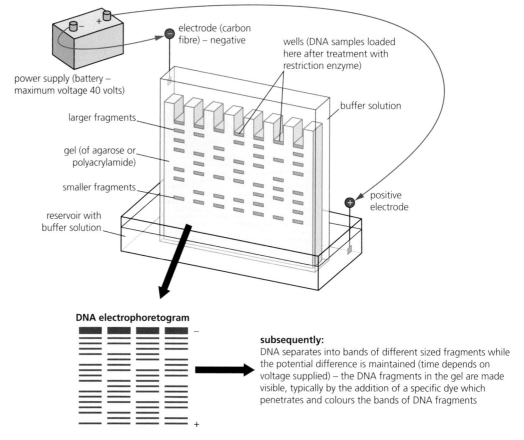

Figure 7.6 Electrophoretic separation of DNA fragments

03_30 Biology for the IB Diploma Second edition
Barking Dog Art

DNA differences used to determine evolutionary relationships

The relatedness of organisms is studied experimentally by investigation of differences in DNA.

1 By DNA hybridisation

The genetic differences between the DNA of various organisms give us data on degrees of divergence in their respective evolutionary histories. By the technique of DNA hybridisation, the matching of DNA samples of different species has enabled the discovery of how closely related particular species are (Figure 7.7).

The degree of similarity of samples of DNA from two organisms is disclosed by measuring the temperature at which they separate. The more distantly related the organisms are, the fewer the bonds (due to base-pairing) that will form between the strands of DNA when mixed. A lower temperature is then required to separate them (Figure 7.7).

DNA hybridisation is a technique that involves matching the DNA of different species, to discover how closely they are related.

DNA extracted from cells and 'cut' into fragments, about 500 bases long

fragments are heated to cause them to become single strands

single strands are mixed with DNA strands from another species, prepared in exactly the same way (therefore comparable)

base pairing causes strands of DNA to align with complementary DNA

the greater the complementarity of the two strands, the more bonds link them together

high complementarity   low complementarity

The closeness of the two DNAs is measured by finding the temperature at which they separate – the fewer bonds formed, the lower the temperature required.

The degree of relatedness of the DNA of **primate species** can be correlated with the estimated number of years since they shared a common ancestor.

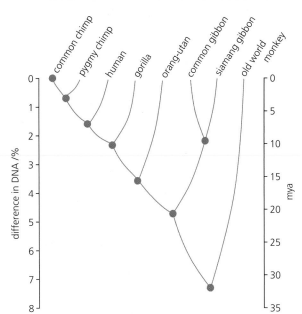

Figure 7.7 Genetic difference between DNA samples and evolutionary relatedness

2 By application of developments in the new discipline of bioinformatics

At the centre of this development is the creation and maintenance of databases concerning nucleic acid sequences (and the proteins derived from them). Already, the genomes of many prokaryotes and eukaryotes have been sequenced, as well as that of humans. This huge volume of data requires organisation, storage and indexing to enable practical use of the subsequent analyses. These tasks involve applied mathematics, informatics, statistics and computer science, and are collectively referred to as **bioinformatics**.

One possible outcome is the use of the GenBank database to determine differences in base sequence of a gene in two species. For example, it is possible to compare the nucleotide sequences of the cytochrome c oxidase gene of humans with that of the Sumatran orang-utan or other species, as shown in Figure 7.8, with a view to determining the degree of relatedness of the two species. This process involves using The National Center for Biotechnology Information web services, which provide access to biomedical and genomic information.

These approaches exploit the fact that organisms that are closely related show fewer differences in the composition of specific nucleic acids (and therefore cell proteins) that they possess. But despite these impressive additional sources of evidence, evolutionary relationships are still only partly understood, so current taxonomy is only partly an evolutionary or **phylogenetic** classification.

Key terms

Bioinformatics The storage, manipulation and analysis of biological information via computer science.

Phylogeny The evolutionary development and history of a species or other group of organisms.

Tip

Learning and understanding the detailed structure of nucleic acids is very useful in many sections of the specification for both AS and A Level.

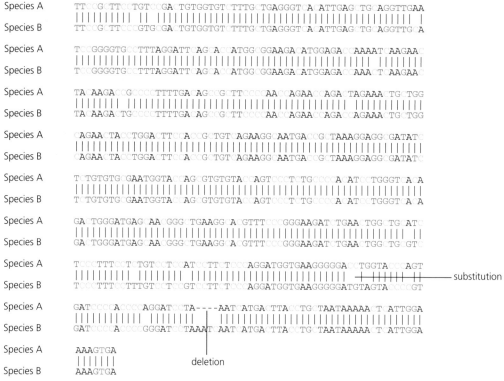

Figure 7.8 Cytochrome oxidase base sequence comparison for two species

Test yourself

1 Explain why using observable features alone may give a false classification of living organisms.

2 Use the diagrams in Figure 7.5 to explain why a restriction enzyme such as *Eco*RI would be used to produce DNA fragments that could be easily attached to other pieces of DNA

3 Which part of the DNA nucleotide gives the molecule an overall negative charge?

4 Which other property of DNA fragments allows them to be separated by the agarose gel?

5 Why will DNA formed from two strands from distantly related organisms separate at a lower temperature?

6 What is the difference between an analogous and a homologous structure?

The five kingdoms

At one time the living world seemed to divide naturally into two kingdoms (Table 7.1).

Table 7.1 Living things divided into two kingdoms

The plants	The animals
Photosynthetic (autotrophic nutrition)	Ingestion of complex food (heterotrophic nutrition)
Mostly rooted (i.e. stationary) organisms	Typically mobile organisms

These two kingdoms grew from the original disciplines of biology, namely botany, the study of plants, and zoology, the study of animals. Fungi and microorganisms were conveniently 'added' to botany! Initially there was only one problem; fungi possessed the typically animal heterotrophic nutrition but were more plant-like in structure.

Later, with the use of the electron microscope came the discovery of the two types of cell structure, namely prokaryotic and eukaryotic. As a result, the bacteria with their prokaryotic cells could no longer be 'plants' since plants have eukaryotic cells. This led to the idea that living things should be divided into five kingdoms (Table 7.2). The evolutionary relationships of the kingdoms are suggested in Figure 7.9.

Table 7.2 The five kingdom classification

Prokaryotae (prokaryotes)	bacteria and cyanobacteria (photosynthetic bacteria), predominately unicellular organisms
Protoctista (protoctists)	eukaryotes, predominately unicellular, and seen as resembling the ancestors of the fungi, plants and animals
Fungi	eukaryotes, predominately multicellular organisms, non-motile, and with heterotrophic nutrition
Plantae (plants)	eukaryotes, multicellular organisms, non-motile, with autotrophic nutrition
Animalia (animals)	eukaryotes, multicellular organisms, motile, with heterotrophic nutrition

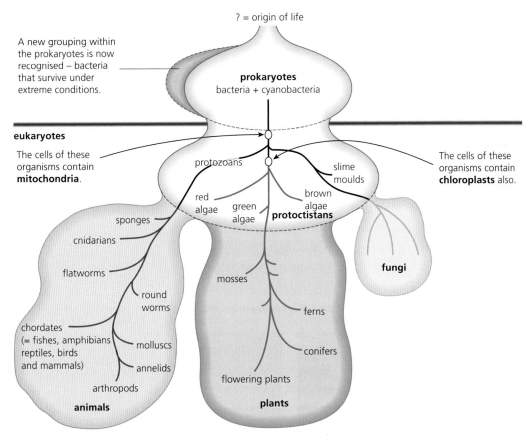

Figure 7.9 Possible evolutionary relationship of the five kingdoms

A new scheme of classification

Then came the discovery of species of bacteria that survive and prosper in extremely hostile environments (the **extremophiles**), such as the 'heat-loving' bacteria found in hot-springs at about 70 °C). Subsequently, extremophiles were found in a wider range of hostile habitats.

Table 7.3 The range of extremophile bacteria

'Salt-loving' bacteria (halophytes)	common in salt lakes and where sea water becomes trapped and concentrated by evaporation and where salt has crystallised
'Alkali-loving' bacteria (alkalinophiles)	survive at above pH 10 – conditions typical of soda lakes
Bacteria that thrive in extremely acidic conditions (acidophiles)	found in conditions of <pH2, such as some sulfur bacteria found in hot, thermal vents
'Heat-loving' bacteria (thermophiles)	occur in hot-springs at about 70 °C; some are adapted to survive at temperatures of 100–115 °C (**hyperthermophilic** prokaryotes)
Bacteria that thrive in sub-zero temperatures	common at temperatures of −10 °C, as in the ice of the poles where salt depresses the freezing point of water

The classification of living organisms into three domains on the basis of their ribosomal RNA

These evolutionary relationships have been established by comparing the sequences of bases (nucleotides) in the ribosomal RNA (rRNA) present in species of each group.

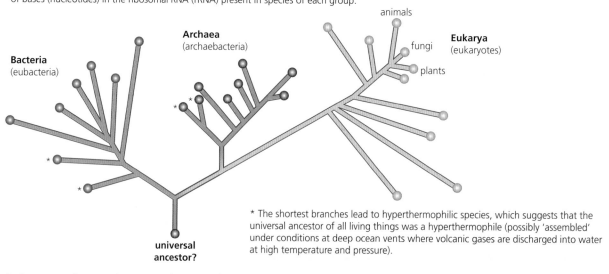

* The shortest branches lead to hyperthermophilic species, which suggests that the universal ancestor of all living things was a hyperthermophile (possibly 'assembled' under conditions at deep ocean vents where volcanic gases are discharged into water at high temperature and pressure).

Archaea were discovered among prokaryotes of extreme and inaccessible habitats. Subsequently, other members of the Archaea were found more widely – in the gut of herbivores and at the bottom of lakes and mountain bogs, for example.

Figure 7.10 Ribosomal RNA and the classification of living organisms

These microorganisms of extreme habitats all have cells that we can identify as prokaryotic. However, the larger RNA molecules present in the ribosomes of extremophiles were discovered to be different from those of previously known bacteria. Further analyses of the biochemistry of extremophiles, in comparison with that of other groups, suggested new evolutionary relationships and led on to a new scheme of classification (Figure 7.10).

Classification into three domains

As a result, we now recognise three major forms of life, called domains. The organisms of each domain share a distinctive, unique pattern of ribosomal RNA and there are other differences, which establish their evolutionary relationships (Table 7.4).

These domains are:

- the **Archaea** (the extremophile prokaryotes)
- the **Eubacteria** (the true bacteria)
- the **Eukaryota** (all eukaryotic cells – the protoctista, fungi, plants and animals).

Incidentally, the Archaea have now been found in an even broader range of habitats than merely extreme environments. Some occur in the oceans and some in fossil fuel deposits deep underground. Some species occur in deep ocean vents, high-temperature habitats such as geysers, in salt pans and in polar environments. Others occur only in anaerobic enclosure such as the guts of termites and of cattle, and at the bottom of ponds, among the rotting plant remains. Here they breakdown organic matter and release methane – with important environmental consequences.

Key term

Domain One of three major forms of life used in the most recent suggestion for classification of living things.

Table 7.4 Biochemical differences between the domains

Biochemical features	Domains		
	Archaea	Eubacteria	Eukaryota
DNA of chromosome(s)	Circular genome	Circular genome	Chromosomes
Bound protein (histone) present in DNA	Present	Absent	Present
Introns in genes	Typically absent	Typically absent	Frequent
Cell wall	Present – not made of peptidoglycan	Present – made of peptidoglycan	Sometimes present – never made of peptidoglycan
Lipids of cell membrane bilayer	Archaeal membranes contain lipids that differ from those of eubacteria and eukaryotes (Figure 7.11)		

Phospholipids of archaeal membranes

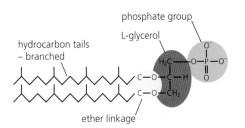

Phospholipids of eubacteria and eukaryote membranes

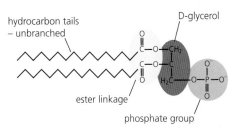

Figure 7.11 Lipid structure of cell membranes in the three domains

The issue of what is a species

On a day-by-day basis, biologists frequently use the term 'species' when they refer to an organism they are studying, within the context of a particular aspect of its biology. For example:

- **ecologists** refer to species as defined by their ecological niche and how they interact with the living and non-living parts of their environment
- **geneticists** refer to species as part of a population whose members have the potential to interbreed and produce viable fertile offspring
- **morphologists** refer to species as defined by common body shape and other structural features by which they are distinguished
- **taxonomists** refer to species as the smallest group of individuals that share a common ancestor – a single 'branch of the tree of life'.

We can see that the term is commonly used on a daily basis, in at least four different contexts, with different meanings and therefore definitions.

Origin of the term 'species'

When Linnaeus devised the binomial system of nomenclature in the eighteenth century there was no problem in defining species. It was believed that each species was derived from the original pair of animals created by God. Since species had been created in this way they were fixed and unchanging.

In fact, present-day living things have arisen by change from pre-existing forms of life. This process has been called 'descent with modification' and 'organic evolution', but perhaps '**speciation**' is better because it emphasises that species change. The fossil record provides evidence that these changes do occur in living things – human fossils alone illustrate this point. We now know that species have evolved, one from another, in the course of the history of life on Earth. The concept of species has been modified – we no longer have a simple definition of a species that is totally accurate in all cases.

Today, as many different characteristics as possible are used in order to define and identify a species. The main characteristics used are:

- morphology and anatomy (external and internal structure)
- cell structure (whether cells are eukaryotic or prokaryotic)
- physiology (blood composition, renal function) and chemical composition (comparisons of nucleic acids and proteins, and the similarities in proteins between organisms, for example).

A **species** can be defined as consisting of organisms of common ancestry that closely resemble each other structurally and biochemically, and which are members of natural populations that are actually or potentially capable of breeding with each other to produce fertile progeny, and which do not interbreed with members of other species.

The last part of this definition cannot be applied to self-fertilising populations or to organisms that reproduce only asexually. Such groups are species because they are very similar to each other morphologically and in all other features.

Having set the scene about the challenge of defining species, we can now agree that on a day-to-day basis the term 'species' is satisfactory and useful provided we think of a species as defined above.

Naming and classifying species – the scale of the task

We have noted there are vast numbers of living things in the world. Up to now, about 2 million species have been described and named. However, until very recently there has been no attempt to produce an international 'library of living things', where new discoveries are automatically checked out (see below). Consequently, some known organisms may have been 'discovered' more than once. Meanwhile, previously unknown species are being discovered all the time. In the UK alone, several hundred new species have been described in the past decade. We might have expected all the wildlife in these islands to be known, since Britain was one of the countries to pioneer the systematic study of plants and animals. Apparently this is not the case; previously unknown organisms are frequently found here, too.

Worldwide, the number of unknown species is estimated at between 3–5 million at the very least, and possibly as high as 100 million. So scientists are not certain just how many different types of organism exist.

Scientists do not agree – the debate continues

This chapter is quite different from others on the same subject. It describes several different ideas about how living things are classified but if you read carefully you will see that all the suggested schemes have drawbacks. The five-kingdom model has a group called Protoctista, whose members are really difficult to define. If you look up a definition of Protoctista you will probably find a description concentrating on features they do not possess rather than features they have in common. You might find something like this: 'The Kingdom Protoctista is defined by exclusion: its members are neither animals (which develop from a blastula), plants (which develop from an embryo), fungi (which lack undulipodia and develop from spores), nor prokaryotes.' In other words it is a collection of organisms that do not necessarily have a great deal in common. Not surprisingly, although this model was widely accepted because it had some clear reasoning, many biologists did not agree.

The discovery of more details of the structure of some bacteria led to the idea that classification should be more strictly based on possible evolutionary history, and hence the very different idea of three domains, with the Eukaryota containing almost all the living things that most people would recognise. Although many biologists accept the logic behind this, once again, many do not agree.

The whole field of classification continues to be hotly debated, as it has for at least 300 years! Even the father of modern classification Carolus Linnaeus (1735) chose to include just plants and animals, ignoring the whole group of single-celled organisms discovered by Anthony van Leeuwenhoek, using the first microscope, some 60 years earlier.

> ### Tip
>
> In exam questions, be prepared to show that you understand that the models of classification are still under debate and that none is perfect. The developments, such as studying DNA from different cells, mean that new evidence is being collected all the time.

Table 7.5 shows some developments of the debate following the introduction of techniques of molecular analysis in the 1970s.

Table 7.5 A recent timeline of the classification debate

Date	Author	Model
1977	Woese	Structure of extremophile RNA suggests they are a separate group. = 6 kingdoms
1990	Woese	More molecular evidence suggests ancient origin of two types of bacteria with all other living things in one other group. = 3 domains
1993	Cavalier-Smith	Other evidence used to dispute the idea that all bacteria and other single-celled protists should be grouped together = 8 kingdoms
1998	Cavalier-Smith	Groups all bacteria together and links all Protozoa = 6 kingdoms

This is a very simplified summary and there have been many other suggestions based on evidence from molecular biology advances since 1998. It is important therefore to understand that all the different models have their strong points and their drawbacks.

Test yourself

7 Give the correct binomial names of five organisms that would be classified as Prokaryotes in the five-kingdom classification.

8 Name the **two** compounds that form a peptidoglycan.

9 Which molecules were first found to be different in Archaea compared with other domains?

10 Who is often described as the father of classification?

11 What was the main problem with defining the group Protoctista?

12 Why are there likely to be many species yet to be discovered?

How do scientists investigate their ideas?

The first thing to understand is that scientific ideas and facts are constantly changing. Some basic principles have stood the test of time, but other ideas have changed dramatically or have needed modification and are still changing. Only 65 years ago biologists did not agree that DNA was the genetic material.

Unfortunately many people think of scientific progress as the story of Archimedes – odd-looking people working away in a strange laboratory and rushing out shouting 'eureka' as they reveal some perfect answer. The truth is that very big steps forward are rare and a great deal of scientific research involves painstakingly repeating investigations to provide reliable evidence for a proposed model, but without this type of research we could never build a solid, reliable body of information.

The process of research begins with an idea or theoretical model. This may be an accepted model or a new approach but it will be only a theory unless there is solid evidence to support it. At this stage it is necessary to make some prediction based on the model, which can be tested experimentally. Designing a reliable investigation to test the prediction requires ingenuity, imagination and extensive background knowledge. To make real advances it is often necessary to design new methods or to use available technology in a novel way. This stage is normally a collaborative effort with a team of other scientists. In this way it is possible to collect evidence to support the model. As more and more evidence is collected and more predictions prove to be correct, then the model will be accepted by most scientists. However, it only takes one well-designed investigation to produce results that contradict the model to undermine it completely.

How do scientists check the validity of investigations?

This is a really important question as many scientific developments are built on the work of others over many years. It is essential that conclusions made in the past and the present are valid and that progress is made based upon reliable information. There are several ways in which evidence is carefully checked and becomes accepted by the scientific community.

1 **Scientific journals**

   All scientists publish full details of their investigations in well-known scientific journals. Their reports must contain full details of their methodology, the original data and an analysis of their findings, following some strict rules. These journals are available to scientists worldwide who can read about the work of others and the latest developments in their field.

2 **Peer review**

   Before a scientific journal will accept work for publication it must be verified by senior scientists in the place where it was carried out. It is then scrutinised by an independent panel of scientists who are experts in the same field. They check the details of the method, the data collected and the validity of conclusions. Peer reviewers often ask for more details or a revision of conclusions before approving its publication. The process of peer review and publication of scientific papers is quite strict and therefore ensures that the information contained in the papers is very reliable. In this way a large body of scientific knowledge and understanding has been built up over many years.

3 **Conferences (symposia)**

   Most important fields of research are carried out by many scientists in several countries. Universities and other institutions often host meetings of scientists from around the world specialising in one particular area of research. At these meetings invited participants often present their latest findings before they have been published. However, the most important function of these meetings is to allow individuals to share ideas, discuss common problems and argue their case where different models are proposed.

Finally it is important to realise that scientists are also human. Debates on the merits of different models can become very heated as proponents defend their ideas.

Tip

Reliability and validity are not the same. Reliability is concerned with collecting data objectively and showing that results are repeatable. Validity is about the design of the investigation, controlling variables effectively and being confident that it is the independent variable that is shown to be affecting the dependent variable.

Tip

This section on how scientists investigate and validate their findings is also relevant to many other parts of the specification so can be tested in any of the exam papers, including those questions testing practical skills.

Exam practice questions

1 Which of the following represents the correct hierarchy of classification?

    A phylum → class → family → order → genus → species

    B phylum → class → order → family → genus → species

    C phylum → class → order → genus → family → species

    D phylum → class → family → genus → order → species   *(1)*

2 Which of the following applies to all members of the same species?

    A have identical external features

    B have the same DNA

    C cannot interbreed with any other species

    D produce fertile offspring with other members of the same species   *(1)*

3 *Halobacterium salinarum* is a bacterium found in very saline environments.

    a) Name two cellular features of this bacterium that could be investigated to show that it is not a eukaryote.   *(2)*

    b) Analysis of the genes for ribosomal RNA (rRNA) is often used to distinguish Archaea from prokaryotes and eukaryotes. Explain why this molecule is particularly useful for this purpose.   *(3)*

    c) Many Archaea are extremophiles, which live in harsh environmental conditions. Some live in hot springs at temperatures of 80 °C or higher.

    What are the major problems faced by cells at these temperatures and how might the modifications of Archaea cell structure help to overcome them?   *(4)*

4 The table shows the base sequence of the same section of DNA taken from the gene for 12S ribosomal RNA in three different animals. All are mammals but the dog and mole are modern placental mammals where the young develop inside the uterus supplied with nutrients through the placenta. Marsupial mammals are largely confined to Australia and are much more primitive, giving birth to tiny underdeveloped young, which are then kept in an external pouch to develop further.

Animal	12S rRNA DNA base sequence
Dog	G G T C C T A G C C T T C C T A T T A G T T T T T A G T A G A C T T A C
Mole	G G T C C C A G C C T T T C T A T T A G C T G T C A G T A A A A T T A C
Marsupial mole	G G T C C T A G C C T T A T T A T T A A T T A T T G C T A G T C C T A C

    a) How many amino acids would be coded by these base sequences?   *(1)*

    b) Count the number of differences in base sequence between:

        i) the mole and the dog   **ii)** the mole and the marsupial mole.   *(2)*

    c) Which two animals are most closely related? Explain your answer.   *(2)*

    d) Explain how the evolutionary history of these animals may account for the relationships between them.   *(4)*

e) i) The strands of DNA from each animal were treated with the restriction enzyme *Hpa*II, which breaks the bond between the bases G–C. How many fragments would be formed from each of the DNA samples shown in the table? *(1)*

ii) Following this enzyme treatment each sample was separated by electrophoresis. Which sample would produce a band on the electrophoresis gel that was closest to the negative electrode? Explain your answer. *(4)*

Stretch and challenge

5 Peptic ulcer disease is a common complaint. Sections of the stomach wall become damaged and the highly acidic contents cause severe pain and can lead to perforation, with the risk of septicaemia. It is also a strong risk factor for stomach cancer.

For many years the main cause was thought to be excess stomach acid. Treatments ranged from simple antacids taken orally to more sophisticated drugs such as hydrogen ion pump inhibitors to limit acid production. Most doctors around the world treated patients in this way and the pharmaceutical industry spent many millions of pounds producing a range of ingenious ways to limit acid production.

This view was challenged by two Australian doctors, Marshall and Warren, between 1980 and 1990. Their story leads from an initial rejection of their research paper to the award of a Nobel prize in 2005. It illustrates that research is a human activity, not always as objective as it might be and subject to many influences. It is also a good example of the role played by peer review, journals and conferences in the process of validation.

You will need to read the story, which is presented as a timeline and can be found by searching for 'Marshall and Warren *Helicobacter* timeline' in a search engine.

Further research into *Helicobacter pylori* will also provide you with interesting background information.

Use this timeline and your own understanding of scientific research to answer the following questions. The abbreviation PUD is used for peptic ulcer disease.

a) What did Marshall and Warren suggest about the role of *Helicobacter pylori* in PUD?

b) Describe two pieces of evidence that suggested a bacterium might be involved in PUD well before Marshall and Warren began their work.

c) Describe the role played by meetings, conferences and congresses held in 1982, 1983, 1984 and 1990.

d) In 1984, Marshall carried out a very unusual demonstration. Why would the scientific community consider this to be of very low validity?

e) In 1994 the patents for the popular drugs used to reduce acid in the stomach ran out. Why would this make drug companies less likely to oppose the introduction of the new antibiotic treatment?

Tip

Question 4 is a very long question for an AS paper but it is a good example of how you will need to follow through several parts of a question that may require knowledge taken from different sections of the specification. These are synoptic questions and a common feature of the full A level examination.

8 Natural selection and biodiversity

Prior knowledge

In this chapter you will need to recall that:

→ individuals of the same species often occur in groups called populations, living in one habitat
→ single populations are linked to others in communities
→ a group of communities with the non-living parts of the habitat form an ecosystem
→ organisms show adaptations to their environment
→ the theory of natural selection was proposed by Charles Darwin in his book *On the Origin of Species*
→ the animals and plants of the Galapagos islands in the Pacific Ocean showed many features that provided Darwin with evidence for his theory
→ the theory of natural selection has been developed using modern biological knowledge
→ evolution by natural selection can lead to the formation of new species
→ there is a worldwide threat to biodiversity
→ there are basic rules for simple genetic crosses
→ individuals can be homozygous or heterozygous with respect to one pair of alleles
→ genes are carried on chromosomes
→ meiosis halves the chromosome number in gamete production
→ conservation efforts are attempting to preserve biodiversity.

Test yourself on prior knowledge

1 Name the title of the book published by Charles Darwin, which first proposed the theory of natural selection.

2 Which ship took Darwin on his famous expedition around the world?

3 Name the group of islands that provided Darwin with some of his most important evidence.

4 How did Darwin come to the conclusion that there was a 'battle for survival' in most species?

5 Explain what is meant by:

   a) continuous variation

   b) discontinuous variation.

   Give **one** example of each.

6 What is an allele?

7 Explain the terms:

   a) homozygous

   b) heterozygous.

8 State the **two** main causes of variation in living organisms.

9 List **two** adaptations that are common to vertebrates living in polar regions.

10 State the name given to a group of members of the same species living in one location.

Niche – a concept central to ecology

Ecology is the study of living things within their environment. It is an essential part of modern biology – understanding the relationships between organisms and their environment is just as important as knowing about the structure and physiology of animals and plants, for example. One of the ideas that ecologists have introduced into biology is that of the ecosystem. An ecosystem is defined as a community of organisms and their surroundings – the environment in which they live. An ecosystem is a basic functional unit of ecology since the organisms that make up a community cannot realistically be considered independently of their physical environment. An example of an ecosystem is woodland.

Within an ecosystem are numerous habitats. The term *habitat* refers to the place where an organism lives. Within a woodland ecosystem, for example, some organisms have a habitat restricted to a small area. An example is a leaf-tissue parasite such as the holly leaf-miner insect, especially at the larval stage, as it is restricted to the interior of the holly leaf. Other species are abundant, for example *Pleurococcus*, a single-celled alga found on all damp surfaces such as most tree trunks and branches. So there is no particularly precise definition of a habitat – but the term is useful.

On the other hand, the term ecological niche is more informative. It defines just how an organism feeds, where it lives and how it behaves in relation to other organisms in its habitat. A niche identifies the precise conditions a species needs.

We can illustrate the value of the niche concept by reference to two common and rather similar sea birds, the cormorant and the shag (Figure 8.1).

> **Key term**
>
> **Ecosystem** A community of organisms and their surroundings – the environment in which they live.

> **Key term**
>
> **Ecological niche** This describes not only where a species lives but all of its activities, such as feeding, its predators and how it interacts with the non-living environment around it.

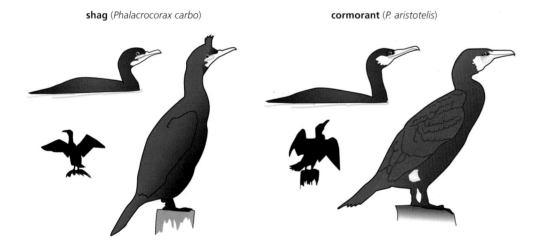

shag (*Phalacrocorax carbo*)   cormorant (*P. aristotelis*)

diet is a key difference in the niches of these otherwise similar birds

prey		% of prey taken by	
		shag	cormorant
surface-swimming prey	sand eels	33	0
	herring	49	1
bottom-feeding prey	flatfish	1	26
	shrimps, prawns	2	33

Figure 8.1 The sea birds cormorant and shag – their niches

Both birds live and feed along the coastline and they rear their young on similar cliffs and rock systems. We can say that they apparently share the same habitat. However, their diet and behaviour differ. The cormorant feeds close to the shore on sea-bed fish, such as flatfish. The shag builds its nest on much narrower cliff ledges. It also feeds further out to sea and captures fish such as sand eels from the upper layers of the waters. Since these birds feed differently and have different behaviour patterns, although they occur in close proximity, they avoid competition. They have different niches.

Adaptation of organisms to their environment

Adaptation is the process by which an organism becomes fitted to its environment. There are countless examples of this process to be observed in all habitats.

Adaptations can be physiological or anatomical. Physiological adaptations are those which are the results of changes to the metabolism of the organism which are advantageous to survival in their particular habitat, such as the production of different algal pigments as described in the following section. Anatomical adaptations are changes to the actual structure of organisms, such as the size of ears in hares and rabbits.

Physiological adaptation

An example of a physiological adaptation is shown in the marine algae known as 'greens', 'browns' and 'reds', which flourish at different zones of the shoreline community.

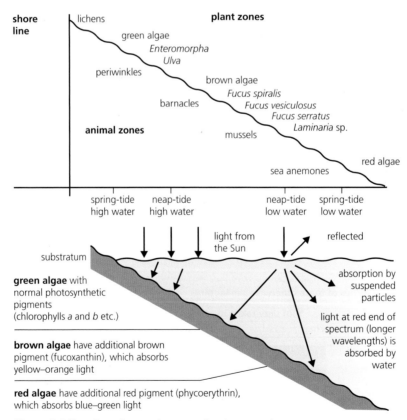

Figure 8.2 Zonation of green, brown and red seaweeds

The colour differences in these seaweeds are due to the particular photosynthetic pigments they contain. These pigments enable algae to absorb and exploit different wavelengths of light. In the marine environment, with increasing depth, progressively more of the higher wavelengths of white light are absorbed or scattered by the sea water and its suspended particles. Consequently, the red algae, equipped to absorb the blue–green light that is transmitted to greater depths, flourish there. Here, the brown and green algae cannot photosynthesise because the wavelengths of light they are adapted to absorb do not reach that depth.

Meanwhile, at lesser depths, red seaweeds are progressively crowded out in competition with the vigorous-growing brown seaweeds and green seaweeds as their particular pigments permit the efficient absorption of the incident light available closer to the surface.

Incidentally, the barnacles *Chthamalus* and *Semibalanus* exhibit differing abilities to endure exposure in the intertidal zone, and this too is an example of physiological adaptation.

Physiological adaptations of extremophiles

In recent years the discovery of unique ecosystems surrounding hydrothermal vents has provided the most remarkable examples of extreme adaptations. These vents are found in the deep ocean on the boundaries of tectonic plates, where there is volcanic activity as the plates forming the Earth's crust move against each other. The water ejected from these vents is highly acidic, contains many toxic sulfides and is at a temperature of over 350 °C and a pressure 250 times greater than that at the surface. The greatest surprise was to find any life at all under such conditions, let alone the variety of worms and crustaceans, such as the giant tube worm *Riftia pachyptila*, which is over 2 m long. There is no light at this depth and the whole ecosystem depends on bacteria using metal sulfides from the vents in chemosynthesis to produce organic compounds on which all the other organisms depend.

> **Key term**
>
> **Chemosynthesis** A method used by some microorganisms to release energy from inorganic molecules. Typically molecules such as ammonia or metal sulfides are oxidised and the energy released used to build organic molecules.

Many of these organisms are still being studied but it is obvious that, in addition to specialised nutrition, they must have remarkable adaptations to thrive in such a niche. Compared with other organisms they have remarkably stable enzymes, membranes and nucleic acids, which enable them to function in extreme temperatures, pressures and pH levels.

Anatomical adaptations

Examples of anatomical adaptations are body structures adapted to regulate heat loss in various mammals. Mammals are described as endotherms, since their body's heat comes from the metabolic reactions of many body organs. Body temperature is largely regulated by varying heat loss from the body. The total heat produced from internal organs largely depends upon the volume of the body, but the amount of heat loss is dependent upon the surface area. As the size of an organism increases, the volume increases more rapidly than the surface area. In other words its surface area-to-volume ratio decreases, reducing the relative heat loss. (We shall look at the concept of surface area-to-volume ratios in Chapter 10.) Consequently, animals in cold regions of the world tend to be large. An example is the polar bear. Smaller animals in colder regions need a high metabolic rate and consequently require a regular and substantial food supply to survive.

Meanwhile, mammals living in hot regions typically have external ears adapted as efficient radiators. These flaps of skin bear little fur (hair provides a heat insulation layer) and are richly supplied with blood capillaries. Heat brought to the external ears from the body interior by warm blood is quickly lost when capillaries here are dilated. A comparison of ear size in hares and rabbits in natural habitats at various latitudes on the North American continent appears to support this (Figure 8.3).

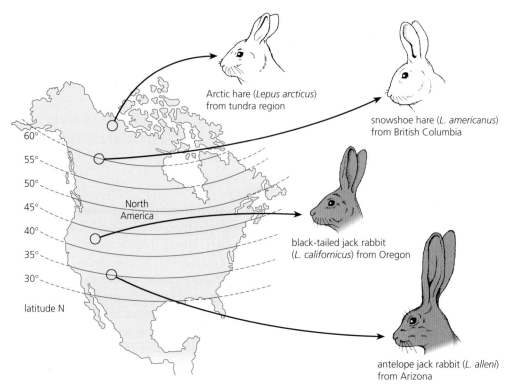

Figure 8.3 External ear sizes of hares in relation to latitude

Behavioural adaptation

Many animals are unable to regulate their body temperature effectively by internal means. They are known as ectotherms and rely on absorbing heat from their surroundings. However, this does not prevent them from occupying a wide variety of niches. Desert environments present particularly challenging problems associated with temperature control. They are extremely hot during the day and often well below freezing at night. Despite this, ectotherms such as lizards and snakes are common desert animals.

To survive in such conditions, ectotherms have adapted their behaviour to avoid large fluctuations in their body temperature (Table 8.1 and Figure 8.4).

Table 8.1 Typical behavioural adaptations of desert ectotherms

Adaptation	Function
Activities limited to morning and evening	Avoids the hottest times of day
Early morning basking on rocks	Raises body temperature quickly after cold nights
Burrowing into sand	Avoids direct sunlight and finds a cooler environment
Seeking deep crevices at night	Rocks retain heat longer during the night

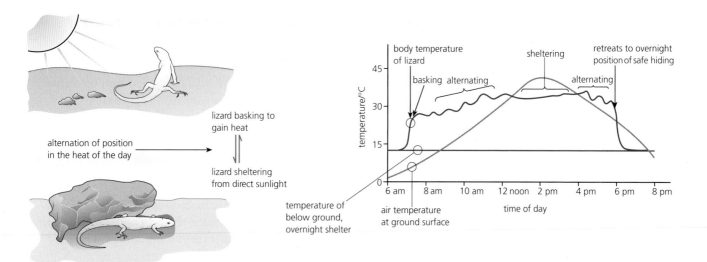

Figure 8.4 Thermoregulation in an ectotherm

Even advanced mammals, such as humans, with many highly developed temperature control mechanisms, have adopted numerous behavioural adaptations to ensure their survival. We build shelters and heat them, we wear clothes, which we change according to the environmental conditions, and we avoid exposure to the most extreme weather conditions.

Migration is a behavioural technique employed by many animals to take advantage of seasonal changes for feeding and breeding. Many birds regularly make incredible annual journeys across the world. The Arctic tern, *Sterna paradisaea*, uses the abundant insect food in northern Europe and the southern Arctic to raise its chicks during the summer. It then embarks upon one of the longest migrations known to avoid the severe winters. During the southern hemisphere summer it is found on the fringes of Antarctica, only to return to the Arctic later in the year. This remarkable feat of endurance and navigation entails a round-trip of over 45 000 miles. The Arctic tern therefore experiences two summers and probably more daylight than any other animal.

Incidentally, the differing feeding habits of the cormorant and shag are also examples of behavioural adaptation.

Test yourself

1 Apart from its exact place in a habitat, name **three** other properties of a species described by the term 'ecological niche'.

2 Why might you expect Peruvian people living in the high Andes to have high red blood cell counts?

3 Why do animals living in polar regions tend to be large?

4 Describe anatomical adaptations to desert life that are typically found in cacti.

5 State **three** ways in which the cormorant and the shag occupy different niches in order to avoid competition.

6 Research the main ways in which it is thought that birds are able to navigate over long distances.

How adaptation is brought about – natural selection

It was Charles Darwin whose careful observations over many years led him to realise what natural process brought about the adaptations of organisms in response to challenging environmental conditions. He coined the term 'natural selection' for this.

Charles Darwin put forward his ideas in 1859, in a book titled *On the Origin of Species*, published by John Murray of Albemarle Street, London. He was proposing a mechanism for the **evolution** of organisms.

By 'evolution', we mean the gradual development of life in geological time. The word evolution is used widely, but in biology it specifically means the processes by which life has been changed from its earliest beginnings to the diversity of organisms we know about today, living and extinct.

Charles Darwin (and nearly everyone else in the scientific community of his time) knew nothing of Mendel's work. Instead, biologists generally subscribed to the concept of 'blending inheritance' when mating occurred (which would only reduce the genetic variation available for natural selection, if it actually occurred). Today we are really talking about 'Neo-Darwinism', which is essentially a restatement of the concepts of evolution by natural selection in terms of Mendelian and post-Mendelian genetics.

The evidence and arguments for natural selection are as follows:

1 **Organisms produce many more offspring than survive to be mature individuals.** Darwin did not coin the phrase 'struggle for existence', but it does sum up the point that the over-production of offspring in the wild leads naturally to competition for resources. Table 8.2 lists the normal rates of production of offspring in some common species but clearly not all of these survive to pass on their genes to the next generation.

Table 8.2 Numbers of offspring produced

Organism	Number of eggs/seeds/young per brood or season
Rabbit	8–12
Great tit	10
Cod	2–20 million
Honey bee (queen)	120 000
Poppy	6000

In fact, in a stable population, a breeding pair gives rise to a single breeding pair of offspring, on average. All their other offspring are casualties of the 'struggle'; many organisms die before they can reproduce.

So, populations do not show rapidly increasing numbers in most habitats, or at least, not for long. Population size is naturally limited by restraints we call 'environmental factors'. These include space, light and the availability of food. The never-ending competition for resources results in the majority of organisms failing to survive

Key term

Evolution The development of new types of living organism from pre-existing types by the accumulation of genetic differences over long periods of time.

and reproduce. In effect, the environment can only support a certain number of organisms, and the number of individuals in a population remains more or less constant over a period of time.

2 **The individuals in a species are not all identical**, but show variations in their characteristics. Today, modern genetics has shown us that there are several ways by which genetic variations arise in gamete formation during meiosis and at fertilisation. You will learn much more about the origins of variation in Chapter 8 of the *Edexcel A level Biology 2* but for AS, the important point is that variation is largely produced during meiosis and it is the essential raw material for selection. You cannot select if you have nothing to choose from!

As described on page 108, genetic variations arise via:

- **random assortment** of paternal and maternal chromosomes in meiosis – this occurs in the process of gamete formation
- **crossing over** of segments of individual maternal and paternal homologous chromosomes that results in new combinations of genes on the chromosomes of the haploid gametes produced by meiosis
- the **random fusion of male and female gametes** in sexual reproduction – this source of variation *was* understood in Darwin's time.

Additionally, variation arises due to **mutations** – either chromosome mutations (page 109) or gene mutations (page 62).

As a result of all these, the individual offspring of parents are not identical. Rather, they show variations in their characteristics.

3 **Natural selection results in offspring with favourable characteristics.** When genetic variation has arisen in organisms:

- the favourable characteristics are expressed in the phenotypes of some of the offspring
- these offspring may be better able to survive and reproduce in a particular environment; of course, other offspring will be less able to compete successfully, survive and reproduce.

Thus natural selection operates, determining the survivors and the genes that are perpetuated in future progeny. In time, this selection process leads to adaptation to the environment, later to new varieties and then to new species.

The operation of natural selection is sometimes summarised in the phrase 'survival of the fittest', although these were not words that Darwin used, at least not initially. To avoid the criticism that 'survival of the fittest' is a circular phrase (how can fitness be judged except in terms of survival?) the term 'fittest' is applied in a particular context. For example, the fittest of the wildebeest of the African savannah (hunted herbivore) may be those with the acutest senses, quickest reflexes and strongest leg muscles for efficient escape from predators. By natural selection, the health and survival of wildebeests is assured.

7 Explain why the rabbit produces only 8–12 offspring per season whereas the cod produces up to 20 million.

8 If a pair of rabbits produced eight offspring every year and their offspring become sexually mature after only 1 year, how many rabbits would there be after 3 years if all the rabbits survived? You may assume that there are equal numbers of males and females.

9 State **three** reasons why the majority of offspring in most populations do not survive to breed.

10 Which forms of genetic variation are simply mixing up alleles into different combinations and which actually change the alleles present?

11 Describe Darwin's role on board *H.M.S. Beagle*.

12 Darwin returned from his voyage in 1836 but his book *On the Origin of Species* was not published until 1859. Why did it take so long?

New evidence for evolution

Evidence from **fossils** (palaeontological evidence) was at one time a main source of information about life forms now extinct. Fossilisation is an extremely rare, chance event; scavengers and bacterial action normally dismember and decompose dead plant and animal structures before they can be fossilised. Of the organisms that have been fossilised, most are never found, recovered or interpreted. Nevertheless, numerous fossils have been uncovered. They include:

- petrified remains (organic matter of the dead organism is replaced by mineral ions)
- moulds (the organic matter decays but the vacated space becomes a mould)
- traces (an impression of a form, such as of a leaf or a footprint)
- preserved, intact whole organisms (trapped in amber, ice or in anaerobic, acidic peat, for example).

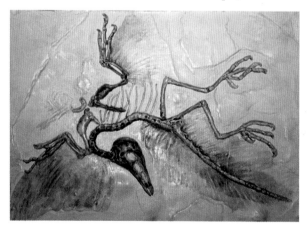

Figure 8.5 A fossil of *Archaeopteryx*, found in 1861

An example can be seen in Figure 8.5.

Additionally, it has sometimes been possible to date quite accurately the rocks surrounding fossils by exploiting the known rates of decay of certain isotopes, including carbon (^{14}C) and the ratio of potassium to argon (^{40}K/^{40}Ar) present in lava deposits. Using the decay rate of ^{14}C gives ages of fossils formed in the last 60 000 years. Using the ratio of ^{40}K/^{40}Ar gives an approximate age of sedimentary rocks (and their fossils) below and above a lava layer from geological time back to the Cambrian period (580 mya), although these are unreliable for the most recent half million years.

Exciting and illuminating fossil finds abound. Two of the most moving are, perhaps, those of the first hominid (a 'southern ape' named Lucy) at Hadar in Ethiopia in 1974, and the footsteps at Laetoli in Tanzania in 1976, found in volcanic ash and dated 3.6 mya – our first record of bipedalism.

Today, studies in comparative physiology and biochemistry are a new tool in the investigation of evolutionary change.

Most living things have DNA as their genetic material. The genetic code is virtually universal. The processes of 'reading' the code and protein synthesis, using RNA and ribosomes, are very similar in prokaryotes and eukaryotes, too. Processes such as respiration involve the same types of steps and similar or identical intermediates and biochemical reactions, similarly catalysed. ATP is the universal energy currency. Among the autotrophic organisms, the biochemistry of photosynthesis is virtually identical as well.

So, early biochemical events in the evolution of life have been 'inherited' widely, as and when forms of life diversified. However, large molecules like nucleic acids and the proteins they code for are subjected to some changes with time, so knowledge of these changes may be an aid to the study of the timings of evolutionary change. It is possible to measure the relatedness of different groups of organisms by the amount of difference between specific molecules such as DNA, proteins and enzyme systems. One aspect of these investigations is **proteomics**. This is the study (qualitative and quantitative) of the proteins coded for by specific genes of the human genome.

Key term

Proteomics The study of the proteins coded for by specific genes found in the human genome.

Immunological studies

The immune reaction provides a mechanism of detecting differences in specific proteins, and therefore (indirectly) their relatedness. Serum is the liquid produced from blood when blood cells and fibrinogen have been removed. Protein molecules present in the serum act as antigens when serum is injected into animals with an immune system that lacks these particular proteins. Typically, a rabbit is used when investigating relatedness to humans. The injected serum triggers the production of antibodies against the injected 'foreign' proteins. Then, fresh serum produced from the treated rabbit's blood (it now contains antibodies against human proteins) is tested against serum from a range of animals. The more closely related the animal is to humans, the greater the precipitation observed. This is illustrated in Figure 8.6.

The precipitation produced by reaction with human serum is taken as 100 percent. For each species, the greater the precipitation, the more recently the species shared a common ancestor with humans. This technique, called comparative serology, has been used by taxonomists to establish phylogenetic links in a number of cases, in both mammals and non-vertebrates.

Immunological studies are a means of detecting differences in specific proteins of species, and therefore (indirectly) their **relatedness**.

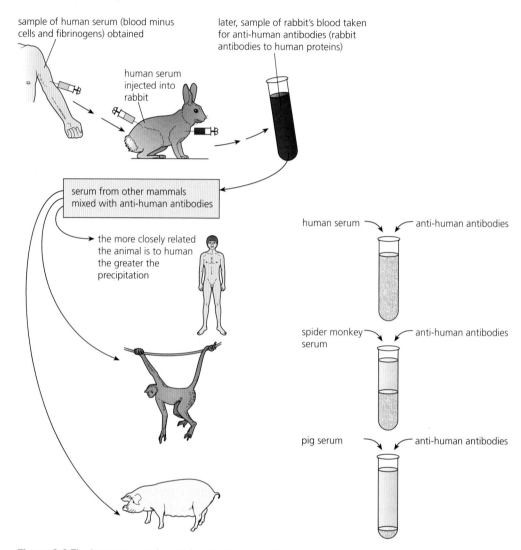

Figure 8.6 The immune reaction and evolutionary relationships

Genetic differences in nucleic acids

The technique of DNA hybridisation involves matching DNA from different species to test the degree of base pairing that occurs (Figure 8.7). This tells us the approximate degree of divergence between closely related groups, such as families within the primates. This data can then be correlated with data on the estimated number of years since they shared a common ancestor.

DNA as a molecular clock

Measurement of changes in DNA from selected species has potential as a molecular clock. DNA in eukaryotic cells occurs in both the chromosomes of the nucleus (99 per cent) and also in the mitochondria. Mitochondrial DNA (mtDNA) is a circular molecule, very short in comparison with nuclear DNA. Cells contain any number of mitochondria, typically between 100 and 1000.

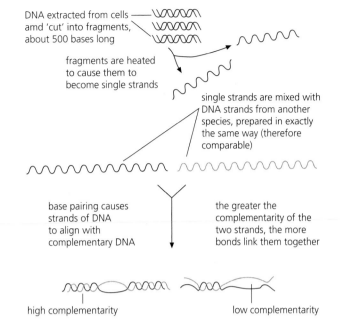

DNA hybridisation is a technique that involves matching the DNA of different species, to discover how closely they are related.

DNA extracted from cells amd 'cut' into fragments, about 500 bases long

fragments are heated to cause them to become single strands

single strands are mixed with DNA strands from another species, prepared in exactly the same way (therefore comparable)

base pairing causes strands of DNA to align with complementary DNA

the greater the complementarity of the two strands, the more bonds link them together

The closeness of the two DNAs is measured by finding the temperature at which they separate – the fewer bonds formed, the lower the temperature required.

high complementarity

low complementarity

Figure 8.7 DNA hybridisation

Mitochondrial DNA has approximately 16 500 base pairs. Mutations occur at a very slow, steady rate in all DNA, but chromosomal DNA has with it enzymes that may repair the changes in some cases. These enzymes are absent from mtDNA.

Thus mtDNA changes 5–10 times faster than chromosomal DNA – involving about 1–2 base changes in every 100 nucleotides per million years. Consequently, the length of time since organisms (belonging to different but related species) have diverged can be estimated by extracting and comparing samples of their mtDNA.

Furthermore, at fertilisation, the sperm contributes a nucleus only (no cytoplasm and therefore no mitochondria). So, all the mitochondria of the zygote come from the egg cell, and there is no mixing of mtDNA genes at fertilisation. All the evidence about relationships from studying differences between samples of mtDNA is easier to interpret in the search for early evidence of evolution.

Ribosomal RNA studies

We have seen in Chapter 7 how ribosomal RNA sequencing has led to new debates as to the main kingdoms and their evolutionary relationship, and a new three-domain model with the Archaea as a separate evolutionary line.

Natural selection and speciation

You will remember that in Chapter 7 we discussed how difficult it can be to define exactly what we mean by a species. The key ideas were a group of organisms that (a) normally interbreed and (b) produce fertile offspring. It is obvious that organisms undergoing natural selection will begin to change, but at what point do they change enough to be recognised as a separate species?

We refer to all of the possible genes and alleles in a population as the **gene pool**.

> **Key term**
>
> **Gene pool** The total variety of genes and alleles present in a sexually reproducing population.

Natural selection operates on this gene pool and can lead to changes in the proportions of genes and alleles present in the population. Mutation can also introduce new forms of genes. The details of exactly how these changes can be measured and monitored are discussed further in Chapter 8 of *Edexcel A level Biology 2*.

How changing gene pools may lead to speciation

Key term

Speciation The name given to the process by which one species may evolve from another.

Species exist almost exclusively as local populations, even though the boundaries to these populations are often rather open and ill-defined. Individuals of local populations tend to resemble each other more closely than they resemble members of other populations. Local populations are very important as they are potential starting points for speciation.

Speciation is much more likely if part of the population is isolated in some way so that the gene pool of each part begins to change in different ways. Even then, many generations may elapse before the composition of the gene pool has changed sufficiently to allow us to call the new individuals a different species. Isolation therefore is very important in this process and can be brought about in different ways.

Key term

Allopatric speciation This occurs when two populations are totally separated from each other by a barrier. This prevents interbreeding and mixing of their gene pools.

Allopatric speciation

If a population is suddenly divided by the appearance of a barrier, resulting in two populations isolated from each other, allopatric speciation occurs. Before separation, individuals shared a common gene pool but after isolation, 'disturbing processes' like natural selection, mutation and random genetic drift may occur independently in both populations, causing them to diverge in their features and characteristics.

Geographic isolation between populations occurs when natural, or human-imposed, barriers arise and sharply restrict movement of individuals (and their spores and gametes, in the case of plants) between the divided populations (Figure 8.8).

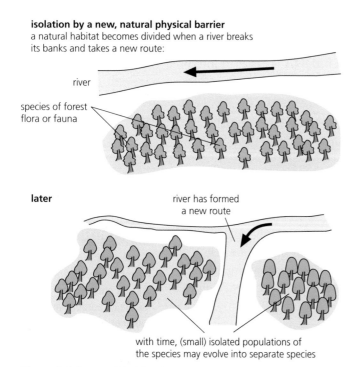

isolation by a new, natural physical barrier
a natural habitat becomes divided when a river breaks its banks and takes a new route:

river

species of forest flora or fauna

later

river has formed a new route

with time, (small) isolated populations of the species may evolve into separate species

Figure 8.8 Geographical barriers

Geographic isolation also arises when motile or mobile species are dispersed to isolated habitats – as, for example, when organisms are accidentally rafted from mainland territories to distant islands. The 2004 tsunami generated examples of this in South East Asia. Violent events of this type have punctuated world geological history with surprising frequency.

Charles Darwin visited the isolated islands of Galapagos, off the coast of South America, during his voyage with *The Beagle* in 1831–36. The islands are 600 miles (970 km) from the South American mainland. Their origin is volcanic – they appeared out of the sea about 16 million years ago, at which point they were of course uninhabited. Today, they have a flora and fauna that relate to mainland species.

Darwin encountered examples of population divergence on the Galapagos islands. For example, the tortoises found on these islands had distinctive shells. With experience, an observer could tell which individual island an animal came from by its appearance, so markedly had the local, isolated populations diverged since their arrival from the mainland. These giant tortoises are certainly unlike any in other parts of the world and their differences could lead to the formation of separate species. In any case they are unlikely to interbreed with those on the other islands.

Sympatric speciation

Once again, the unique habitats of the Galapagos islands provide examples of sympatric speciation in action (Figure 8.9). The iguana lizard had no mammal competition when it arrived on the Galapagos islands. It became the dominant form of vertebrate life and was extremely abundant when Darwin visited. By then, two species were present, one terrestrial and the other fully adapted to marine life. The latter is assumed to have evolved locally as a result of pressure from overcrowding and competition for food on the islands (both species are vegetarian), which drove some members of the population out of the terrestrial habitat.

Key term

Sympatric speciation
This occurs when two populations are still able to mix freely in the same area but some individuals accumulate changes in their gene pools, which are sufficient to prevent interbreeding.

Many organisms (e.g. insects and birds) may have flown or been carried on wind currents to the Galapagos from the mainland. Mammals are most unlikely to have survived drifting there on a natural raft over this distance, but many large reptiles can survive long periods without food or water.

immigrant travel to the Galapagos

The **giant iguana lizards** on the Galapagos Islands became dominant vertebrates, and today are two distinct species, one still terrestrial, the other marine, with webbed feet and a laterally flattened tail (like the caudal fin of a fish).

The Galapagos islands
Today the tortoise population of each island is distinctive and identifiable.

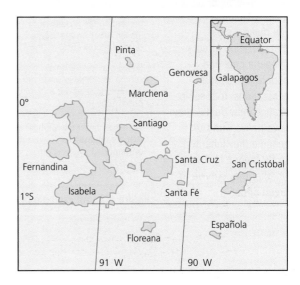

Figure 8.9 The Galapagos islands and species divergence

Test yourself

13 Give **two** reasons why fossils are very rare.

14 Which **three** cellular molecules are found in almost all living things?

Where populations co-exist in the same area, speciation can occur where changes brought about by natural selection result in reproductive isolation.

Reproductive isolation mechanisms occasionally develop that are strong enough to prevent interbreeding between members of small, isolated populations that have diverged genetically, if only slightly, as a result of their isolation. Cases of reproductive isolation are likely to be less consistently effective than geographic separation in bringing about complete isolation in the early stages. Examples can be seen in Table 8.3.

Table 8.3 Types of reproductive isolation

Mismatch of genitalia	This can make successful copulation impossible.
Barriers to fertilisation	Changes in the uterus of animals or the stigma of plants can prevent gametes from meeting (Figure 8.10).
Formation of sterile hybrids	Horses can successfully mate with asses (donkeys) but their offspring (mules) will be sterile. This is because the horse and ass have different chromosome numbers of 60 and 66 respectively, but the hybrid mule has 63. This makes it impossible for the mule's chromosomes to form pairs in meiosis and therefore produce viable gametes.
Behavioural changes	Many animals have elaborate courtship displays where even small changes will result in rejection by a potential mate. Many birds will only choose mates with the correct song; the Galapagos finches are known to select partners with the correct beak size and shape.
Temporal changes	Where such factors as available food supply cause changes in the timing of gamete production, fertilisation will be prevented.
Ecological changes	Where two populations occupy different parts of the habitat, they are unlikely to meet. The separation of marine and terrestrial iguanas, explained above, is a good example.

Tip

When answering questions, make sure you make it clear that selection affects the gene pool of a population in future generations. It does not change individuals – they have to live with the genome they inherited, so for them it is simply whether they survive to breed or not.

Incompatibility in flowering plants refers to physiological mechanisms that may make fertilisation impossible by preventing the growth of pollen tubes on the stigma or through the style.

Pollen that lodges on a stigma 'germinates' and attempts to send out a pollen tube that may eventually reach the embryo sac. Growth of pollen tubes that are opposed or unsupported by the stigma tissue fails.

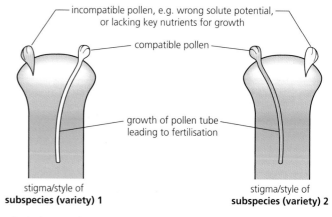

Figure 8.10 An example of reproductive isolation in a flowering plant

Types of selection

Natural selection operates on individuals, or rather on their phenotypes. Phenotypes are the product of a particular combination of alleles, interacting with the effects of the environment of the organism.

Consequently, natural selection causes changes to the composition of gene pools. However, the effects of these changes vary. We can recognise different types of selection.

- **Stabilising selection** occurs where environmental conditions are stable and largely unchanging. It does not lead to evolution, but rather it maintains favourable characteristics that enable a species to be successful, and the alleles responsible for them, and eliminates variants and abnormalities that are useless or harmful. Probably most populations undergo stabilising selections. The example in Figure 8.11 comes from human birth records on babies born between 1935 and 1946 in London. It shows there is an optimum birth weight for babies, and those with birth weights heavier or lighter are at a selective disadvantage.

Key term

Stabilising selection
This occurs where conditions are favourable and not changing, so pressures to change in one way are less than the advantage in remaining the same.

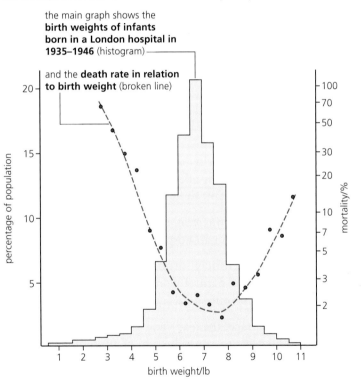

The birth weight of humans is influenced by **environmental factors** (e.g. maternal nutrition, smoking habits, etc.) and by **inheritance** (about 50%).

When more babies than average die at very low and very high birth weights, this obviously affects the gene pool because it tends to eliminate genes for low and high birth weights.

the main graph shows the **birth weights of infants born in a London hospital in 1935–1946** (histogram)

and the **death rate in relation to birth weight** (broken line)

The data are an example of continuous variation. The 'middleness' or central tendency of this type of data is expressed in three ways:

1 mode (modal value) – the most frequent value in a set of values

2 median – the middle value of a set of values where these are arranged in ascending order

3 mean (average) – the sum of the individual values, divided by the number of values

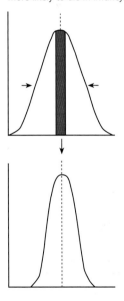

This is an example of **stabilising selection** in that the values (weights) at the extremes of a continuous variation are at a selective disadvantage. This means that infants of these birth weights are more likely to die in infancy.

Figure 8.11 Birth weight and infant mortality, a case study in stabilising selection

- **Directional selection** is associated with changing environmental conditions. In these situations, the majority of an existing form of an organism may no longer be best suited to the environment. Some unusual or abnormal forms of the population may have a selective advantage.
  An example of directional selection is the development of resistance to an antibiotic by bacteria. Certain bacteria cause disease, and patients with bacterial infections are frequently treated with an antibiotic to help them overcome the infection.

Key term

Directional selection
This occurs when environments are changing and there is a clear advantage in the population changing in one particular direction.

Test yourself

15 Why should the hybrid DNA of more closely related organisms need a higher temperature to separate the strands than the hybrid DNA of less closely related species?

16 Give **two** reasons why analysis of mitochondrial DNA (mtDNA) will yield more information about evolutionary history than chromosomal DNA.

17 State the main difference between sympatric and allopatric speciation.

Key term

Pathogen An infectious agent that causes an illness or disease in its host. The term is generally applied to microorganisms but includes such things as viruses.

Antibiotics are very widely used. In a large population of a species of bacteria, some may carry a gene for resistance to the antibiotic in question. Sometimes such a gene arises by spontaneous mutation. Alternatively the gene is acquired in a form of sexual reproduction between bacteria of different populations.

A 'resistant' bacterium has no selective advantage in the absence of the antibiotic and must compete for resources with non-resistant bacteria. But when the antibiotic is present, most bacteria of the population are killed off. Resistant bacteria remain and create the future population, all of which now carry the gene for resistance to the antibiotic (Figure 8.12). The gene pool has been changed abruptly.

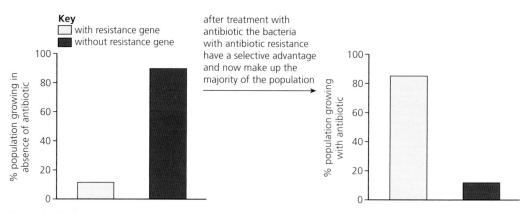

Figure 8.12 Directional selection

An evolutionary race

The discovery of antibiotics in the 1940s and 1950s was hailed as the end of many diseases caused by bacteria. For a time this seemed to be a reasonable assumption as new types of antibiotics were discovered and were very effective. Unfortunately this ignored the effect of evolution and how widespread use of antibiotics would mean that selection pressures were increased and along with it the rate of evolutionary change. First, increased doses of antibiotics were needed to bring about the same effect, and then strains of bacteria that were entirely resistant to their effect began to appear, until today there is a major problem with bacteria that are resistant to most of the common families of antibiotics.

Methicillin resistant *Staphylococcus aureus* (**MRSA**) is today well known as a 'superbug' causing serious problems in all hospitals.

No biologist would be entirely surprised by these developments as similar examples can be found throughout the natural world. It is generally referred to as an '**evolutionary race**' between the host or medical science and the pathogens.

As in any prolonged race, the lead can change many times. It would be true to say that in the past 50 years the introduction of antibiotics means that medical science has gained a large advantage. But that lead is now under serious threat as it becomes much more difficult to find new antibiotics while the majority of pathogens are rapidly developing multiple resistance to those in use.

Biodiversity

Species richness and biodiversity

Species richness is defined as the total number of different species within a given area or community. To produce this information, a precise listing of all the different types of organism is required. However, the abundance of each species present is not required. As such, species richness is not a complete measure of biodiversity of a habitat. The diversity of species present in a habitat can be measured by applying the formula known as the Simpson Diversity Index:

$$\text{diversity} = \frac{N(N-1)}{\Sigma n(n-1)}$$

where N = total number of organisms of all species found
and n = number of individuals of each species.

Key term

Species richness The total number of different species within a given area or community.

Measuring genetic variability

In a population – a group of individuals of a species living close together and able to interbreed – the alleles of the genes located in the reproductive cells of those individuals make up a **gene pool**. A sample of the alleles of the gene pool will contribute to the genomes (gene sets of individuals) of the next generation, and so on, from generation to generation.

The size of an interbreeding population has a direct impact on the genetic diversity of the individuals. A very small population can be described as an **inbreeding** group – the individuals are closely related. In fact, the smaller the population, the more closely related the offspring will be. The important genetic consequence of inbreeding is that it leads to **homozygosity** – at more and more of the loci there will be identical alleles. There is progressively less variation in the population. While the individuals of that population may initially be well adapted, in the face of environmental changes the population is less able to adapt. The genetic fitness of a population is compromised.

It is in these latter cases of small, isolated populations that genetic variability is most critical. In the cases of populations of endangered species, the question is whether there is sufficient genetic diversity to allow the population to adapt to future changes in the environment and so survive.

This is a practical problem faced by modern zoos, which have taken on the role of attempting to conserve genetic variation within endangered species via captive breeding programmes. It is also an issue for endangered organisms in the wild, where population numbers have been reduced to small, isolated groups in former strongholds. This problem can be illustrated by examining current conservation research being undertaken in the tropical rainforests of Sabah in Malaysia, Borneo.

A conservation case study based on genetic diversity analyses

Orang-utans (*Pongo pygmaeus*), together with gorillas and chimpanzees, are great apes. Most of their features (including their large cranium and well-developed brains, elongated arms and highly developed muscles) are common to humans too, although they are not our direct ancestors. Orang-utans are the largest ape species after the gorilla – females typically weigh in excess of 35 kg and males up to 80 kg. A standing male may be up to 150 cm high. They are arboreal and active by day, with a diet largely consisting of fruits and shoots. They may be solitary, or live in pairs or very small

Table 8.4 Estimated orang-utan numbers in Sabah, Borneo, over the past 100 years or so

Year	Number of orang-utans
1900	310 000
1980	25 000
2003	13 000

family groups, sleeping in tree 'nests' and moving on from day to day. The young (birth weight about 1.5 kg) remain with their mother for up to 5 years of parental care. They reach maturity at about 10 years, and their total life expectancy is 30 years or more. Humans are their only predator, either directly (through poaching) or indirectly (through logging and forest clearing for agriculture).

Sabah, situated on the north-eastern part of the island of Borneo, has an orang-utan population that has been in decline for the past 100 years (Table 8.4). This collapse was triggered by massive deforestation that began in the 1890s and accelerated in the 1950s and 1970s.

Currently, the orang-utans of Sabah are threatened with extinction in the near future due to the continuing logging of the forest trees, clearing of whole forests for oil palm plantations and illegal killings. Today's population estimates are based on ground surveys and are confirmed by aerial surveys (nest counts) made by helicopter. While Sabah is the main stronghold of these primates in North Borneo, the present estimated population is of only 11 000 individuals, distributed in the remaining pockets of rainforest, as shown in Figure 8.13. More than half live outside protected areas, in forests still frequently disturbed by selective logging activities.

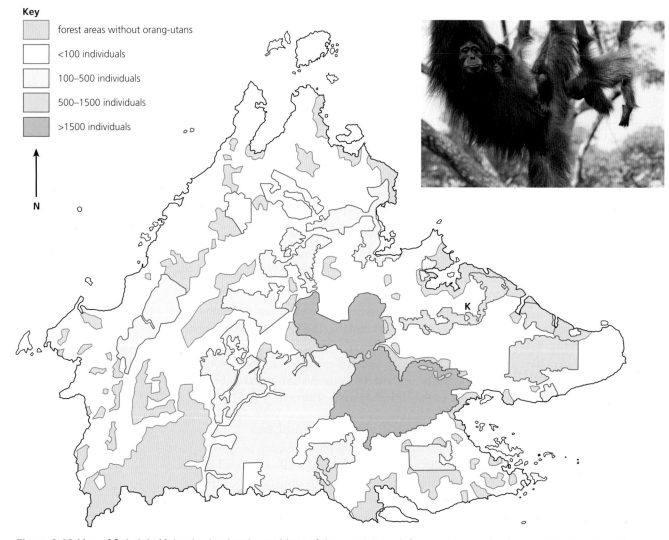

Key

- forest areas without orang-utans
- <100 individuals
- 100–500 individuals
- 500–1500 individuals
- >1500 individuals

N

Figure 8.13 Map of Sabah in Malaysia showing the positions of the remaining rainforest, colour coded to show the densities of orang-utans in each; the Kinabatangan Wildlife Sanctuary is marked **K**

A team of conservation biologists, led by Professor Mike Bruford and Dr Benoit Goossens of the Biodiversity and Ecological Processes Group, School of Biosciences, Cardiff University, is investigating the genetic diversity of these populations in an attempt to devise sustainable schemes that will effectively support remaining orang-utan populations. Their data are being collected from animals living in Kinabatangan Wildlife Sanctuary (Figure 8.11).

By collecting hair and faeces found at fresh nest sites, they are able to extract DNA to create genetic profiles. Genetic markers called microsatellites are applied to the DNA samples, which distinguish, among other things, heterozygous genes from homozygous genes. They serve as tools to evaluate inbreeding levels, the genetic structure and past history of populations, and to assess both gene flow between populations and effective population size.

The results establish that, in Kinabatangan Sanctuary, the orang-utans do not cross the rivers but do move freely through the forest areas on either side. Studies indicate that, if nothing is done about the existing sanctuary provision here, then this situation will almost certainly lead to extinction. If an elaborate series of wildlife corridors is set up between remaining areas of forest, survival can hopefully be assured.

Since these measures are probably more expensive than local economies can sustain, the team has devised a programme of transfers of individuals between sites (referred to as translocations), together with a modest programme of corridor establishment. This proposal, known as the Kinabatangan Management Plan, is being prepared for presentation to the agencies and authorities who must find the necessary funding for whichever measures they choose to support.

The case of the vanishing rainforests

Rainforests cover almost 2 per cent of the Earth's land surface, but they provide habitats for almost 50 per cent of all living species. It has been predicted that if all non-vertebrates occurring in a single cubic metre of tropical rainforest soil were collected for identification, there would be present at least one completely previously unknown species. Tropical rainforests contain the greatest diversity of life of any of the world's biomes.

Key term

Biome A very large area of interconnected ecosystems of the Earth with similar climatic and geographic conditions.

Sadly, tropical rainforests are being rapidly destroyed. Satellite imaging of the Earth's surface provides the evidence that this is so – where no other reliable sources of information are available. The world's three remaining tropical forests of real size are in South America (around the Amazon Basin), in West Africa (around the Congo Basin) and in the Far East (particularly, but not exclusively, on the islands of Indonesia).

The current rate of destruction is estimated to be about one hectare ($100\,m \times 100\,m$ – a little larger than a football pitch) every second. This means that each year an area larger than the British Isles (31 million hectares) is cleared. While extinction is a natural process, this current rate is on a scale equivalent to that at the time of the extinction of the dinosaurs (an event 65 million years ago at the Cretaceous–Tertiary boundary). Table 8.5 states several reasons for conservation of rainforest ecosystems.

New species evolve, but other species (less suited to their environment, perhaps) become extinct, as much of the fossil record throughout geological time indicates. One example of a well-documented extinction is that of the dodo bird.

Table 8.5 Why conserve rainforest ecosystems?

Ecological reasons	Most species of living things are not distributed widely, but instead are restricted to a narrow range of the Earth's surface. In fact, the majority of species living today occur in the tropics. So when tropical rainforests are destroyed, the only habitat of a huge range of plants is lost, and with them very many of the vertebrates and non-vertebrates dependent upon them.
	In effect, the rainforests are critically important 'outdoor laboratories' where we learn about the range of life that has evolved, the majority of which consists of organisms as yet unknown.
	The soils under rainforests are mostly poor soils that cannot support an alternative ecosystem for very long. To destroy rainforest is to remove the most productive biome on the Earth's surface – in terms of converting the Sun's energy to biomass.
	If destroyed but later left to regenerate, only species that have not become extinct may return. Re-grown rainforest will be deprived of its variety of life.
Economic reasons	The whole range of living things is functionally a gene pool resource, and when a species becomes extinct its genes are permanently lost. The destruction of rainforests decreases our genetic heritage more dramatically than the destruction of any other biome. As a consequence future genetic engineers and plant and animal breeders are deprived of a potential source of genes. Many new drugs and other natural products, in some form or another, are manufactured by plants. The discovery of new, useful substances often starts with rare, exotic or recently discovered species.
	7 million km^2 of humid tropical forests have so far been cleared – about half of the original forest present before clearance programmes started. Much is 'cleared' to make way for the production of crops for food (as illustrated in the article in Figure 8.14). Yet only 2 million km^2 have remained in agricultural production. Mostly, the soil does not sustain continued cropping.
	Rainforest is a continuing resource of hardwood timber, which if selectively logged can be productive, but when cleared as forest, is lost as a source of timber in the future.
	Trees help stabilise land and prevent disastrous flooding downriver. Huge areas of productive land are washed away once mountain rainforest has been removed.
	Trees in general are carbon dioxide 'sinks' that help reduce global warming. Without these trees, other ways of reducing atmospheric CO_2 must be found.
Aesthetic reasons	These habitats are beautiful, exhilarating and inspirational places to visit. They are part of the inheritance of future generations, which should be secured for future people's enjoyment, too.
Ethical reasons	This biome is home to many forest peoples who have a right to traditional ways of life.
	Similarly, many higher mammals, including relatively close 'relatives' of *Homo*, live exclusively in these habitats. The needs of all primates must be respected.

The IUCN Red List of Threatened Species

Currently, the rate of extinctions is exceptionally high. Environmentalists seek the survival of endangered species by initiating and maintaining local, national and international action. For example, the **International Union for the Conservation of Nature (IUCN)**, working with appropriate local organisations, publishes a series of Regional Red Lists. These assess the risk of extinction to species within countries and regions. The lists are based upon criteria relevant to all species and all regions of the world. They convey the urgency of conservation issues to the international community and to policy makers. The aim is to stimulate action to combat loss of endangered species and of the habitats that support them.

Practical conservation – what does it entail?

Conservation involves applying the principles of ecology to manage the environment so that, despite human activities, a balance is maintained. The aims of conservation are to preserve and promote habitats and wildlife, and to ensure natural resources are used in a way that provides a sustainable yield. Conservation is an active process, not simply a case of preservation, and there are many different approaches to it. Practical conservation involves:

- the designation and maintenance of representative habitats as nature reserves
- preservation of endangered species and their genetic diversity through the maintenance of botanical and zoological gardens (with their captive breeding programmes) and the establishment of viable seed banks.

Table 8.6 What active management of nature reserves involves

Continuous monitoring of the reserve so that causes of change are understood, change may be anticipated, and measures taken early enough to adjust conditions without disruption, should this be necessary.

Maintenance of effective boundaries and the limiting of unhelpful human interference. The enthusiastic involvement of the local human community sends out messages about the purposes of conservation (a local 'education' programme, in effect) and that everyone has a part to play.

Measures to facilitate the successful completion of life cycles of any endangered species for which the reserve is 'home', together with supportive conditions for vulnerable and rare species.

Restocking and re-introductions of once-common species from stocks produced by captive breeding programmes of zoological and botanical gardens.

Conservation by promotion of nature reserves

Nature reserves comprise carefully selected land set aside for restricted access and controlled use, to allow the local maintenance of biodiversity. This is not a new idea; the New Forest in southern Britain was set aside for hunting by royalty over 900 years ago. An incidental effect was to produce a sanctuary for wildlife.

Today, this solution to extinction pressures on wildlife includes the setting up and maintenance of areas of special scientific interest as nature reserves, of our National Parks (the first National Park was Yellowstone, set up in North America in 1872) and of the African game parks, which have been more recently established. In total, these sites represent habitats of many different descriptions, in many countries around the world. Some of the conservation work they achieve may be carried out by volunteers.

In a nature reserve, the area enclosed is important – a tiny area may be too small to be effective. However, the actual dimensions of an effective reserve vary with species size and life style of the majority of the threatened species it is designed to protect. Also, for a given reserve there is an 'edge effect'. A compact reserve with minimal perimeter is less effective than one with an extensive perimeter interface with its surroundings.

The use to which the surrounding area is put is important, too; if it is managed sympathetically, it may indirectly support the reserve's wildlife. Another feature is geographical isolation – reserves positioned at great distances from other protected areas are less effective than reserves in closer proximity. Also, it has been found that connecting corridors of land are advantageous (Figure 8.14). In agricultural areas these may simply take the form of hedgerows protected from contact with pesticide treatments that nearby crops receive.

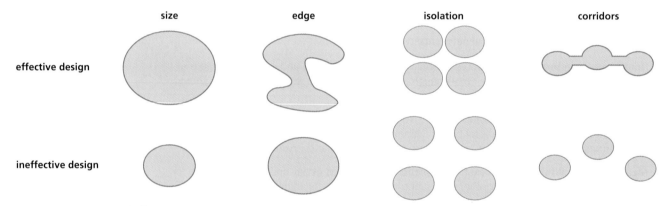

Figure 8.14 Features of effective nature reserves

Ex-situ conservation – an appraisal
Zoological (and botanical) gardens and their captive breeding programmes
Endangered species typically have very low population numbers and are in serious danger of becoming extinct. For some species whose numbers have dwindled drastically, captive breeding may be their last hope of survival.

Today, many zoos cooperate to manage individuals of the same species, held in different zoos, as a single population. A 'stud book' – a computerised database of genetic and demographic data – has been compiled for many of these species. This provides the basis for the recommendation and conduct of crosses designed to preserve the gene pool and avoid the problems of inbreeding.

Animals are shipped between zoos, or the technique of artificial insemination is employed. Some species can be very hard to breed in captivity, while with others there has been a high success rate.

Other successes include individuals bred in captivity that have been released and have survived in the wild. Examples include red wolves, Andean condors, bald eagles and golden lion tamarins.

Critics of the process fear that genetic diversity may have dwindled so much already that a species cannot be regenerated, and suggest that the work concentrates on a few, highly attractive species, and that at great cost (with funds diverted from more effective habitat conservation) it gives a false sense that extinction problems are being solved.

The building up of seed banks

Storing seeds in seed banks is an inexpensive and space-efficient method of *ex-situ* conservation (Figure 8.15). The natural dormancy of seeds allows for their suspended preservation for long periods, typically in conditions of low humidity and low temperature. The steps, following collection and preparation are:

- seed drying (to below 7 per cent water)
- packaging (in moisture-proof containers)
- storage (at a temperature of $-188\,°C$)
- periodic germination tests
- re-storage or replacement.

Figure 8.15 Wellcome Trust Millenium Building, Wakehurst Place – home of the Millenium Seed Bank (an initiative of the Royal Botanical Gardens, Kew)

The advantages of all these approaches are reviewed in Table 8.7.

Table 8.7 Pros and cons of *in-situ* and *ex-situ* conservation

In-situ conservation – terrestrial and aquatic nature reserves	*Ex-situ* conservation – captive breeding programmes of zoological and botanical gardens, and seed banks
Habitats that are already rare are especially vulnerable to natural disaster – rare habitats themselves are easily lost if a range of examples are not preserved as nature reserves.	Originally, zoos were collections of largely unfamiliar animals kept for curiosity, with little concern for any stress caused, but now captive breeding programmes make good use of these resources.
When a habitat disappears the whole community is lost, threatening to increase total numbers of endangered species.	Captive breeding maintains the genetic stock of rare and endangered species.
A refuge for endangered wildlife allows these species to lead natural lives in a familiar environment for which they are adapted, and be a part of their normal food chains.	The genetic problems arising from individual zoos having very limited numbers to act as parents is overcome by inter-zoo cooperation (and artificial insemination in some cases).
The biota of a reserve may be monitored for early warning of any further deterioration in numbers of a threatened species, so that remedial steps can be taken.	Animals in zoos tend to have significantly longer life-expectancies, and are available to participate in breeding programmes for much longer than wild animals.
The offspring of endangered species are nurtured in their natural environment and gain all the experiences this normally brings, including the acquisition of skills from parents and peers around them.	Captive breeding programmes, for most species they are applied to, have been highly successful, although the young do not grow up in the 'wild', so there is less opportunity to observe and learn from parents and peers.
There is an established tradition of maintaining reserves and protected areas in various parts of the world, so there is much experience to share on how to manage them successfully.	Captive breeding programmes generate healthy individuals in good numbers for attempts at re-introduction of endangered species to natural habitats – a particularly challenging process, given that natural predators abound in these locations.
Nature reserves are popular sites for the public to visit (in approved ways), thereby maintaining public awareness of the environmental crisis due to extinctions, and individual responsibilities that arise from it.	Zoos and botanical gardens are accessible sites for the public to visit (often sited in urban settings where many may have access), contributing effectively to public education on the environmental crisis.
Reserves are ideal venues to which to return endangered individuals that are the product of captive breeding programmes – providing realistic conditions for re-adaptation to their habitat, where progress can be monitored.	Seed banks are a convenient and efficient way of maintaining genetic material of endangered plants, which make use of the ways in which seeds survive long periods in nature.

Test yourself

18 Explain why homozygosity will prevent natural selection.

19 State the difference between *in-situ* and *ex-situ* conservation.

20 Describe the purpose of stud books in captive breeding programmes.

Exam practice questions

1 A population is best described as:

   A a group of organisms occupying the same habitat

   B a group of organisms of the same species occupying the same habitat

   C a group of organisms of the same species occupying the same ecosystem

   D a group of organisms of the same genus occupying the same ecosystem *(1)*

2 Which of the following would promote the development of antibiotic-resistant strains of bacteria?

   A Failing to complete a course of antibiotic treatment

   B Taking very high doses of antibiotics

   C Treatment with more than one antibiotic

   D Treatment with other drugs but not antibiotics *(1)*

3 **a)** Describe how seeds are prepared and stored by seed banks. *(3)*

   b) Explain the importance of seed banks in maintaining biodiversity. *(3)*

4 Three weed species appeared in a vegetable plot left as bare soil for 1 year. Individual plants were counted and the results are shown in the table.

Species	Number of individual plants
Groundsel	45
Shepherd's purse	40
Dandelion	10
Total	95

   a) Calculate the Simpson Diversity Index for this habitat. *(3)*

   b) Explain why using a measure such as a diversity index would be more useful than using species richness to compare the diversity of this habitat with others. *(3)*

> **Tip**
>
> You will find the formula you require in the diversity section of this chapter. In an exam you may be required to select the right formula from a data sheet or you may be given it in the question).

5 **a)** Use the information in this chapter and Figure 8.11 to explain why the current organisation of the Kinabatangan Sanctuary will almost certainly lead to the extinction of orang-utans even though this is a protected reserve. *(4)*

   b) Changing the organisation of the Kinabatangan Sanctuary and conserving the orang-utan population in the future is an expensive process. Discuss the ethical and economic arguments that could be made to justify this investment. *(5)*

6 The kakapo (*Stripops habroptila*) is the world's largest parrot, weighing up to 2 kg. It cannot fly but can climb trees well and walk long distances. It feeds on a variety of fruits, seeds and roots and is nocturnal. If in danger, it

will stand perfectly still for long periods. When breeding every 2–4 years it builds its nest on the ground and the males compete for 'calling posts' from where they emit a loud booming sound at night to attract females. Prior to the nineteenth century, kakapos were a very successful species, colonising all the islands of New Zealand, where it is thought their main predator was the (now extinct) giant eagle.

a) Use the information provided to explain how the unique adaptations shown by the kakapo made it a successful bird species in its niche before 1800. *(4)*

b) After the end of the nineteenth century, the giant eagle was extinct and New Zealand was rapidly colonised by European settlers. They brought with them domesticated cats as well as rats, and began to farm large areas of the land and hunt kakapos as an easily available source of meat. In 2012 there were only 125 birds remaining on a few isolated islands.

Explain why the changes after the nineteenth century have resulted in the near-extinction of the kakapo. *(3)*

c) At the present time conservationists are making an effort to save the kakapo from extinction. All the known kakapos have been collected and released onto a small, isolated island free from predators.

What are the advantages and disadvantages of this strategy to prevent extinction? *(4)*

Stretch and challenge

7 Humans have taken such control of their environment that they have eliminated natural selection pressures and therefore will cease evolving.

To what extent would you agree or disagree with this statement?

8 Areas designated National Parks in the UK are subject to strict controls but they also provide exceptional recreational opportunities in a small, highly populated island.

To what extent are the roles of National Parks in conservation of habitats and biodiversity in conflict with pressures for recreational spaces and industrial development? Does this mean that we must choose one or the other?

Tips

Parts (a) and (b) of this question are aimed at AO2, which is designed to test your ability to apply knowledge and understanding to a new situation.

The command word 'explain' is important here. You cannot gain credit for merely repeating information from the stem of the question. You must explain clearly how the changes named would have their effect on the population.

Part (c) is designed to test AO3, which assesses your ability to use your knowledge to consider evidence or argument and come to a reasoned conclusion.

Tip

It would be best to approach Question 8 with some research-specific examples of projects being undertaken in one or more National Parks with a view to managing this conflict. Use your research evidence to come to some conclusion.

9 Cell transport mechanisms

Prior knowledge

In this chapter you will need to recall that:

→ cell surface membranes are found on the surface of plant and animal cells and around cell organelles

→ plant and animal cells have cytoplasm bound by a membrane; in addition, plant cells are covered with a cell wall

→ cell surface membranes have a common structure formed mainly from a lipoprotein bilayer

→ diffusion is the movement of molecules from a region of high concentration to a region of lower concentration

→ osmosis is the movement of water molecules from a dilute to a more concentrated solution across a partially permeable membrane

→ both osmosis and diffusion will continue until there is no longer a concentration difference

→ both diffusion and osmosis are driven by the kinetic energy of the molecules concerned; there is no external energy input needed

→ where substances need to be moved against a concentration gradient then energy will be needed in the form of ATP

→ ATP is the universal form of chemical energy used in all cells.

Test yourself on prior knowledge

1 What type of lipids are found in cell surface membranes?

2 What are the **two** main types of protein present in the membrane?

3 What is the most common chemical constituent of a plant cell wall?

4 What name is given to the difference in concentrations across a membrane?

5 Which has a greater concentration of water molecules, a dilute solution or a concentrated solution?

6 Which molecule is formed when energy is released from ATP?

7 Why will diffusion 'stop' when the two concentrations are equal, even though there is still a great deal of random movement of the particles?

8 What is the only molecule that moves during osmosis?

The structure of the cell surface membrane

The cell surface membrane is the structure that maintains the integrity of the cell (it holds the cell's contents together). It is also the barrier across which all substances entering and leaving the cell must pass. So the membrane's properties, based on its chemical composition and structure, are all important in the operation of the cell.

The cell surface membrane is made almost entirely of protein and lipid, together with a small and variable amount of carbohydrate. In Figure 9.1, a model of the molecular structure of the cell surface membrane, known as the **fluid mosaic model**, is illustrated. The cell surface membrane is described as a *mosaic* because the proteins are clearly scattered about in this pattern, and *fluid* because the components (lipids and proteins) are able to move past each other in a linear plane.

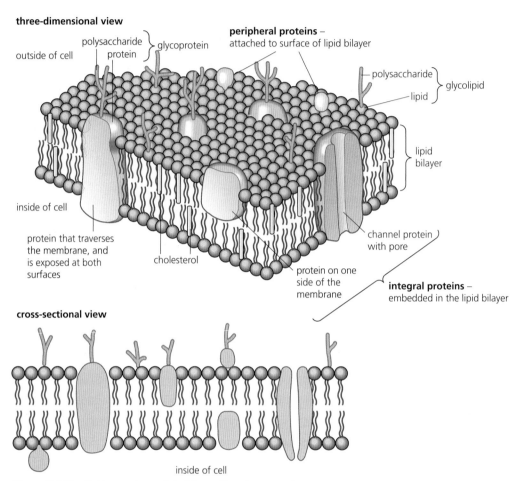

Figure 9.1 The fluid mosaic model of the cell surface membrane

Evidence for the fluid mosaic model

It was in 1972 that two cytologists, S. J. Singer and G. L. Nicolson, proposed the fluid mosaic model of membrane structure. The model was built up from a body of evidence that had accumulated over a period of time, from studies of cell structure (cytology), cell biochemistry and cell behaviour (cell physiology). This is another good example of the way in which important advances are made in science, which was discussed in Chapter 7. This evidence was in the form of ten important observations.

1 Cell contents are observed to flow out when the cell surface is ruptured, as illustrated in Figure 9.2 in a damaged red blood cell. This confirms the presence of a physical barrier around the cytoplasm that is, under normal circumstances, well able to contain and protect the cell contents.

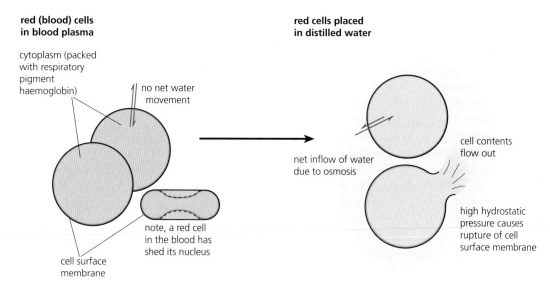

red (blood) cells
in blood plasma

cytoplasm (packed
with respiratory
pigment
haemoglobin)

no net water
movement

note, a red cell
in the blood has
shed its nucleus

cell surface
membrane

red cells placed
in distilled water

net inflow of water
due to osmosis

cell contents
flow out

high hydrostatic
pressure causes
rupture of cell
surface membrane

Figure 9.2 A red cell with a damaged cell surface membrane

2 Water-soluble compounds enter cells less readily than compounds that dissolve in lipids (these will be non-polar compounds and hydrophobic substances). This implies that lipids are a major component of the cell surface membrane.

3 Lipids obtained from cell membranes consist of a type of compound known as a **phospholipid**. The chemical structure of a phospholipid is shown in Figure 1.19 in Chapter 1.

A phospholipid has a 'head' composed of a glycerol group, to which is attached one ionised phosphate group. This latter part of the molecule has **hydrophilic** (water-loving) properties. For example, **hydrogen bonds** readily form between the phosphate head and water molecules.

The remainder of the phospholipid comprises two long, fatty acid residues consisting of hydrocarbon chains. These 'tails' have **hydrophobic** (water-hating) properties. So phospholipids are unusual in being partly hydrophilic and partly hydrophobic.

4 The behaviour of phospholipids when added to water was predicted from this structure – and is demonstrated in practice. With a small quantity of phospholipid in contact with water, these molecules form a monolayer that floats with the hydrocarbon tails exposed above the water (see Figure 1.20 in Chapter 1). When more phospholipid is available, the molecules arrange themselves as a bilayer, with the hydrocarbon tails facing together. This latter is the situation in the cell surface membrane model. Furthermore, in the lipid bilayer, attractions between the hydrophobic hydrocarbon tails on the inside, and between the hydrophilic glycerol/phosphate heads and the surrounding water molecules on the outside, result in a stable, strong barrier.

5 Chemical analysis of cell surface membranes has also shown that, although a significant proportion of lipid is present, there is insufficient in total to cover the whole of the cell surface in a bilayer. Furthermore, protein is also present as a major component. The proteins of cell surface membranes are globular proteins (see Chapter 2).

Key terms

Integral proteins Proteins present in the cell surface membrane that are partially or totally buried within the lipid bilayer.

Peripheral proteins Proteins present in the cell surface membrane that are superficially attached to the lipid bilayer.

6 Work on the extraction of protein from cell surface membranes indicated that, while some occur on the external surfaces and are easily extracted, other proteins occur buried within or across the lipid bilayer. These are difficult to extract.

7 Electron micrographs (EMs) of cell surface membrane fragments, which had by chance split down the midline, showed that some proteins occur buried within or across the lipid bilayer (Figure 9.3). Proteins that occur partially or fully buried in the lipid bilayer are described as integral proteins. Those that are superficially attached on either surface of the lipid bilayer are known as peripheral proteins. The roles of membrane proteins have also been investigated, and these are diverse. Membrane proteins may be channels for transport of metabolites, or enzymes or carriers; others may be receptors or antigens.

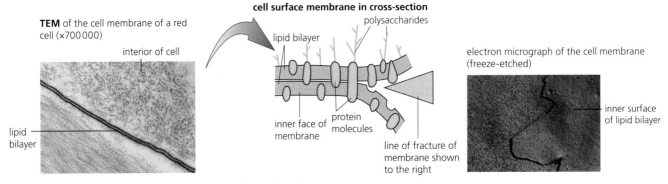

Figure 9.3 Cell surface membrane structure; evidence from the electron microscope

8 Experiments in which specific components of membranes are 'tagged' by reaction with marker chemicals (typically fluorescent dyes) show that the component molecules within membranes are continually on the move. The membrane's structure can truly be described as 'fluid'.

9 Lipid bilayers have been found to contain molecules of a rather unusual lipid, in addition to phospholipids. This lipid is known as **cholesterol**. Cholesterol has the effect of disturbing the close-packing of the phospholipids, thereby increasing the flexibility of the membrane.

10 On the outer surface of the cell, antenna-like carbohydrate molecules form complexes with certain of the membrane proteins (forming **glycoproteins**) and lipids (**glycolipids**). The functions of these complexes have since been shown to be cell–cell recognition, or as receptor sites for chemical signals. Others are involved in the binding of cells into tissues.

Energy transfers in cells

In this chapter and several others you will meet ideas about how cells use and transfer energy. At this level it is really important to describe these processes in a scientifically accurate way. To do this we must keep to one of the most important rules in science. It may sound daunting but it is the first law of thermodynamics and is very straightforward. Put simply, it says that **we cannot create or destroy energy**. Energy can only be transferred from one form to another and the amount of energy you start with is always the same as the amount of energy at the end.

The role of ATP

Energy made available within the cytoplasm may be transferred to a molecule called **adenosine triphosphate** (**ATP**). This substance occurs in all cells at a concentration of $0.5–2.5\,mg\,cm^{-3}$. It is a relatively small, soluble organic molecule – **a nucleotide** – with an unusual feature. It carries three phosphate groups linked together in a linear sequence.

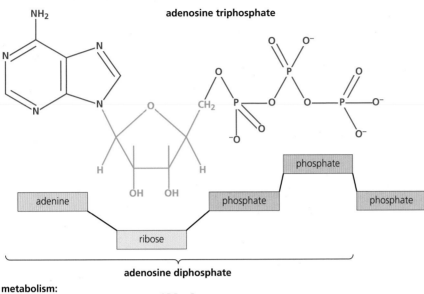

> ### Tip
> At AS and A level, you must watch your language very carefully. In examinations, phrases such as 'energy is made in respiration' or 'energy is used up in muscles' are simply wrong and will gain you no credit.

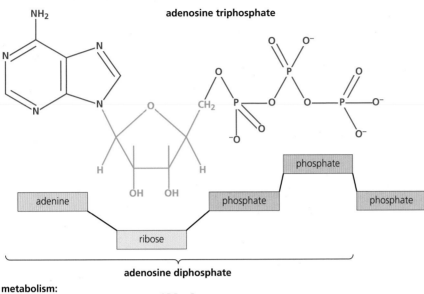

Figure 9.4 The structure and role of ATP

ATP is formed from adenosine diphosphate (ADP) and a phosphate ion (P_i) by transfer of energy from other reactions. ATP is referred to as 'energy currency' because, like money, it can be used for different purposes, and it is constantly recycled. ATP contains a considerable amount of chemical energy locked up in its structure. What makes ATP special as a reservoir of **stored chemical energy** is its role as a common intermediate between energy-yielding reactions and energy-requiring reactions and processes.

Energy-requiring reactions include the synthesis of cellulose from glucose, the synthesis of proteins from amino acids and the contraction of muscle fibres.

The free energy available in ATP is approximately $30–34\,kJ\,mol^{-1}$, made available in the presence of a specific enzyme. Some of this energy is lost as heat in a reaction, but much free energy is made available to do useful work – more than sufficient to drive a typical energy-requiring reaction of metabolism.

Sometimes ATP reacts with water (a hydrolysis reaction) and is converted back to ADP and P$_i$. Direct hydrolysis of the terminal phosphate groups like this happens in muscle contraction, for example. Therefore ADP is exactly the same molecule as ATP in Figure 9.4 but with one less phosphate group attached to it.

Mostly, ATP reacts with other metabolites and forms **phosphorylated** intermediates, making them more reactive in the process. The phosphate groups are released later, so both ADP and P$_i$ become available for re-use as metabolism continues. These are very good examples of energy transfers not 'making' or 'using up' energy.

In summary, ATP is a molecule universal to all living things; it is the source of energy for chemical change in cells, tissues and organisms. The important features of ATP are that it can:

- move easily within cells, by facilitated diffusion
- take part in many steps in cellular respiration and in very many reactions of metabolism
- transfer energy in relatively small amounts, sufficient to drive individual reactions.

Test yourself

1 State which part of a phospholipid is:
   a) hydrophobic
   b) hydrophilic.
2 State what type of protein is found in the cell surface membrane structure.
3 Give the name for proteins buried within the phospholipid bilayer.
4 What other lipid is found in the cell surface membrane structure?
5 State the function of glycoproteins found on the surface of the cell surface membrane.
6 What type of chemical molecule is ATP?
7 What does the first law of thermodynamics state about the nature of energy?
8 Name the type of reaction involved in converting ATP to ADP + P$_i$.

Movement across the cell surface membrane

There is continuous and rapid movement of many substances across the cell surface membrane of living cells. Water, respiratory gases (O_2 and CO_2), nutrients such as glucose, essential ions and excretory products are always moving across these membranes, either entering or leaving the cell. Cells may secrete substances such as hormones and enzymes, and they may receive growth substances and certain hormones.

Plant cells secrete the chemicals that make up their walls through their cell membranes, and assemble and maintain the wall outside the membrane. Certain mammalian cells secrete structural proteins such as collagen, in a form that can be assembled outside the cells in the production of connective tissues, for example.

In addition, the cell surface membrane is where the cell is identified by surrounding cells and organisms. For example, protein receptor sites are recognised by hormones, neurotransmitter substances from nerve cells and other chemicals sent from other cells. Figure 9.5 is a summary of this movement, and also identifies the possible mechanisms of transport across membranes.

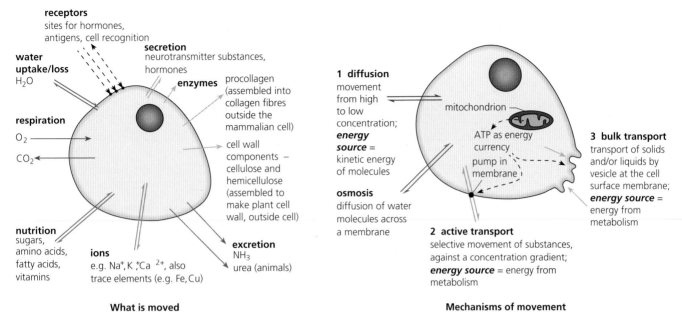

receptors
sites for hormones,
antigens, cell recognition

water uptake/loss
H_2O

secretion
neurotransmitter substances,
hormones

enzymes

procollagen
(assembled into
collagen fibres
outside the
mammalian cell)

respiration
O_2
CO_2

cell wall
components –
cellulose and
hemicellulose
(assembled to
make plant cell
wall, outside cell)

nutrition
sugars,
amino acids,
fatty acids,
vitamins

ions
e.g. Na^+, K^+, Ca^{2+}, also
trace elements (e.g. Fe, Cu)

excretion
NH_3
urea (animals)

1 diffusion
movement
from high
to low
concentration;
energy source =
kinetic energy
of molecules

osmosis
diffusion of water
molecules across
a membrane

mitochondrion

ATP as energy
currency

pump in
membrane

2 active transport
selective movement of substances,
against a concentration gradient;
energy source = energy from
metabolism

3 bulk transport
transport of solids
and/or liquids by
vesicle at the cell
surface membrane;
energy source =
energy from
metabolism

What is moved

Mechanisms of movement

Figure 9.5 Movements across the cell surface membrane

1 Movement by diffusion

The atoms, molecules and ions of fluids (liquids and gases) undergo continuous random movements. Given time, these movements result in the complete mixing and even distribution of the components of a gas mixture, and of the atoms, molecules and ions in a solution. So, for example, from a solution we are able to take a tiny random sample and analyse it to find the concentration of dissolved substances in the whole solution – because any sample has the same composition as the whole. Similarly, every breath we take has the same amount of oxygen, nitrogen and carbon dioxide as the atmosphere has as a whole. This process is called **diffusion**.

Where a difference in concentration has arisen between areas in a gas or liquid, random movements carry molecules from a region of high concentration to a region of low concentration. As a result, the particles become evenly dispersed. The energy for diffusion comes from the **kinetic energy** of molecules.

Key terms

Diffusion The free passage of molecules (and atoms and ions) from a region of their high concentration to a region of low concentration until they are evenly distributed.

Kinetic energy The energy possessed by a particle because it is in continuous motion.

Diffusion in cells

Diffusion across cell surface membranes (Figure 9.6) occurs where:

- the cell surface membrane is fully permeable to the solute – the lipid bilayer of the cell surface membrane is permeable to non-polar substances, including steroids and glycerol, and also oxygen and carbon dioxide in solution, all of which diffuse quickly via this route
- the pores in the membrane are large enough for a solute to pass through. Water diffuses across the cell surface membrane via the protein-lined pores of the membrane (**channel proteins**), and via tiny spaces between the phospholipid molecules. This latter movement occurs easily where the fluid-mosaic membrane contains phospholipids with unsaturated hydrocarbon tails, for here the hydrocarbon tails are spaced more widely. The membrane is consequently especially 'leaky' to water, for example.

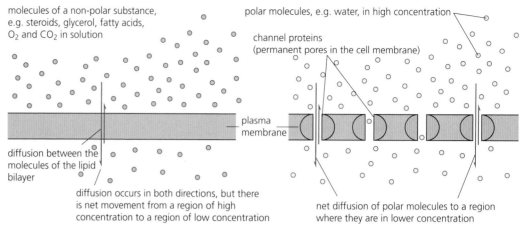

molecules of a non-polar substance, e.g. steroids, glycerol, fatty acids, O_2 and CO_2 in solution

polar molecules, e.g. water, in high concentration

channel proteins (permanent pores in the cell membrane)

plasma membrane

diffusion between the molecules of the lipid bilayer

diffusion occurs in both directions, but there is net movement from a region of high concentration to a region of low concentration

net diffusion of polar molecules to a region where they are in lower concentration

Figure 9.6 Diffusion across the cell surface membrane

Key term

Betalains A group of plant pigments found in certain plants such as beetroot.

Core practical 5

Investigate the effect of temperature on beetroot membrane permeability

Background information

The cells of beetroot contain an intensely red, water-soluble pigment called betalain in their interior – it is found in the large central vacuole of each cell. The pigment may escape in sufficient quantities to be detected in the aqueous medium around the tissue. This occurs if harmful external conditions are applied to beetroot tissue. For example, let us consider the effect of externally applied heat energy. One effect of heat is to denature proteins, and this applies to the proteins of membranes as well as those elsewhere in cells. Once membrane proteins have been denatured, the integrity of the lipid bilayer of the cell surface membranes may also be compromised. Escape of the vacuole contents will indicate this has happened. So, in effect, we can experimentally investigate the approximate temperature at which membrane proteins are seriously denatured.

It is also possible to adapt this technique to investigate the effect of chemical substances (for example, strong solutions of ions, or organic solvents such as alcohol) on the permeability of membranes.

Carrying out the investigation

Aim: To determine the temperature at which the beetroot cell surface membrane is denatured by heat.

Risk assessment: Good laboratory practice is sufficient to avoid a hazard. Wear eye protection when performing this experiment.

1. The first step is to prepare washed beetroot tissue cylinders, about 3 cm long and 0.5 cm in diameter (Figure 9.7). You should carefully cut ten cylinders.
2. Submerge one cylinder in a water bath at 70 °C for 1 minute. Then withdraw it and place it in 15 cm³ of distilled water in a test tube (labelled 'treatment at 70 °C') at room temperature for 15 minutes. After this, you should remove the tissue cylinder and discard it.
3. Cool the water bath to 65 °C and repeat the process with a second cylinder – heat-treat it, allow it to stand in a tube of fresh distilled water (labelled 'treatment at 65 °C') for 15 minutes, and then discard it.
4. Repeat, using heat treatments with water that is 5 °C cooler each time. Make 25 °C the lowest temperature treatment.
5. The distilled water in the test tubes becomes coloured by any pigment that has escaped from the tissue cylinders as a result of heat treatment. You can measure the pigment loss from the tissue into the test tube solutions using a colorimeter containing a complementary colour filter (so, for a red solution, a blue filter is required). Set the scale to zero using the solvent (distilled water) only.

Questions

Read through the method carefully.

1. If the pigment betalain leaks from the central vacuole of a plant cell, how many membranes must it cross to reach the surrounding water? What are the names of these membranes?
2. Why is it necessary to wash the cylinders in water before heat treating them?

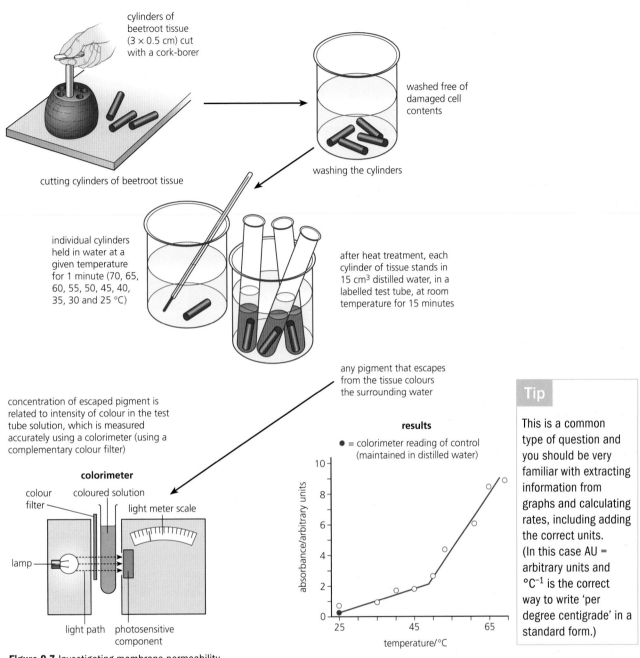

cylinders of beetroot tissue (3 × 0.5 cm) cut with a cork-borer

cutting cylinders of beetroot tissue

washed free of damaged cell contents

washing the cylinders

individual cylinders held in water at a given temperature for 1 minute (70, 65, 60, 55, 50, 45, 40, 35, 30 and 25 °C)

after heat treatment, each cylinder of tissue stands in 15 cm³ distilled water, in a labelled test tube, at room temperature for 15 minutes

any pigment that escapes from the tissue colours the surrounding water

concentration of escaped pigment is related to intensity of colour in the test tube solution, which is measured accurately using a colorimeter (using a complementary colour filter)

colorimeter

colour filter
coloured solution
light meter scale

lamp

light path   photosensitive component

results

● = colorimeter reading of control (maintained in distilled water)

y-axis: absorbance/arbitrary units (0 to 10)
x-axis: temperature/°C (25, 45, 65)

Figure 9.7 Investigating membrane permeability

Tip

This is a common type of question and you should be very familiar with extracting information from graphs and calculating rates, including adding the correct units. (In this case AU = arbitrary units and °C⁻¹ is the correct way to write 'per degree centigrade' in a standard form.)

3 The method also suggests that the cylinders are kept in distilled water at room temperature for exactly 15 minutes. Would this be regarded as a good description of variable control?

4 Look at the graph in Figure 9.7. Is this showing absorption or transmission measured by the colorimeter? Explain your answer.

5 If the membrane proteins do not denature before 40 °C, why should the graph rise between 25 °C and 40 °C as shown in Figure 9.7?

6 Use the graph in Figure 9.7 to compare the rate of increase of absorption with temperature between 25–45 °C and 50–65 °C.

Facilitated diffusion

In facilitated diffusion, a substance that otherwise is unable to diffuse across the cell surface membrane does so as a result of its effect on particular molecules present in the membrane. These latter molecules, made of globular protein, form into pores or channels large enough for diffusion – and close up again when that substance is no longer present (Figure 9.8). In facilitated diffusion, the energy comes from the kinetic energy of the molecules involved, as is the case in all forms of diffusion. Energy from metabolism is not required. An important example of facilitated diffusion is the movement of ADP into and out of mitochondria.

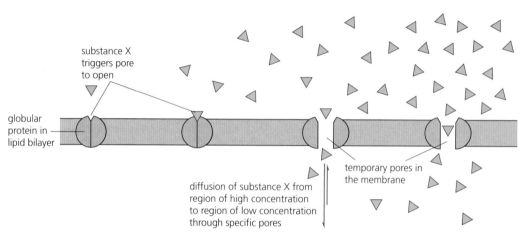

substance X triggers pore to open

globular protein in lipid bilayer

diffusion of substance X from region of high concentration to region of low concentration through specific pores

temporary pores in the membrane

Figure 9.8 Facilitated diffusion

Osmosis – a special case of diffusion

Osmosis is a special case of diffusion. It is the diffusion of water molecules across a membrane that is permeable to water. Since water makes up 70–90 per cent of living cells and cell membranes are **partially permeable** membranes, osmosis is very important in biology.

Dissolved substances attract a group of polar water molecules around them. The forces holding water molecules in this way are weak chemical bonds, including **hydrogen bonds**. Consequently, the tendency for random movement by these dissolved substances and their surrounding water molecules is restricted. Organic substances like sugars, amino acids, polypeptides and proteins, and inorganic ions like Na^+, K^+, Cl^- and NO_3^-, have this effect on the water molecules around them.

The stronger the solution (that is, the more solute dissolved per volume of water), the larger the number of water molecules that are slowed down and held almost stationary. So, in a very concentrated solution, many more of the water molecules have restricted movement than in a dilute solution. On the other hand, in pure water, all of the water molecules are free to move about randomly, and do so.

When a solution is separated from water (or a more dilute solution) by a membrane permeable to water molecules (such as the cell surface membrane), water molecules that are free to move tend to diffuse across the membrane, while dissolved molecules and their groups of water molecules move very much less, if at all. So there is a net flow of water into a concentrated solution, from water or a weaker solution, across the membrane. The membrane is described as partially permeable.

Key term

Osmosis The net movement of water molecules (solvent), from a region of high concentration of water molecules to a region of lower concentration of water molecules, across a partially permeable membrane.

Test yourself

9 What is the driving force that causes movements by diffusion?

10 Name **two** substances that can pass freely through the lipid bilayer of cell surface membranes.

11 By what route do substances pass through the cell surface membrane by facilitated diffusion?

12 Which important cellular compounds pass across mitochondrial membranes by facilitated diffusion?

Osmosis and plant cells

Figure 9.2 shows what happens when water enters an animal cell by osmosis. As the cell expands the membrane is stretched until it bursts. Whilst this demonstrates just how important it is for animal body fluid concentrations to be strictly controlled, the situation in plant cells is very different. This is caused by the differences between animal and plant cell structure. Plant cells have similar membranes but they are covered with a strong cell wall, often made of cellulose, which only expands a little.

The best way to understand the principles involved is to think of the plant cell as a bicycle tyre containing an inner tube. If you remove the tyre and pump up the inner tube it is quite easy to get it to expand like a balloon and eventually burst. However, with the tyre replaced, pumping up the inner tube becomes progressively harder and harder as the pressure inside builds, until you can no longer force more air inside. The tyre is acting exactly like the plant cell wall. As water enters the plant cell the **turgor pressure (P)** inside the cell builds up, because the plant cell wall exerts an equal and opposite pressure, until it equals the pressure of the water entering by osmosis. At this point, water neither enters nor leaves and the cell is said to be **fully turgid**. A fully turgid cell is quite hard and solid and is a vital means of support to the plant. In young plants this is often the only means of support but is also very important to adult plants.

The opposite happens when water leaves the plant cell by osmosis. The pressure inside the cell drops and eventually the outer membrane (**plasmalemma**) shrinks away from the cell wall (just as if you take all the air out of an inner tube). This process of breaking away from the cell wall is called **plasmolysis**. At this point the cell loses all its firmness and becomes soft, and is said to be **flaccid**. The effect of this can be seen in Figure 9.9. As the plant is starved of water the turgor pressure in the cells falls and the plant loses support and wilts.

Water potential

So far we have used a general description of water movements into and out of plant cells, but we need to express what is happening in more scientific terms. This is done by use of the term **water potential** (given the Greek letter **psi** – ψ).

The units of water potential are those of pressure, Pascals (Pa), or more usually **kilopascals (kPa)**.

(1 Pascal = 1 Newton per square metre)

This is where we need to think logically and it leads us to using negative numbers, which can be confusing.

Key terms

Turgor pressure (P) The pressure inside plant cells caused by water entering the cell.

Fully turgid Plant cells are said to be fully turgid when the turgor pressure inside is so high that it prevents further entry of water by osmosis.

Plasmalemma The name given to the cell surface membrane surrounding the outside of the cytoplasm in plant and animal cells.

Plasmolysis This occurs as water leaves a plant cell by osmosis, causing the cytoplasm to shrink away from contact with the cell wall.

Flaccid Plant cells are said to be flaccid when water is withdrawn by osmosis and they lose their firmness.

Water potential A measure of the tendency for water to pass from one place to another. In biology this often means the tendency of water to move into or out of a cell.

a plant cell with adequate supply of water

in this state, all cell walls exert pressure on the surrounding cell walls and the tissue is turgid – fully supported

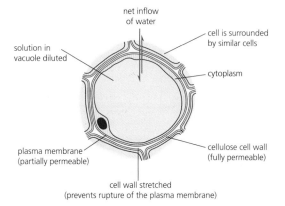

net inflow of water

cell is surrounded by similar cells

solution in vacuole diluted

cytoplasm

plasma membrane (partially permeable)

cellulose cell wall (fully permeable)

cell wall stretched (prevents rupture of the plasma membrane)

b wilting

In the cells of the epidermis of the leaf stalk of rhubarb (*Rheum rhaponticum*) the solutions in the vacuoles are coloured. They can be seen under the microscope without staining.

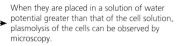

When they are placed in a solution of water potential greater than that of the cell solution, plasmolysis of the cells can be observed by microscopy.

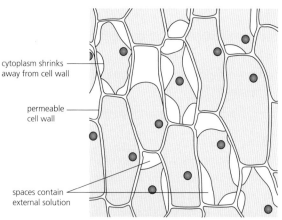

cytoplasm shrinks away from cell wall

permeable cell wall

spaces contain external solution

Figure 9.9 Turgidity supports plant tissue

Start by looking again at the explanation of osmosis as a special case of diffusion, at the beginning of this section.

- Water will always move from a region where there is more water (dilute solution) to a region where there is less water (concentrated solution).
- Making a solution more concentrated (adding solutes) means the water molecules are less free to move.
- So when there are no dissolved solutes (pure water) then the water molecules have their greatest potential to move.
- This means that pure water has the highest possible water potential.
- In other words, any solution must have a *lower* water potential than pure water.
- However, if every other solution is more negative then we are left with only one conclusion. **The water potential of pure water is zero.**

Negative values of water potential in increasingly concentrated solutions:

$$0 \longrightarrow -10\,\text{kPa} \longrightarrow -20\,\text{kPa} \longrightarrow -30\,\text{kPa}$$
(pure water)

more concentrated solutions = more negative water potential

The greater negative water potential of a solution is caused by the solutes dissolved in it and is called its osmotic potential (π), which, because it is a solution, will always be negative.

There can be a positive pressure if some force is causing water to leave the solution or cell.

For example, the turgor pressure exerted by the plant cell wall as the cell takes in water tends to push water out of the cell and therefore would be labelled positive (+) Remember, we have already stated that the osmotic potential of the cell contents would tend to draw water into the cell and this is negative, therefore it is a logical step to assume that the turgor pressure tending to force water out will be positive.

Overall, therefore, we have the following relationship, which you will be expected to learn:

water potential = turgor pressure + osmotic potential

$$\psi = P + \pi$$

> **Key term**
>
> **Osmotic potential (π)** The increased water potential of a solution caused by the solutes dissolved in it. Sometimes referred to as solute potential, it will always have a negative value.

Example

1 Plant tissue has been immersed in pure water overnight. Cells in the tissue are found to have an osmotic potential of −600 kPa. What will be their turgor pressure?

$$\psi = P + \pi$$

2 Two adjacent animal cells, M and N, are part of a compact tissue. They have the following values:

	Cell M	Cell N
Osmotic potential (π)	−580 kPa	−640 kPa
Turgor pressure (P)	+410 kPa	+420 kPa

In which direction will water flow between these cells?

$$\psi = P + \pi$$

Answers

1 If the plant cell is fully turgid, water will not enter or leave so WP = 0 (there is no tendency for water to enter).

Hence 0 = P − 600

and P = +600 kPa

Not surprisingly, our answer shows that turgor pressure will be equal and opposite to the osmotic potential, which explains why there is no net water movement.

2 Water potential of cell M = +410 − 580 = −170 kPa

Water potential of cell N = +420 − 640 = −220 kPa

Water will flow from a less negative to a more negative water potential. So water flows from cell M to cell N.

13 Explain why water molecules will be attracted to ions in a solution.

14 Name the membrane that surrounds the vacuole in a plant cell.

15 Which structure in a plant cell will cause the build up of turgor pressure as water enters by osmosis?

16 State the units of water potential.

17 Name the compound that has the highest possible water potential.

Core practical 6

Determining the water potential of a plant tissue

Background information

In order to determine the water potential of the cells in a plant tissue you will be expected to understand the terms 'water potential', 'osmotic potential' and 'turgor pressure'.

The principle of this investigation is to place plant tissue in a range of concentrations of an external solution and to measure changes caused by osmosis. The most accurate way to measure the changes caused by osmosis is to measure the change in mass of the tissue as it gains or loses water.

The theory behind water potentials tells us that when the water potential of the cell contents and the osmotic potential of the external solution are the same, then the overall water potential will be zero and at this point there will be no gain or loss in mass of the tissue left in the solution. This is what we are attempting to find in this investigation.

Carrying out the investigation

Risk Assessment: There are no significant risks in this procedure provided normal laboratory rules are observed.

1 Take one large potato and cut six cylinders, 4 cm long, using the same cork borer (at least 0.5 cm in diameter). Check that none of the cylinders has any peel attached. Blot each cylinder carefully to remove any excess liquid and weigh it. You will need to be as accurate as possible and it is advisable to use a balance capable of measurements to at least 0.01 g.

2 Label five large boiling tubes, one for each of the following: distilled water, 0.25 M, 0.5 M, 0.75 M and 1.0 M. It is useful to add more concentrations within this range if time or apparatus permit.

3 Make up 30 cm^3 of these concentrations of sucrose solution. Place one cylinder of potato in each boiling tube and add 30 cm^3 of the correct appropriate solution to each tube. (The actual volume may vary according to the boiling tube you are using but it is important that the volume is the same for each and as much as you can fit into your chosen container.)

4 Leave all the boiling tubes and their contents in a cool place for at least 2 hours (but not longer than 24 hours). After this time, carefully pour out the liquid and blot the cylinders very carefully as some may be quite soft. Reweigh the dried cylinders and record the new mass, again taking care to match each one with their initial mass.

5 Calculate the percentage change in mass for each cylinder and plot a graph of percentage change on the vertical axis and osmotic potential of the solution along the horizontal axis, using Table 9.1.

Table 9.1 Osmotic potential of different concentrations of sucrose solution

Concentration of sucrose/M	Osmotic potential of the solution/kPa
0	0
0.25	−680
0.5	−1450
0.75	−2370
1.0	−3510

Use your graph to identify the osmotic potential of the sucrose solution at which the potato tissue neither gains nor loses mass. This will be equal to the water potential of the potato tissue.

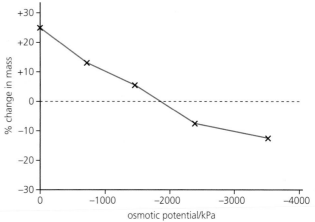

Figure 9.10 Graph of typical results from this investigation

Questions

1 You could take the opportunity here to use a 1 M sucrose solution to make your own dilutions to match the values required. Exactly 15 cm³ of 1 M solution made up to 30 cm³ with distilled water would give you a 0.5 M solution. How could you make up the other concentrations?
2 Why is it essential to use the same potato?
3 Why should you use as large a volume as possible of the sucrose solution?
4 Why should you leave the cylinders in solution for at least 2 hours?
5 How do you calculate percentage change?
6 Why is change in mass measured as a percentage?
7 Should the graph be drawn as a line of 'best fit'?

Tip

Using the final value and not the initial value is a very common error in calculating percentage change. If you wish to know the change then you must compare it with what you started with.

2 Movement by active transport

You have seen that diffusion is due to random movements of molecules and occurs spontaneously, from a high to a low concentration. However, many of the substances required by cells have to be absorbed from a weak external concentration and taken up into cells that contain a higher concentration. Uptake against a concentration gradient cannot occur by simple diffusion as it requires a source of energy to drive it. This type of uptake is known as **active transport**.

● **Active transport can occur against a concentration gradient** – that is, from a region of low concentration to a region of higher concentration. The cytoplasm of a cell normally holds some reserves of substances valuable in metabolism, like nitrate ions in plant cells or calcium ions in muscle fibres. The reserves of useful molecules and ions do not escape; the cell membrane retains them inside the cell. Yet when more of these or other useful molecules or ions become available for uptake, they are actively absorbed into the cells. This happens even though the concentration outside is lower than inside.

● **Active uptake is highly selective**. For example, in a situation where potassium ions (K^+) and chloride ions (Cl^-) are available to an animal cell, K^+ ions are more likely to be absorbed. Similarly, where sodium ions (Na^+) and nitrate ions (NO_3^-) are available to plant cells, NO_3^- ions will be absorbed more rapidly. This is often important in ensuring that the needs of the cell are met and that unwanted ions are excluded.

Key term

Active transport The movement of substances across a cell surface membrane against a concentration gradient, using energy in the form of ATP.

- **Active transport involves special molecules of the membrane**. These molecules pick up particular ions and molecules and transport them to the other side of the membrane, where they are then released. These **carrier proteins** are globular proteins that span the lipid bilayer (Figure 9.1). Movements by these carrier proteins require reaction with ATP; this reaction supplies metabolic energy to the process. Most of these proteins are specific to particular ions and molecules and this is the way selective transport is brought about. If the carrier for a particular substance is not present, the substance will not be transported.

Active transport is a feature of most living cells. You meet examples of active transport in the active uptake of ions by plant roots, in the mammalian gut where absorption occurs, in the kidney tubules where urine is formed, and in nerve fibres where an impulse is propagated.

The carrier proteins of cell surface membranes are of different types. Some transport a particular molecule or ion in one direction (Figure 9.11) while others transport two substances (like Na^+ and K^+) in opposite directions (Figure 9.12). Occasionally, two substances are transported in the same direction, for example Na^+ and glucose during the absorption of glucose in the small intestine.

Many ion channels in the carrier proteins have controlled opening and closing. These are called **gated ion channels**. Those which are controlled by small potential differences are called **voltage-gated ion channels**, while those that are sensitive to chemical signals are called **ligand-gated ion channels**.

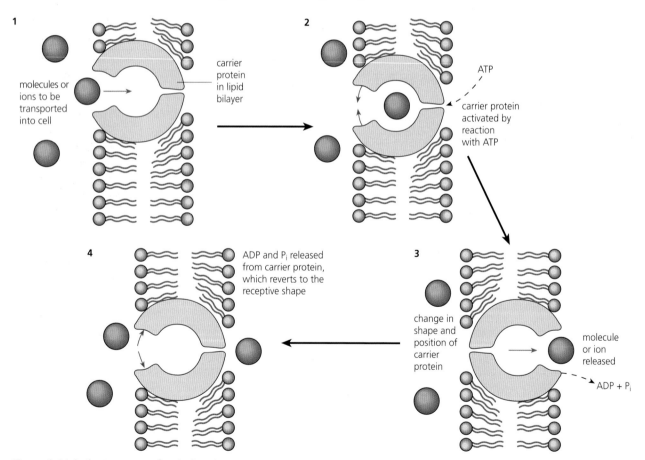

Figure 9.11 Active transport of a single substance

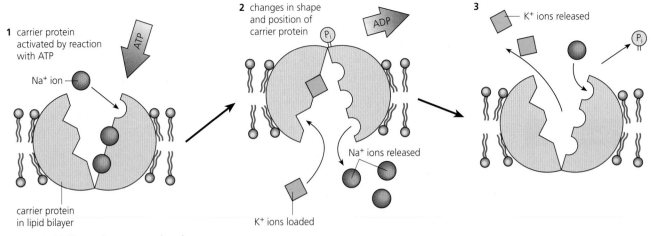

Figure 9.12 The sodium–potassium ion pump

3 Movement by bulk transport

Another mechanism of transport across the cell surface membrane is known as **bulk transport**. It occurs through the movement of vesicles of matter (solids or liquids) across the membrane, by processes known generally as cytosis. Uptake is called endocytosis and export is exocytosis (Figure 9.13).

The strength and flexibility of the fluid mosaic membrane makes this activity possible. Energy from metabolism (ATP) is also required to bring it about. For example, when solid matter is being taken in (phagocytosis), part of the cell surface membrane at the point where the vesicle forms is pulled inwards and the surrounding cell surface membrane and cytoplasm bulge out. The matter thus becomes enclosed in a small vesicle.

In the human body, there are a huge number of phagocytic cells, called the macrophages. They engulf the debris of damaged or dying cells and dispose of it (phagocytosis means 'cell eating'). For example, we break down about 2×10^{11} red blood cells each day, which are ingested and disposed of by macrophages.

Key terms

Vesicles Membrane-bound cell organelles containing liquid or solid particles.

Cytosis The bulk transport of materials across cell membranes contained in vesicles.

Endocytosis Movement of materials into cells.

Exocytosis Movement of materials out of cells.

Phagocytosis This occurs when cells (phagocytes) use their membranes to surround external particles to form vesicles within their own cytoplasm.

Macrophages Large white blood cells that are able to engulf cell debris and foreign particles by the process of phagocytosis.

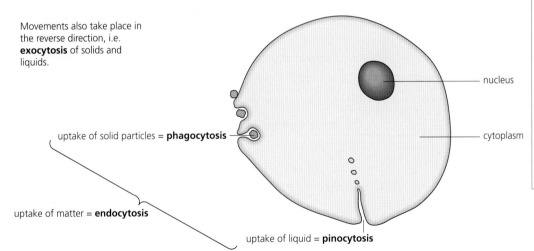

Figure 9.13 Transport by cytosis

Key term

Pinocytosis Movement of materials in liquid form into and out of cells.

Test yourself

18 Name the source of energy needed to transport substances against a concentration gradient.

19 State which structural part of the cell surface membrane is required for active transport.

20 NO_3^- ions are often transported into plant roots by active transport. State why these ions are so important to a plant.

21 Very large molecules and solid particles cannot be transported through the cell surface membrane. Name the process that allows them to be taken up by a cell.

22 State the name of the process by which macrophages engulf red blood cells.

Cells produce many highly active substances such as enzymes and hormones. Simply releasing these into the cytoplasm would obviously cause major disruption. To overcome this problem many compounds are only assembled into their active form inside the membranes of the Golgi body. From there they are transported in bulk inside membrane-bound vesicles to be released outside the cell by exocytosis.

Bulk transport of fluids is referred to as pinocytosis.

Bulk transport is one way in which very large molecules, which are not able to be moved in other ways, can be transported into the cell.

Properties of transported materials

You have seen that substances can be transported across membranes in different ways. The method by which they are transported will vary according to the nature of the substance itself.

The most important features of the molecules, atoms or ions to be transported are:

- the size of the particle – obviously very large molecules such as some proteins will be simply too big to pass through carrier channels or between the molecules of the membrane itself, so bulk transport will be the only pathway. In general, smaller particles are transported more quickly than larger particles
- the solubility of the particle – almost all transport takes place in solution, so particles with limited solubility will only be transported slowly. When substances dissolve they dissociate into charged ions. Not only does this make them smaller but the ions are also much more mobile. Substances that dissolve easily in lipids will obviously pass through the phospholipid layer very easily
- the charge present – the structure of the cell surface membrane makes it difficult for charged particles to pass through. Electrostatic attraction or repulsion will prevent free movements. All ions are charged atoms and polar molecules have weak charges on their structure (see Chapter 1 for the structure of a water molecule). For this reason most ions and some other charged particles pass through the membrane using specialised protein channels.

Some common substances transported into and out of cells and their properties are listed in Table 9.2.

Table 9.2 Properties and membrane transport of some common cellular substances

Substances	Properties	Mode of membrane transport
O_2, CO_2, N_2	Small non-polar molecules	Direct diffusion
H_2O	Small polar molecule	Facilitated diffusion through special protein channels (aquaporins) and sometimes directly through the phospholipid bilayer by osmosis
Glucose	Large polar molecule	Facilitated diffusion using specialised carrier proteins
Glycerol, fatty acids, steroids; vitamins A, D and E	Lipid–soluble molecules	Direct diffusion
Ions, e.g. Na^+, Cl^-, NO_3^-	Small charged atoms	Active transport using carrier proteins; passive movements can occur through protein channels when the concentration gradient is favourable
ATP, ADP	Larger polar molecules	Move in and out of mitochondria by facilitated diffusion

Exam practice questions

1 Cholesterol molecules in the phospholipid bilayer increase the flexibility of the cell surface membrane. The main reason for this is because:

   A cholesterol molecules carry a negative charge

   B cholesterol molecules disturb the tightly packed phospholipids

   C cholesterol molecules penetrate all the way through the membrane

   D cholesterol molecules form additional pores in the membrane *(1)*

2 The formation of ATP from ADP + P$_i$ is:

   A a condensation reaction

   B an oxidation reaction

   C a reduction reaction

   D a hydrolysis reaction

3 Explain how very large molecules such as proteins in solution can be taken across the cell surface membrane into a cell. *(4)*

Tip

Question 3 is a very straightforward question asking you to recall accurately some basic specification content (AO1). However, it is a good exercise in writing an accurate description to gain all the marks available – something that will be vital if you are to achieve higher grades.

4 The diagram shows a typical plant cell that has been immersed in a concentrated glucose solution for several hours.

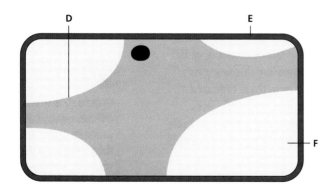

   a) Name the parts of the cell labelled D and E. *(2)*

   b) Name the substance present in the space labelled F. *(1)*

   c) Explain why a plant with many cells in the condition shown in the diagram will begin to wilt. *(4)*

Tip

Questions 1 and 2 are simple recall questions (AO1). There will be several of these in a question paper. Do think carefully before selecting your answer. If you are not sure, at least try to eliminate the possible answers that you know are wrong before selecting an answer. There will be clear instructions on how to correct your choice if you change your mind but, above all, do not select two possible answers, as this is always marked wrong.

Tip

Question 4 is another example of AO1 but with the need to show you understand the principles involved. Part (b) often catches students out.

5 Use your knowledge of the fluid mosaic model of cell surface membrane structure to explain the following observations.

    a) When extracting proteins from a cell membrane, one group of proteins was shown to be easily extracted but a second group was much more difficult to extract. *(3)*

    b) Analysis of the total phospholipid content of a single cell membrane showed that there was much more than required for a single layer surrounding the cell but less than that required to form a complete double layer. *(3)*

> **Tip**
>
> Question 5 is a typical question that asks you to apply your knowledge (AO2). Although the answers are covered in this book it would be an application question, as the specification only refers to the actual structure, so this is actually asking you to apply your knowledge of the structure to these observations.

> **Tip**
>
> Question 6 is an application question (AO2) where you are asked to use your knowledge of the water potential equation to perform the calculation. It is important to check for simple arithmetical errors and to set out your calculations clearly. If you simply write down an answer without clear working, one slip will lose you all of the marks.

6 A small cylinder of plant tissue is placed in a sodium chloride solution.

    a) If the plant cells have a turgor pressure of $+250\,kPa$ and an osmotic potential of $-650\,kPa$, calculate their water potential if placed in a sodium chloride solution of osmotic potential $-1245\,kPa$. *(4)*

    b) Explain why a phospholipid bilayer will inhibit the passage of water molecules, yet they are able to move freely across the cell surface membrane. *(3)*

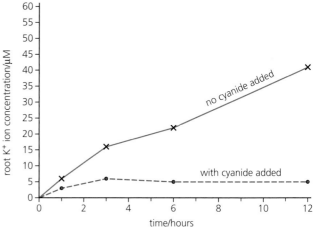

7 The uptake of K^+ ions by barley roots was investigated by placing fresh root sections in a solution containing radioactively labelled K^+ ions. Samples of roots were then taken out of the solution over a period of 12 hours. The concentration of K^+ ions in each sample was then determined by measuring the amount of radioactivity they contained. This investigation was then repeated with potassium cyanide added to the K^+ ion solution. Cyanide ions inhibit an enzyme in mitochondria called cytochrome c oxidase, which prevents them from carrying out their main function. The results of this investigation are shown in the graph.

    a) Calculate the rate of uptake of K^+ ions in both solutions between 3 and 12 hours. *(3)*

    b) Explain the difference in the rate of K^+ ion uptake by the roots in the different solutions. *(3)*

    c) Explain the difference between the K^+ ion concentration found in the roots placed in the cyanide-containing solution, after 3 hours and 12 hours. *(2)*

> **Tip**
>
> Question 7 is a more difficult question as it asks you to do several things. Calculating rates from a graph is a common type of question, which you should be comfortable with, but remember to take the values from the graph as accurately as the printed illustration will allow. Linking ATP production to active transport is an obvious AO2 application question, which can rely on information from several parts of the specification (synoptic).

8 A student investigated the effect of temperature on beetroot cell membranes. She cut a cylinder of beetroot, washed it in distilled water and blotted it carefully. She then placed the cylinder in a water bath at 70 °C for 1 minute. Then she took the cylinder out of the water bath and placed it in 15 cm³ of distilled water in a test tube at 25 °C for 15 minutes. After this, she removed the cylinder.

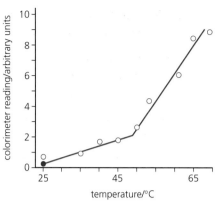

She then measured the amount of red coloured pigment in the distilled water using a colorimeter. The colorimeter was set to read absorbance using a blue filter.

She then repeated the procedure nine times, using identical cylinders from the same beetroot with the temperature of treatment being reduced by 5 °C each time. The results of this investigation are shown in the graph.

a) Explain why a blue coloured filter would be used in the colorimeter to measure the absorbance in this investigation. *(2)*

b) Explain how the temperature treatment of each cylinder could be changed to give more reliable data. *(2)*

c) The student concluded that the results of this investigation 'showed that the proteins in the beetroot membranes denatured above 50 °C'. Discuss the validity of this conclusion. *(4)*

Stretch and challenge

9 Fish are aquatic animals that can be found living in a wide range of habitats, from freshwater streams containing almost no ions to seawater containing high concentrations of ions, especially Na^+ and Cl^-. Many fish have impermeable skins with scales but all have gills, which means that a very large surface area is in contact with the water and separated from it by thin cell surface membranes.

a) Use your knowledge of membranes and transport to explain the problems faced by fish living:

i) in freshwater

ii) in seawater.

b) Consider ways in which these fish might overcome the problems they face. Begin by thinking of possible ideas yourself and then research the actual physiology of some fish. You may come across remarkable examples such as salmon or eels, which spend much of their lives in the open ocean but return to freshwater rivers to breed.

Tip

Question 8 illustrates two important principles. You must understand the science behind the core practicals, not just regard them as a recipe. Secondly this question is a part of AO3, which asks you to be able to suggest amendments to experimental design and make judgements about exactly what conclusions can (and cannot) be drawn from experimental data.

Gas exchange and transport

Prior knowledge

In this chapter you will need to recall that:

→ living organisms exist in a wide range of sizes from minute unicellular forms to very large mammals
→ gas exchange occurs mainly by diffusion
→ plants have much lower metabolic rates and therefore lower rates of gas exchange
→ living organisms show adaptations for gas exchange depending upon their size and environment
→ gas exchange surfaces need to have a large surface area and thin membranes
→ larger organisms need to couple their gas exchange surfaces to a transport system
→ mammals use haemoglobin in their bloodstream to transport oxygen
→ haemoglobin is contained within red blood cells
→ the gas exchange surfaces in fish are called gills, and in mammals, lungs
→ most gas exchange in plants takes place through stomata in leaves.

Test yourself on prior knowledge

1 Explain why gas exchange surfaces need to have thin membranes.
2 Why do some animals have pigments such as haemoglobin in their blood?
3 State the correct name for a red blood cell.
4 Explain why plants have lower metabolic rates than animals.
5 Most plants close many of their stomata at night. Explain why.
6 State which gas is likely to diffuse out of leaves at night.
7 Explain why most stomata are usually found on the under-surfaces of leaves.
8 Describe where, in an active mammal, oxygen is most likely to pass from the blood into the tissues.
9 State which of the following contains most oxygen per cm^3 at the same temperature and pressure:

 A water

 B air

Size and surface area

The size and shape of an organism influence its method of gas exchange. The amount of gas an organism needs to exchange is largely proportional to its volume (the bulk of respiring cells), but the amount of exchange that can occur is proportional to its surface area over which diffusion takes place. This relationship is very important as it determines important features of living things.

Consider what happens as a cell increases in size. If we take the simple example of a cell being cube–shaped, as shown in Figure 10.1, with a side of length, l, then its volume is l^3 and its surface area is $6 \times l^2$. This is really important mathematically because as l increases, the volume increases as l³ and the surface area increases as l². In terms of gas exchange, the volume (demand) increases a lot more than the surface area (supply). As a result it does not take much of an increase in size before this fact becomes a limiting factor (it limits further increases in size).

In an organism that consists of a single small cell, the surface area is large in relation to the amount of cytoplasm it contains. Here, the surface of the cell is sufficient for efficient gas exchange because the sites where respiration occurs in the cytoplasm are never very far from the surface of the cell. The **surface-area-to-volume ratio** is very high for single-celled organisms, and this makes for efficient gas exchange.

The geometric 'organisms' in Figure 10.1 illustrate how the surface-area-to-volume ratio changes as the size of an organism increases. Increasing size lowers the surface area per unit of volume of the whole structure – that is, the larger the object, the smaller its surface-area-to-volume ratio.

This is another example of how ratios can be very useful when comparing different organisms. For example, a blue whale is about 30 m long and weighs about 140 tonnes, so it would be rather pointless to say that it is bigger than a dormouse, which is about 70 mm long without its tail and weighs about 17 g. However, a dormouse has a surface-area-to-volume ratio over 4 times greater than that of the blue whale. This tells us many things about the different problems faced by these animals.

It is obviously advantageous for single cells to remain small, but if organisms are to exploit different habitats then they need to increase in size and show cell specialisation to increase efficiency. Therefore multicellular organisms need special adaptations to overcome the limitations of larger size and so we see a wide range of respiratory surfaces.

One way of increasing surface area is to change shape. A thin and flat shape – such as that of the leaves of a plant, the fronds of seaweed and the body of a flatworm – has a larger surface-area-to-volume ratio and therefore gas exchange is extremely efficient (Figure 10.2).

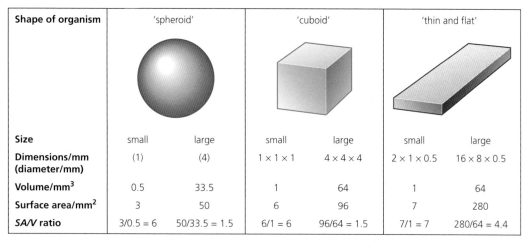

Shape of organism	'spheroid'		'cuboid'		'thin and flat'	
Size	small	large	small	large	small	large
Dimensions/mm (diameter/mm)	(1)	(4)	1 × 1 × 1	4 × 4 × 4	2 × 1 × 0.5	16 × 8 × 0.5
Volume/mm³	0.5	33.5	1	64	1	64
Surface area/mm²	3	50	6	96	7	280
SA/V ratio	3/0.5 = 6	50/33.5 = 1.5	6/1 = 6	96/64 = 1.5	7/1 = 7	280/64 = 4.4

Figure 10.1 Size, shape and surface-area-to-volume ratios

Amoeba, a large, single-celled animal (protozoan) living in pond water and feeding on the tiny protozoa around it. Food is taken into food vacuoles. Gases are exchanged over the whole body surface.

Size = about 400 μm

Ulva, the sea lettuce, an anchored or free-floating seaweed. It floats near the surface of water and photosynthesises in the light. Gases are exchanged over the whole body surface.

Size = about 5–15 cm long, about 30–35 μm thick

Dugesia tigrina, a free-living flatworm found in ponds under stones or leaves or gliding over the mud. It feeds on smaller animals and fish eggs. It is a very thin animal that exchanges gases over the whole body surface.

Size = about 20 mm

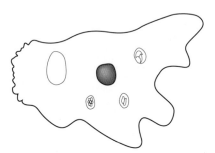
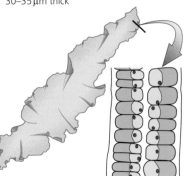
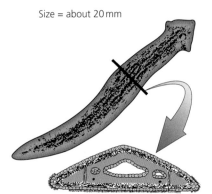

Figure 10.2 Organisms in which gas exchange takes place through their external surface

There are many different ways of adapting respiratory surfaces to provide sufficient diffusion to meet the needs of the organism, but all follow the same pattern:

- The surfaces have a large area.
- The surfaces have thin membranes.
- There must be some means of maintaining a diffusion gradient across the membranes.
- Where demands are high then the surfaces must be linked to a transport system.

Test yourself

1 Name **two** ways in which organisms can increase their surface area.

2 Copy and complete the following table for cells that are perfect spheres.
   (Volume (V) = (4/3) πr^3; surface area (SA) = $4\pi r^2$)

Radius/mm	Surface area/mm²	Volume/mm³	SA:V ratio
1	12.57	4.19	3
2			
3			
4			

3 State the mathematical reason for the trend in *SA:V* ratio shown in the table in Question 2.

4 Explain why thin membranes and a high diffusion gradient are essential features of a respiratory surface.

Gas exchange

Gaseous exchange in insects

Insects are an amazing group of animals. There are more species of insect, and they are more numerous, than any other specialised multicellular group. There are many reasons for their success, but one major factor is that they possess a tough external **exoskeleton**. This has been adapted for rapid flight as well as agile movement on land. The insect exoskeleton is made up of **chitin**, which is a specialised polysaccharide usually combined with other compounds to make a composite material of great strength and flexibility. However, chitin is impermeable to oxygen, and so forms a barrier to gas exchange.

In order to provide sufficient oxygen to active tissues such as flight muscles, insects have a branching network of fine tubes known as tracheae. Tracheae are supported by rings of chitin, which prevent them collapsing when the pressure changes yet still allow them to be flexible. This is exactly the same function as rings of cartilage in the mammalian trachea (and the strengthening rings in a vacuum cleaner hose).

Key term

Tracheae A series of fine tubes that carry gases to and from active organs in the body of an insect.

layout of the tracheal system (air sacs not shown)

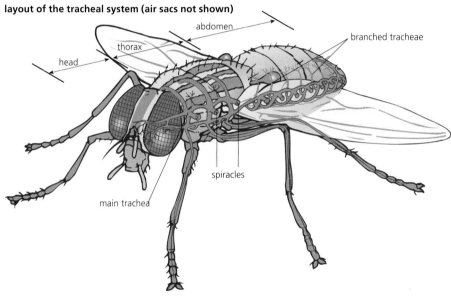

part of the tracheal system (drawing)

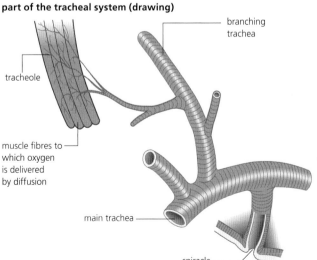

part of the tracheal system (photomicrograph)

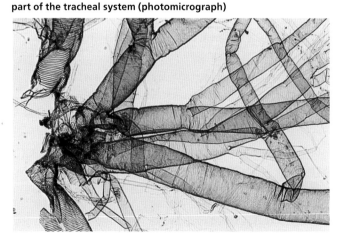

Figure 10.3 The tracheal system of a typical insect

Key term

Spiracles Valves found on the thorax and abdomen of insects through which gaseous exchange takes place.

Larger tracheae form the main pathway along the length of the insect body with much smaller branches called **tracheoles** leading directly into the tissues. Tracheae are connected to the outside atmosphere by valves called spiracles, which can often be seen as a series of eight small holes on each side of the abdominal segments and two more pairs on the thorax. Spiracles can be opened and closed by means of small muscles. They often have fine hairs, which prevent the entry of small particles that could block the tracheoles.

Gas exchange along the tracheae is largely brought about by **diffusion**. Tracheoles, ending in the tissues, have very thin walls to allow diffusion into and out of the cells. Remember that diffusion relies on concentration gradients so each gas must be considered separately. Hence, high levels of carbon dioxide produced by active tissues mean that there will be a large gradient between the tracheoles and the very low levels of carbon dioxide in the external atmosphere. Exactly the opposite applies for oxygen, so the two gases will move in different directions down their respective concentration gradients.

Simple diffusion along the tubes can be insufficient to supply all the needs of the insect tissues and it is thought that movements of the abdomen, especially during flight, compress and expand the tubes, flushing air through them much faster. Opening and closing of certain spiracles also helps to ensure a flow of fresh air through the main tracheae. Many insects are highly adapted to life in very dry environments. The presence of many fine tubes leading to the atmosphere will obviously be a potential source of water loss. Closing spiracles at times when gas exchange is slow helps to conserve water.

Core practical 7

Dissect an insect to show the structure of the gas exchange system

Ethical issues

All dissections raise ethical issues. At the very least it would be expected that any biologist would ensure that any animal was treated with due care and respect.

The law places very strict guidelines on the use of higher animals in any investigation but this does not apply to invertebrates. There is considerable debate on why this should be the case and you need to research the different sides of the debate.

Most of all you will be expected to understand what is meant by an ethical issue. The main point to grasp here is that, although at the centre of the debate might be the question of what is the right thing to do, different people will hold perfectly valid but opposing ethical viewpoints.

To take two extremes, you might wish to argue that killing any living thing is wrong. On the other hand you might wish to argue that killing any other living thing can be justified as long as it is not a human.

Obviously both opinions can have a reasoned argument but both are full of complications. Does not killing any living thing include all plants or harmful insects? Does killing any other living thing apart from humans mean we are free to slaughter anything we choose, including threatened species, and who decides where we draw the line? So many ethical issues are very difficult to resolve.

So, you are free to take your own ethical stand but you will be expected to show that you have reasoned arguments for doing so and that you understand the reasoning of others who may take a different view.

Carrying out the investigation

Risk assessment: In all dissections, hygiene is important to minimise any risk of infection. Animal materials must be obtained from a biological supplier or premises licenced with the local authority. Gloves are not necessary, but hands should be washed thoroughly after the activity. Note: some people may be allergic to insect cuticle. Safe use of dissection instruments will be demonstrated by your teacher. Any remains from the dissection should be disposed of according to the instructions you are given by your teacher.

Equipment: The dissection can be carried out successfully with a minimum of equipment, but is easier with some simple options:

- a pair of the smallest, sharpest dissection scissors available
- a pair of fine forceps (tweezers)
- a dissection dish
- microscope slides and cover slips
- light microscope + a binocular microscope (useful but not essential)
- a hand lens or magnifying glass
- preserved locust or similar insect (the larger the better)

It is important for this dissection to secure the insect carefully. A classic dissection dish has a thick layer of heavy wax in the bottom to facilitate this. A simple economical alternative is to use a large ice-cream container. The walls of this need to be trimmed to about 7–8 cm high. In the bottom of this use waterproof glue to stick a piece of polystyrene or the flower arranging material 'oasis', about 1 cm thick. This will form a platform for the dissection and allow the insect to be covered with water. You may need to trim the sides of the container further to ensure that, when covered with water, the dissection is easy to access.

Dead houseflies or blowflies can be used but their scale makes the dissection more difficult.

1 Hold the insect with the forceps and examine it carefully using a hand lens. Identify the following:
   a) the three main body sections – head, thorax and abdomen
   b) spiracles – these vary slightly in different insects but are often small ovals on the rear of most abdominal segments. Thoracic spiracles are more difficult to identify; in the locust they are found close to where the legs join the abdomen
   Whilst not part of the respiratory system, other features such as compound eyes and the mouthparts are often well worth a close examination.
2 Secure the insect to the bottom of the dissecting dish using plain straight pins with a row of visible spiracles pointing upwards. Use pins to fix the wings upwards so that the side of the abdomen is not covered (Figure 10.4). Then add just enough water to cover it.

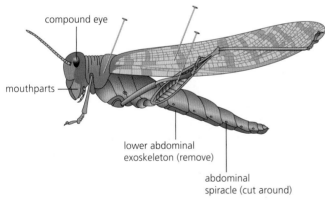

compound eye

mouthparts

lower abdominal exoskeleton (remove)

abdominal spiracle (cut around)

Figure 10.4 Locust dissection

3 Hold the insect firmly with the forceps and cut through the exoskeleton around one spiracle.
4 Use the forceps to take hold of the spiracle and the small piece of exoskeleton and pull gently to remove it from the abdomen. TAKE CARE not to pull too hard. If the spiracle and a small piece of underlying tissue do not come away easily then use the scissors to cut around it to free it.
5 Mount the spiracle and the tissue on a microscope slide with a drop of water and add a cover slip. Take care to hold the cover slip with one edge against the water drop and lower slowly to avoid trapping air bubbles.
6 Examine the slide carefully under medium power of the microscope and look for tracheae with rings of chitin and any visible features of the spiracle to add to your previous observations. The photomicrograph in Figure 10.3 will help you to identify tracheae.
7 Select a suitable part of your slide and draw a simple illustration of what you see. Use single clear, sharp, pencil lines without shading. Try to ensure that the proportions of your drawing are correct and show important features rather than attempting to draw an artistic 'picture'. Add a scale to your drawing to show the approximate size of the tubes.
8 Finally, turn the insect over on its back and relocate it with pins. You may need to cut off the wings to do this. Insert the scissors under one side of the last abdominal segment. Keep the points of the scissors pointing upwards towards you to avoid cutting into the tissues beneath. Cut the exoskeleton in a line down one side toward the head. Repeat this on the other side of the abdomen so that you can carefully remove a whole strip of the exoskeleton under the abdomen.
9 You will now see lots of white fat surrounding the gut along with other organs. Do not disturb this too much but use a magnifying glass to try to identify the tracheal network. It is a double line of fine empty tubes which, unlike other organs, look shiny because they contain air. Compare the pattern of these tubes with the photomicrograph in Figure 10.3.

Test yourself

5 Name the valves linking the tracheae to the atmosphere.
6 Name the structures that prevent tracheae collapsing when the pressure inside is reduced.
7 State **two** ways in which insects can increase the rate of gas exchange through tracheae.
8 The largest insects in the world are only about 10 cm long. What features of their gas exchange system may limit their overall size?

Gaseous exchange in a bony fish

Fish and other aquatic organisms face different problems of gas exchange from those using air. First of all, water contains about 25 times less oxygen by volume than air and this gets less the warmer the water becomes. Secondly water is almost 800 times denser than air, so is much more difficult to move around. Despite this, highly mobile fish such as trout and salmon have adaptations that enable them to extract enough oxygen from water in order to sustain a high metabolic rate.

Fish obtain oxygen from water by means of **internal gills**. The structure of the gills of a bony fish such as a herring is shown in Figures 10.5 and 10.6. You can see that bony fish have four pairs of gills, supported by a bony arch. Each gill has two rows of **gill filaments** arranged in a V-shape. Filaments are very thin structures carrying rows of thin-walled **gill plates** on both surfaces. A tough muscular flap of skin, the operculum, protects the gills and is partly responsible for maintaining a continuous flow of water over them. The space inside the mouth containing the gills is called the buccal cavity.

Key terms

Operculum A tough muscular flap covering the gills of bony fish.

Buccal cavity The space inside the mouth of fish and other animals.

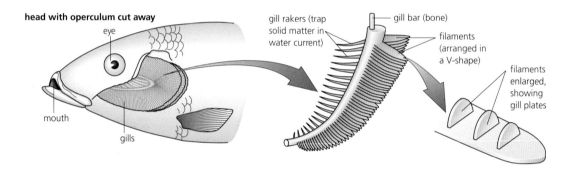

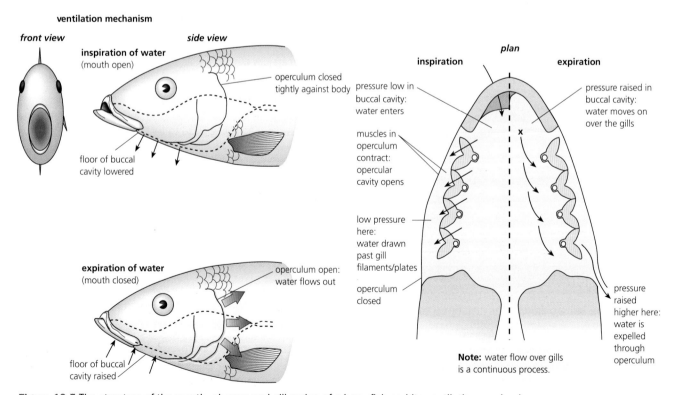

Figure 10.5 The structure of the mouth, pharynx and gill region of a bony fish and its ventilation mechanism

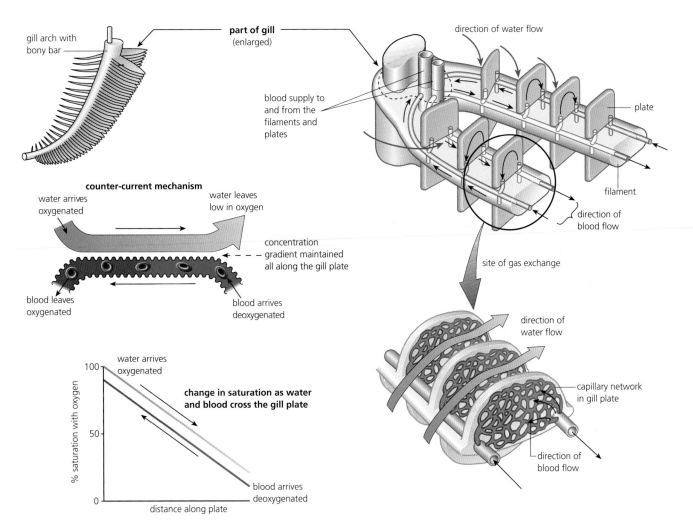

Figure 10.6 Gill structure of a bony fish and gas exchange by counter-current mechanism

Ventilation of the gills

The flow of water across the gills is maintained by changes of water pressure. As the mouth opens, the floor of the buccal cavity is lowered and the operculum is closed tight against the body. This increase in volume decreases the pressure, compared with the outside, so water flows in.

The contracting muscles in the operculum cause it to bulge outwards, increasing the volume of the opercular cavity. This decreases the pressure in this area and water flows across the gills. To keep up this flow, the mouth closes and the floor of the buccal cavity is raised, forcing more water across the gills. Finally this increases the pressure behind the operculum until it exceeds that of the water outside, forcing the operculum to open and allowing water to flow out.

The result of this opening and closing of the mouth is a continuous flow of water across the gills.

Gaseous exchange in the gills

The gill filaments and gill plates have very thin walls and together they form a very large surface area. The capillaries themselves have walls that are only one-cell thick and carry blood with large numbers of red cells. These cells contain **haemoglobin**, which

Key term

Haemoglobin A red iron-containing protein found in red blood cells, which combines with oxygen.

Key term

Counter-current mechanism An arrangement where fluids are caused to flow close to each other in opposite directions to maximise exchange between them.

enables the blood to carry oxygen very efficiently. Further details of this process are discussed later in this chapter.

As water passes over the gills, it flows in the opposite direction to the blood. This is an example of a **counter-current mechanism**, which allows the fish to remove 80–90 per cent of the dissolved oxygen from the water. The principle of the counter-current mechanism is that it maintains a concentration gradient along the whole length of the blood–water boundary. If blood and water flowed in the same direction they would quickly reach the same concentration and therefore only 50 percent of the available oxygen would be transferred. (We will meet this idea again in *Edexcel A level Biology 2* when we consider kidney function.)

Tip

It is useful to bear in mind different applications of the same principle, such as the counter-current mechanism, as they can form the basis of synoptic questions.

Test yourself

9 Describe the features of the gills that give them a large surface area.

10 Explain how the pressure of water in the buccal cavity is increased to force it across the gills.

11 State how much oxygen by volume would be contained in $1\,cm^3$ of pure water under standard conditions.

12 State the advantage of the counter-current flow of water and blood in the gills.

Gaseous exchange in mammals

Almost all mammals are extremely active animals with a very high demand for oxygen. They maintain a constant body temperature, which is also a high energy-demand activity. The organs of gaseous exchange in mammals are the **lungs**. Lungs are extremely efficient. The structure of the human thorax, which houses the lungs, is shown in Figure 10.7.

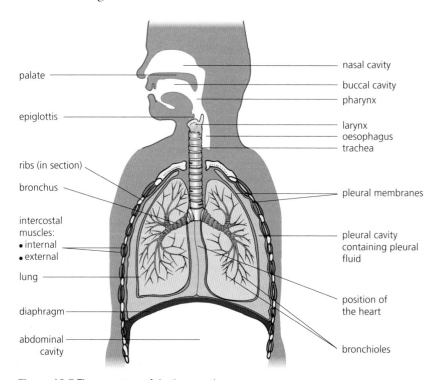

Figure 10.7 The structure of the human thorax

Lungs are housed in the **thorax**, an air-tight chamber formed by the rib-cage and its muscles (**intercostal muscles**), with a domed floor, which is the diaphragm. The diaphragm is a sheet of muscle attached to the body wall at the base of the rib-cage, separating thorax from abdomen. The internal surfaces of the thorax are lined by the pleural membranes, which secrete and maintain the pleural fluid. Pleural fluid is a lubricating liquid derived from blood plasma that protects the lungs from friction during breathing movements.

The lungs connect with the pharynx at the rear of the mouth by the **trachea**. Air reaches the trachea from the mouth and nostrils, passing through the larynx ('voice box'). Entry into the larynx is via a slit-like opening, the glottis. Directly above this is a cartilaginous flap, the **epiglottis**. Glottis and epiglottis work to prevent the entry of food into the trachea. The trachea initially runs beside the oesophagus (food pipe). Incomplete rings of cartilage in the trachea wall prevent collapse under pressure from a large bolus (ball of food) passing down the oesophagus.

The trachea then divides into two **bronchi**, one to each lung. Within the lungs the bronchi divide into smaller **bronchioles**. The finest bronchioles end in air sacs (alveoli). The walls of bronchi and larger bronchioles contain smooth muscle, and are also supported by rings or tiny plates of cartilage, preventing collapse that might be triggered by the sudden reduction in pressure that occurs with powerful inspirations of air.

Ventilation of the lungs

Air is drawn into the alveoli when the air pressure in the lungs is lower than atmospheric pressure, and it is forced out when pressure is higher than atmospheric pressure. Since the thorax is an air-tight chamber, pressure changes in the lungs occur when the volume of the thorax changes. The ventilation mechanism of the lungs is shown in Figure 10.8.

Key terms

Diaphragm A muscular sheet at the bottom of the thorax.

Pleural membranes Double membranes surrounding the lungs, which are lubricated with pleural fluid to prevent friction during breathing.

Alveoli Tiny air sacs within the lungs where gas exchange takes place.

Tip

Be careful to use accurate language when describing ventilation. It is the active movement of muscles that causes a change in lung volume, then the pressure changes and finally air flows in or out from high to low pressure. Be careful to keep to this sequence.

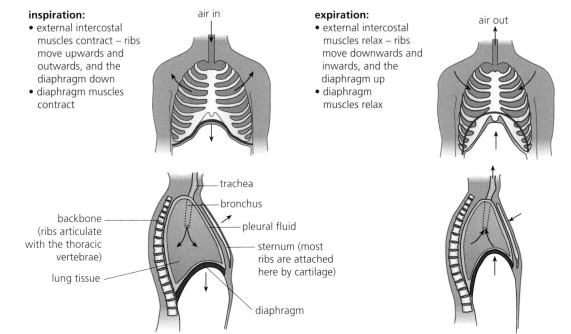

inspiration:
- external intercostal muscles contract – ribs move upwards and outwards, and the diaphragm down
- diaphragm muscles contract

air in

expiration:
- external intercostal muscles relax – ribs move downwards and inwards, and the diaphragm up
- diaphragm muscles relax

air out

trachea
bronchus
backbone (ribs articulate with the thoracic vertebrae)
pleural fluid
sternum (most ribs are attached here by cartilage)
lung tissue
diaphragm

volume of the thorax (and therefore of the lungs) increases; pressure is reduced below atmospheric pressure and air flows in

volume of the thorax (and therefore of the lungs) decreases; pressure is increased above atmospheric pressure and air flows out

Figure 10.8 The ventilation mechanism of human lungs

Squamous epithelium
A single layer of
flattened cells forming
an outer covering.

Endothelium A single
layer of cells forming an
inner lining.

Alveolar structure and gas exchange

There are some 700 million alveoli present in a pair of human lungs, providing a surface area of about $70\,m^2$. This is an area 30–40 times greater than that of the body's external skin. The wall of an alveolus is made up of **squamous epithelium**, which has a single layer of cells about $0.1\,\mu m$ thick. Lying very close is a network of capillaries whose walls are made up of a single layer of thin **endothelial** cells. This means the total thickness of the walls separating air and blood is only $4–5\,\mu m$. The capillaries are extremely narrow – just wide enough for red blood cells to squeeze through (Figure 10.9).

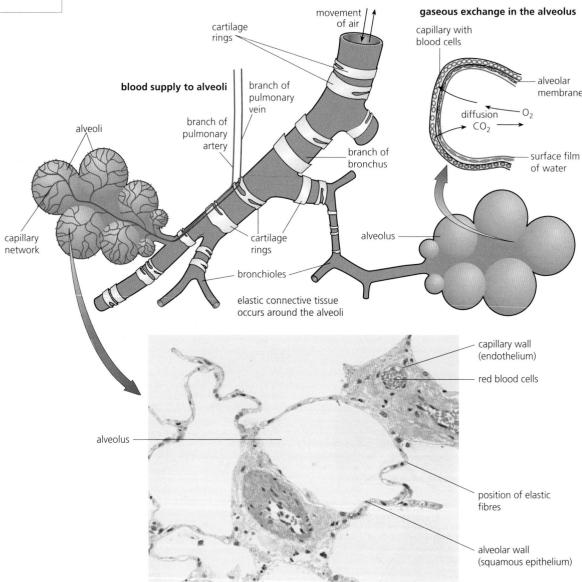

Figure 10.9 Blood supply and gaseous exchange in the alveolus

Blood arriving in the lungs is low in oxygen (it has a lower *partial pressure* of oxygen than the alveolar air, see Table 10.1) but high in carbon dioxide (it has a higher *partial pressure* of carbon dioxide than alveolar air). As blood flows past the alveoli, gaseous exchange occurs by diffusion. Oxygen dissolves in the surface film of water then diffuses across into the blood plasma and finally into the red blood cells. Here, it combines with haemoglobin to form oxyhaemoglobin. (We shall discuss the details of this process later in this chapter.) At the same time carbon dioxide diffuses from the blood into the alveolus.

Table 10.1 The composition of air in the lungs

	Approximate percentage of each gas present/%		
	Inspired air	**Alveolar air**	**Expired air**
Oxygen	20	14	16
Carbon dioxide	0.04	5	4.0
Nitrogen	79	81	79
Water vapour	variable	saturated	saturated

Test yourself

13 Explain how food is prevented from entering the trachea.

14 Name the **two** sets of muscles that are used to increase the volume of the thorax during inspiration.

15 Suggest why a wound penetrating the rib cage would cause the lungs to collapse.

16 If atmospheric pressure at sea level is 102 kPa and 19% of the air is oxygen, what will be the partial pressure of oxygen at sea level?

17 Name the **two** tissues that a molecule of oxygen must cross to pass from an alveolus to the inside of a capillary.

Tip

It is important to remember that plants, like animals, need oxygen to carry out cellular respiration in order to form ATP at all times. Only when light is present do they photosynthesise. At this time the volume of oxygen given off by photosynthesis quickly exceeds that taken up by respiration and so there is a net diffusion of oxygen into the atmosphere. So gaseous exchange varies with the light available in any 24 hour period.

Gas exchange in flowering plants

In contrast to most animals, plants have low metabolic rates and much lower rates of gaseous exchange. In consequence they do not show such advanced adaptations. The main reasons for this are that plants do not move and do not maintain a high body temperature.

Another difference is that plants exchange different gases at different times of day.

Most gaseous exchange in plants is by diffusion directly into and out of the tissues. Oxygen for respiration in root tissues comes directly from the soil through the permeable cell walls and membranes. Where soil oxygen levels are limited, for example by waterlogging, then respiration and subsequent growth are severely restricted.

Above ground, plant leaves and stems are often covered with cuticle or a bark to prevent water loss. This also prevents diffusion of gases. To overcome this, stems often have tiny patches of very loosely packed cells called **lenticels** (Figure 10.10). The loose packaging of cells means that there are many air spaces, which allow direct diffusion to and from the tissues beneath.

In plants the highest rates of gas exchange occur in leaves, to supply carbon dioxide for photosynthesis. The supply of water is also essential for this process, so leaves are adapted to prevent excessive water-loss and to allow rapid diffusion. Layers of waterproof cuticle cover the leaf but the epidermis of the underside of leaves usually contains many thousands of pores called **stomata** (singular stoma – Figure 10.11).

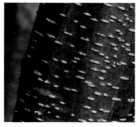

Figure 10.10 Lenticels in a woody stem

Key terms

Lenticels Areas of loosely packed cells forming pores on the surface of plant stems.

Stomata Pores on the leaf surface allowing diffusion of gases into and out of the leaf.

Key term

Spongy mesophyll
Loosely packed inner leaf tissue.

These pores are surrounded by a pair of bean-shaped **guard cells**, which can alter the size of the stoma by changes in their turgor pressure. When turgid, their curved shape means the stoma opens, and when they lose turgor (become flaccid) and straighten, the pore closes. This allows the leaf to maximise gas exchange in the light when photosynthesising, and to limit water loss when less gas exchange is needed in the dark.

Carbon dioxide diffusing through open stomata enters the air spaces of the spongy mesophyll inside the leaf. From here it dissolves in the **water films** surrounding the spongy mesophyll cells and can then freely diffuse through the cell surface membrane into the photosynthesising tissues. Oxygen will diffuse outwards by the same route.

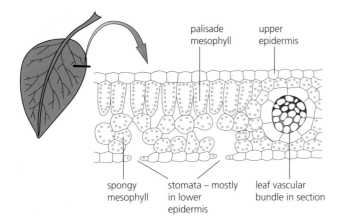

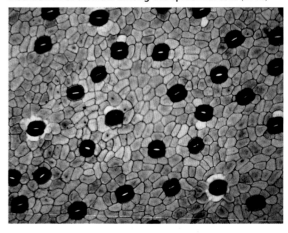

photomicrograph of lower surface of leaf – showing distribution of stomata among the epidermal cells (×100)

Figure 10.11 Distribution of stomata in a typical leaf

Test yourself

18 Suggest why gas exchange organs in plants are less specialised than in animals.

19 Name the actual gas exchange surface in a leaf.

20 State which gases will be exchanged in plant roots during daylight hours.

Transport of gases in the blood

You have seen that active animals need specialised organs to ensure that gas exchange is increased to meet their demands. However, once gas exchange has taken place it is vital that the oxygen is delivered and carbon dioxide removed from all of the tissues of the body in an equally efficient manner. This requires a specialised system of loading and unloading the gases in the blood and rapidly transporting the blood to all parts of the body. This section deals with gas transport by the blood. We shall look at the design of the circulatory system and other functions of the blood cells in the next chapter.

Haemoglobin

Oxygen and carbon dioxide have only limited solubility in the liquid blood plasma, so a much more efficient mechanism is required for their transport. In many animals, including mammals and bony fish, the red pigment **haemoglobin** is used to increase the volume of oxygen carried by the blood. Haemoglobin is a conjugated protein made up of four interlocking sub-units. Each of these is composed of a large globular protein with an iron-containing **haem group** attached (Figure 10.12).

The oxygen molecules are carried by haemoglobin attached to the haem groups. Therefore one haemoglobin molecule can carry four oxygen molecules to form **oxyhaemoglobin**.

$$\underset{\text{haemoglobin}}{Hb} + \underset{\text{oxygen}}{4O_2} \rightleftharpoons \underset{\text{oxyhaemoglobin}}{Hb(O_2)_4} + \underset{\text{hydrogen ions}}{H^+}$$

Haemoglobin is contained inside the red blood cells (**erythrocytes**), not in the blood plasma. Erythrocytes are very strange cells. They are constantly manufactured in the red bone marrow. Mammalian erythrocytes have no nucleus but their cytoplasm is rich in haemoglobin. They are biconcave disc-shaped and have typical cell surface membranes, which gives them a large, thin surface area ideal for transfer of gases by diffusion and the maximum volume of haemoglobin. (see Figure 11.2, Chapter 11). Blood gains its thick, red appearance because the erythrocytes are so numerous (about 5×10^6 per mm^3), which means that about one quarter of all the cells in our bodies are erythrocytes.

Key terms

Conjugated protein
A protein that contains non-polypeptide chemical groups in its structure.

Erythrocytes Mammalian red blood cells.

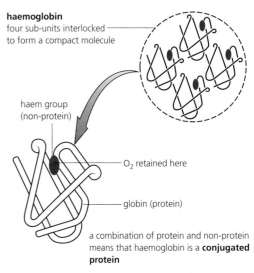

Figure 10.12 The structure of haemoglobin

Partial pressures

Air is a mixture of several different gases. The total pressure of the air is made up of the sum of the pressures of the individual gases. So, if atmospheric pressure is, say 100 kPa, this does not mean that all the gases within it are at a pressure of 100 kPa. To calculate the individual pressures of the gases in the mixture we simply use their relative abundance in the mixture. Therefore if air is 20 per cent oxygen its partial pressure would be 20 per cent of $100 = pO_2 = 20$ kPa. To show this is part of a mixture we use the term partial pressure (*p*). So in this case the partial pressure of the oxygen is 20 kPa. We will often refer to partial pressures in gas exchange because there is almost always a mixture of gases involved and we wish to consider each one individually.

The law of partial pressures states that in a mixture of ideal gases, each gas has a partial pressure that is the pressure that the gas would have if it alone occupied the volume. The total pressure of a gas mixture is the sum of the partial pressures of the individual gases in the mixture.

Key term

Partial pressure (*p*)
The pressure due to one gas in a mixture of gases. The total pressure of a mixture of gases is made of the sum of partial pressures due to each gas.

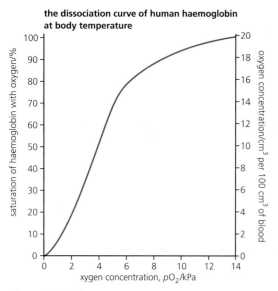

the dissociation curve of human haemoglobin at body temperature

saturation of haemoglobin with oxygen/%

oxygen concentration/cm³ per 100 cm³ of blood

xygen concentration, pO_2/kPa

Figure 10.13 Oxygen dissociation curve for human haemoglobin at body temperature

Oxygen dissociation curves

As oxygen molecules combine with haemoglobin they become attached one by one to the four haem groups. The more oxygen that is available, the more haem groups are filled. If we draw a graph of the amount of oxygen taken up by the haemoglobin compared with the concentration of oxygen around it (expressed as its partial pressure in kPa), we might expect it to be a straight line (directly proportional). This is not the case because as an oxygen molecule becomes attached it makes it easier for the next two to be taken in, but the final oxygen is more difficult to add, so the graph shows a distinctive elongated S-shape (Figure 10.13). This type of graph is known as an **oxygen dissociation curve**.

The graph shows that this results in the ideal situation. Where external oxygen concentrations are low, oxyhaemoglobin will become less saturated and oxygen will be released; and vice-versa where oxygen concentrations are high.

Key terms

Oxygen dissociation curve A graph showing the percentage saturation of haemoglobin at different external concentrations of oxygen.

Bohr effect The reduction of the oxygen-carrying capacity of haemoglobin caused by increasing concentrations of carbon dioxide.

Bohr effect

Most oxygen is required by the most active cells. These cells will also produce higher concentrations of carbon dioxide. High levels of carbon dioxide cause oxyhaemoglobin to release more oxygen. This is known as the **Bohr effect**. The oxygen dissociation curve is shifted to the left (Figure 10.14), which gives a boost to the oxygen available at just the point where it is needed most.

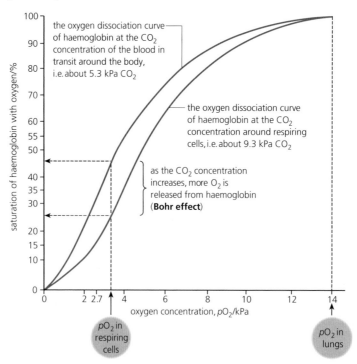

saturation of haemoglobin with oxygen/%

the oxygen dissociation curve of haemoglobin at the CO_2 concentration of the blood in transit around the body, i.e. about 5.3 kPa CO_2

the oxygen dissociation curve of haemoglobin at the CO_2 concentration around respiring cells, i.e. about 9.3 kPa CO_2

as the CO_2 concentration increases, more O_2 is released from haemoglobin (**Bohr effect**)

oxygen concentration, pO_2/kPa

pO_2 in respiring cells

pO_2 in lungs

Figure 10.14 Bohr effect – how carbon dioxide affects oxyhaemoglobin

Myoglobin

Whilst it is vital to have pigments such as haemoglobin in the blood because they have a strong affinity for oxygen, it also creates a problem. How do we get haemoglobin to give up all of its oxygen to other tissues? The obvious answer is to ensure that other pigments are present in the tissues that have a higher affinity for oxygen than haemoglobin. In mammals, working muscles have a high oxygen demand and to ensure that this is met they contain a pigment called **myoglobin**. The higher affinity for oxygen will be shown by an oxygen dissociation curve for myoglobin being further to the left than that of haemoglobin (Figure 10.15).

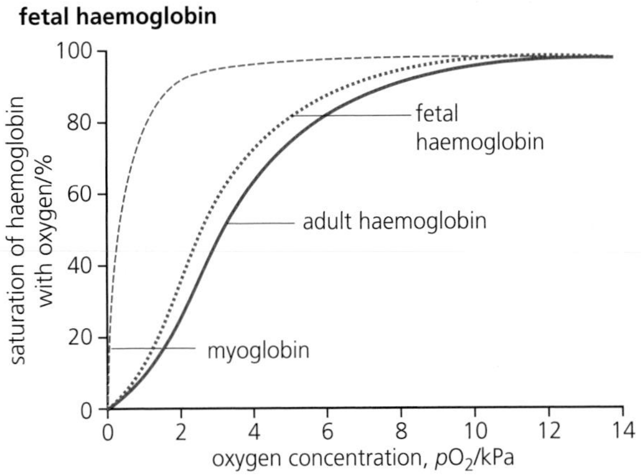

To achieve this higher affinity, myoglobin molecules have a modified structure. Like haemoglobin, myoglobin is a conjugated globular protein and carries oxygen attached to a haem group. Unlike haemoglobin it has only one haem group attached to a complex globular protein, just like the single sub-unit shown in Figure 10.12. As a result myoglobin becomes fully saturated at about pO_2 40 kPa whilst haemoglobin requires about pO_2 80 kPa.

Figure 10.15 Oxygen dissociation curve for adult haemoglobin, myoglobin and fetal haemoglobin

Not only does myoglobin assist in the transfer of oxygen from haemoglobin to muscles but its stronger affinity for oxygen means that it is normally fully saturated so forms a useful reservoir of oxygen within the muscles, enabling them to keep working longer when oxygen demand exceeds supply.

> ### Tip
>
> Many questions use different dissociation curves so it is very useful to remember that, in general, a curve further to the left means that at any given oxygen concentration it will have a higher saturation and therefore a better affinity for oxygen. A curve further to the right means a lower affinity for oxygen.

Fetal haemoglobin

A very similar problem arises in the placenta of pregnant mammals. How can oxygen be transferred across the placenta from the maternal blood to the fetal circulation? In this case the use of pigments such as myoglobin would be unsuitable as their very strong affinity for oxygen would not allow sufficient exchange between fetal blood and the growing tissues.

The answer is provided by the presence of **fetal haemoglobin**. During the development of the embryo the genes coding for haemoglobin are expressed in varying ways. The result of this is that the haemoglobin proteins undergo subtle changes during development that affect their properties. Fetal haemoglobin has a slightly greater affinity for oxygen than adult haemoglobin (again its dissociation curve is slightly to the left) but the change is just enough to transfer oxygen from maternal haemoglobin but not enough to affect its normal working within the fetal circulation (Figure 10.15).

Haemoglobin and carbon dioxide transport

Most carbon dioxide is transported in the plasma and red blood cells as hydrogen carbonate ions:

$$CO_2 + H_2O \xrightarrow{\text{carbonic anhydrase}} HCO_3^- + H^+$$

The presence of the enzyme carbonic anhydrase inside red blood cells means that greater volumes of carbon dioxide can be transported than simply relying upon limited solubility in the plasma. The exchange of carbon dioxide is therefore much more efficient.

A much smaller amount of carbon dioxide (around 10 per cent) is combined directly with the amino groups of the polypeptide chains of haemoglobin inside red blood cells to form carbaminohaemoglobin.

$$\text{carbon dioxide} + \text{haemoglobin} \rightarrow \text{carbaminohaemoglobin} + H^+$$

Both of these reactions release H^+ ions. Haemoglobin can also take up some of these to buffer the blood and prevent it becoming too acidic.

Test yourself

21 Name **two** differences between haemoglobin and myoglobin.

22 State exactly where haemoglobin is found in the blood.

23 Suggest the effect an increase in carbon dioxide concentration would have on the ability of haemoglobin to carry oxygen.

24 Explain why myoglobin in muscles is normally fully saturated with oxygen.

25 State the difference in oxygen affinity between fetal haemoglobin and adult haemoglobin.

Exam practice questions

1 The reason why the surface-area-to-volume ratio of organisms decreases as they become larger is:

   A Surface area increases with the cube of the dimension but volume increases with the square of the dimension.

   B Surface area increases logarithmically but volume increases linearly.

   C Surface area increases with the square of the dimension but volume increases with the cube of the dimension.

   D Surface area increases linearly but volume increases logarithmically. *(1)*

2 Structures that allow gas exchange to take place in plant stems are called:

   A stomata

   B spiracles

   C lenticels

   D micropyles *(1)*

3 Explain how a continuous flow of water is maintained across the gills of a bony fish. *(5)*

Tip

Question 3 is a very simple recall question but it is useful practice in writing concisely and accurately to achieve full marks.

4 The graph below shows the oxygen dissociation curves of fetal and adult haemoglobin.

   a) The curves for fetal haemoglobin and adult haemoglobin show a similar shape as the partial pressure of the oxygen increases. Explain how the structure of a haemoglobin molecule can account for the pattern of oxygen uptake shown by these curves. *(3)*

   b) Explain what is meant by the term partial pressure. *(2)*

   c) i) What is the difference between the oxygen saturation of fetal haemoglobin and adult haemoglobin when the partial pressure of oxygen is 4 kPa? *(1)*

      ii) Explain the importance of this difference to the development of the fetus in a mammal. *(3)*

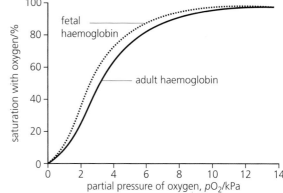

Tip

Dissociation curves are very common in examination questions. They may be directly linked to the specification, as in Question 4, or they may show curves from different animals you have not met before. In this case there is some straight recall (AO1) but also an element of applying your knowledge for AO2 in part a) and in part c) ii).

Tip

Question 4 asks you to apply your knowledge of gaseous exchange in the blood and expects you to have some other scientific knowledge necessary to understand this topic. In both AS and A level examinations you will meet synoptic questions that test your knowledge of two or more parts of the specification or your practical skills. This question begins to test your ability to draw together different scientific concepts.

5 The diagram shows two of the chemical reactions that take place inside red blood cells within a capillary close to actively respiring tissues.

H^+ = hydrogen ions; HbO_2 = oxyhaemoglobin; HHb = haemoglobin; HCO_3^- = hydrogen carbonate ions

a) Describe two ways in which a human red blood cell is adapted for gas exchange. *(2)*

b) i) Haemoglobin acts as a chemical buffer in the blood. What is meant by a chemical buffer? *(2)*

ii) Use the information in the diagram to explain how haemoglobin can act as a buffer in blood. *(3)*

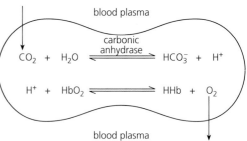

c) The binding of hydrogen ions to oxyhaemoglobin causes small changes in the shape of the oxygen binding sites. Explain how this may bring about the Bohr effect as carbon dioxide concentration increases. *(3)*

Stretch and challenge

6 Haemoglobin is a very common respiratory pigment used for the transport of oxygen. It can be found in a wide range of animals including vertebrates, but it is not the only pigment found in living organisms, as shown in the table below.

Respiratory pigment	Occurrence	Colour change (deoxygenated → oxygenated)
Haemoglobin	Most vertebrates	purple red → bright red
Haemocyanin	Crustaceans (crabs and lobsters)	colourless → blue
Chlorocruorins	Marine annelid worms	green → red

The chemical structure of the sub units making up these pigments is shown here. (Note that most carbon atoms have been omitted to make them clearer.)

Haemocyanin

Haemoglobin

Chlorocruorin

a) Use your knowledge of the structure of haemoglobin to describe the similarities and differences between these pigments.

b) i) What might the structure of these molecules indicate about their origins in evolution?

   ii) The diagram on the right shows the chemical structure of chlorophyll a. What are the similarities and differences between chlorophyll a and haemoglobin? Is this further evidence for evolutionary ancestry?

c) Haemoglobin in humans is always found inside red blood cells. Other pigments are normally found dissolved in the plasma. How does this arrangement make the functioning of haemoglobin more efficient?

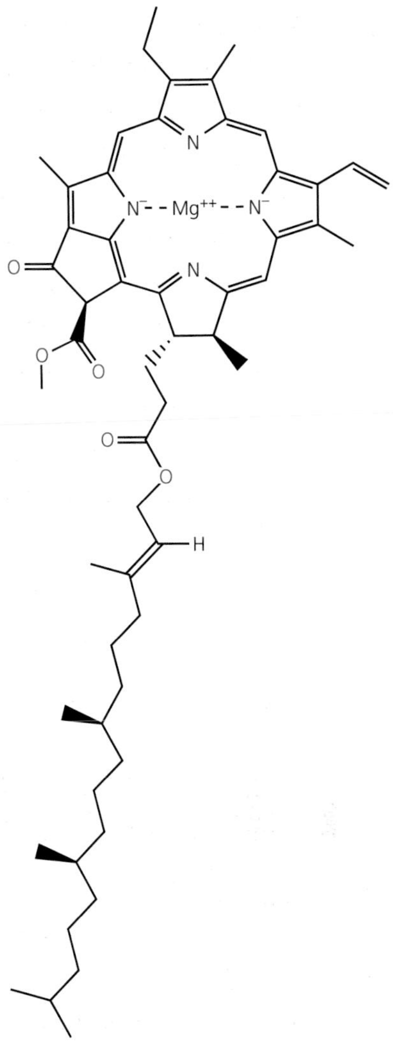

Mammalian circulation

Prior knowledge

In this chapter you will need to recall that:

→ the human heart has four chambers and is divided into two by the septum

→ the left side of the heart carries oxygenated blood and the right side deoxygenated blood

→ the heart is made up of a specialised tissue called cardiac muscle

→ ventricles have thick muscular walls to pump blood to the lungs from the right ventricle and around the body from the left ventricle

→ atria have much thinner walls and pump blood to the ventricles

→ valves between atria and ventricles and at the base of main arteries ensure a one-way flow of blood through the heart

→ arteries have thicker elastic walls and carry high-pressure blood away from the heart

→ veins have thinner walls and contain valves; they carry low-pressure blood back to the heart

→ capillaries have very thin walls, only one-cell thick, and form a network linking arteries and veins

→ blood transports nutrients around the body and defends the body against pathogens by clotting and by destroying foreign bodies using the immune system

→ blood is made up of several types of white cell, red cells and platelets suspended in a fluid called plasma.

Test yourself on prior knowledge

1 Name the main artery carrying blood from the left ventricle.

2 State the type of valves found at the base of main arteries in the heart.

3 The muscular wall of the left ventricle is much thicker than that of the right ventricle. Explain why.

4 Name the only artery that carries deoxygenated blood.

5 Explain why veins need valves but arteries don't.

6 Explain why the walls of capillaries are only one cell thick.

7 Why do arteries need elastic walls?

8 State which component of the blood is concerned with blood clotting.

9 What is the general name for all white blood cells?

10 State the name given to the arteries supplying cardiac muscle with nutrients.

Circulatory systems

Living cells require a supply of water and nutrients such as glucose and amino acids, and most need oxygen. The waste products of cellular metabolism have to be removed, too. In single-celled organisms and very small organisms, internal distances are small, so here movements of nutrients can occur efficiently by diffusion, as discussed in Chapter 10.

In more active organisms, cells need to be supplied with nutrients at a much faster rate, so there is a greater need for an efficient internal transport system. Larger animals have a **blood circulatory system** that links the parts of the body and makes resources available where they are required.

Internal transport systems at work are examples of mass flow. In mass flow, fluid moves in response to a pressure gradient, flowing from a region of high pressure to regions of lower pressure. Any suspended and dissolved substances present in the fluid are carried along in the same direction.

Key term

Mass flow A system of transport that uses a fluid, which is moved by a pressure gradient. Substances to be transported are suspended or dissolved in the fluid and all move in one direction.

Advantages of a double circulation

Mammals have a **closed circulation** in which blood is pumped by a powerful, muscular heart and circulated in a continuous system of tubes – the **arteries**, **veins** and **capillaries** – under pressure. The heart has four chambers and is divided into right and left sides by the **septum**. Blood flows from the right side of the heart to the lungs, where it is oxygenated, and then back to the left side of the heart. From here it is pumped around the rest of the body and back to the right side of the heart. As the blood passes twice through the heart in every single circulation of the body, this is called a **double circulation**.

The circulatory system of mammals is shown in Figure 11.1, alongside an alternative system found in fish. Fish also have a closed circulation, where blood flows only once through the heart in every circulation of the body, a condition known as a **single circulation**. This means that blood emerging from the gills (see Chapter 10) has only a low pressure and therefore flows around the body much more slowly.

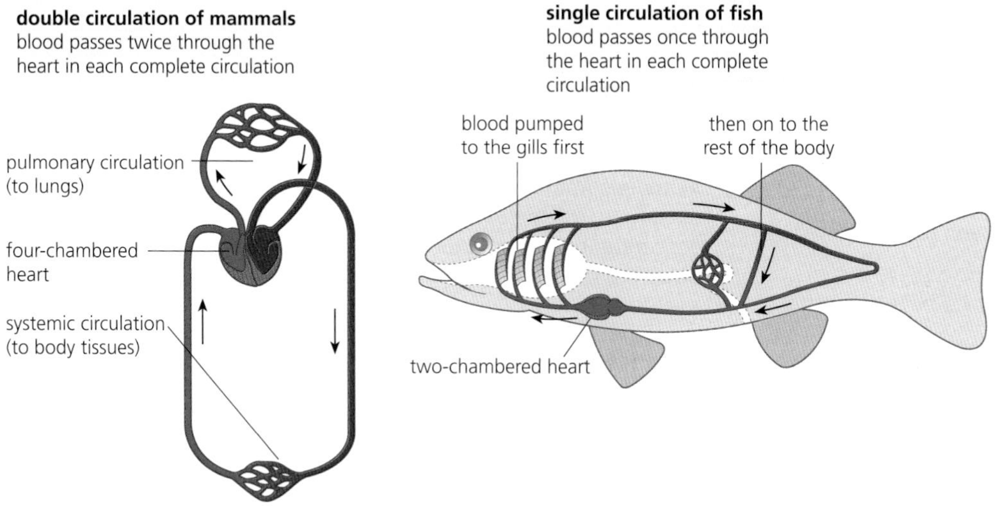

double circulation of mammals
blood passes twice through the heart in each complete circulation

pulmonary circulation (to lungs)

four-chambered heart

systemic circulation (to body tissues)

single circulation of fish
blood passes once through the heart in each complete circulation

blood pumped to the gills first

then on to the rest of the body

two-chambered heart

Figure 11.1 Single and double circulation

Figure 11.1 shows oxygenated blood as red and deoxygenated blood as blue. To overcome the problem of low pressure, a double circulatory system returns blood to the heart. This poses the danger that oxygenated and deoxygenated blood will mix, so blood pumped around the body will contain far less oxygen. The division of the heart into two separate halves prevents this happening so only fully saturated blood from the lungs enters the systemic circulation.

It becomes clear that the major advantages of the mammalian circulation are that:

- oxygenated blood is delivered at high pressure to all body tissues
- oxygenated blood reaches the respiring tissues undiluted by deoxygenated blood.

In discussing the mammalian blood circulation, we will take the human circulation as the example.

Test yourself

1 Why would a fish heart not be divided into two complete halves?
2 Suggest why fish survive and flourish, even though they have a less-efficient single circulation.
3 Give **two** features of a mass-flow transport system.

The transport medium – the blood

Blood is a special tissue consisting of a liquid medium called **plasma**, in which are suspended red cells (**erythrocytes**), white cells (**leucocytes**) and **platelets** (Figure 11.2). The plasma is the medium for exchange of substances between cells and tissues; the red cells are involved in transport of respiratory gases; and the white cells are adapted to combat infection. The roles of the components of blood are summarised in Table 11.1. In addition to transport, the blood functions as an important defence mechanism and the source of tissue fluid and lymph, which we will consider in more detail later in this chapter.

Table 11.1 The components of the blood and their roles

Component	Role
Plasma (Note: 'serum' is plasma from which all cells and the soluble protein fibrinogen have been removed.)	Transport of: • nutrients from gut or liver to all cells • excretory products, e.g. urea from the liver to the kidneys • hormones from the endocrine glands to all tissues and organs • dissolved proteins that have roles including regulating the osmotic concentration (water potential) of the blood • dissolved proteins that are antibodies • heat to all tissues.
Red cells	Transport of: • oxygen from the lungs to respiring cells • carbon dioxide from respiring cells to the lungs (also carried in the plasma).
White cells	**Lymphocytes** have major roles in the immune system, including forming antibodies. **Phagocytes** such as **monocytes** and **neutrophils** ingest bacteria or cell fragments. **Eosinphils** are identified by taking up the red stain eosin and stimulate allergic responses and histamine production.
Platelets	Involved in the blood clotting mechanism.

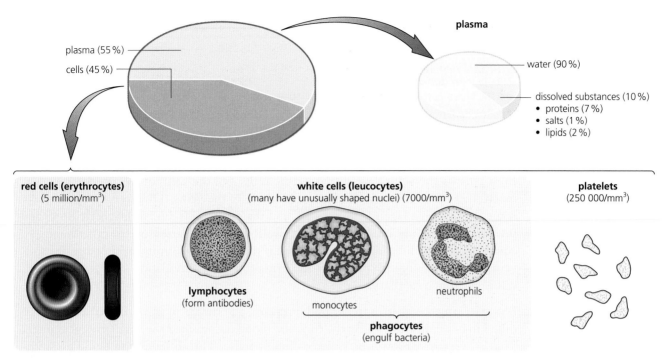

Figure 11.2 The composition of the blood

The plumbing of the circulatory system – arteries, veins and capillaries

There are three types of vessel in the circulatory system:

- **arteries**, which carry blood away from the heart
- **veins**, which carry blood back to the heart
- **capillaries**, which are fine networks of tiny tubes linking arteries and veins.

Figure 11.3 shows an artery, vein and capillary vessel in section, and details of the wall structure of these three vessels.

Figure 11.3 The structure of the walls of arteries (× 20), veins (× 20) and capillaries (× 4000)

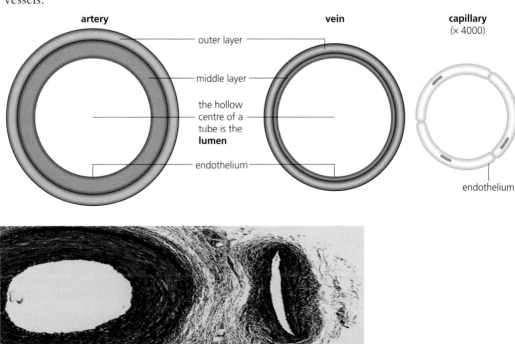

Table 11.2 Differences between arteries, veins and capillaries

	Artery	Capillary	Vein
Outer layer (tunica externa) collagen fibres	Present	Absent	Present
Middle layer (tunica media) elastic fibres and smooth (involuntary muscle)	Thick layer	Absent	Thin layer
Endothelium (tunica intima) pavement epithelium	Present	Present	Present
Valves	Absent	Absent	Present

Tip

The elastic recoil of arteries shows another example of the need to write accurately about energy that we met in Chapter 10. Notice that, once again, we avoid talking about energy being 'used up' or 'lost' and concentrate on how it is transferred from one form to another.

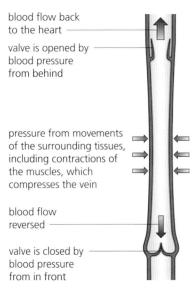

blood flow back to the heart

valve is opened by blood pressure from behind

pressure from movements of the surrounding tissues, including contractions of the muscles, which compresses the vein

blood flow reversed

valve is closed by blood pressure from in front

Figure 11.4 The valves in veins

Key terms

Pulmonary circulation
The pathways of arteries, veins and capillaries carrying blood from the heart to the lungs and back to the heart.

Systemic circulation
The pathways of arteries, veins and capillaries carrying blood from the heart to all body tissues and back to the heart.

Both arteries and veins have strong, elastic walls, but the walls of the arteries are very much thicker and much more elastic than those of the veins. The strength of the walls comes from the collagen fibres present, and the elasticity is due to the elastic fibres and involuntary (smooth) muscle fibres. The walls of the capillaries, on the other hand, consist of endothelium only (endothelium is the innermost lining layer of arteries and veins). Capillaries branch profusely and bring the blood circulation close to cells – no cell is far from a capillary.

Blood leaving the heart is under high pressure, and travels in waves or **pulses**, following each heart beat. By the time the blood has reached the capillaries, it is under very much lower pressure, without a pulse. This difference in blood pressure accounts for the differences in the walls of arteries and veins.

The importance of elastic arteries

As the ventricles of the heart contract, blood is forced into arteries at very high pressure but the ventricles must then relax to refill with blood. The result of this could be that the blood would flow in a stop–start fashion. Obviously, this would be very inefficient. However, elastic fibres and smooth muscle in the walls of arteries are stretched when the blood is at high pressure. Therefore some of the energy in the blood is stored as potential energy in the elastic walls of the arteries. When the pressure of the blood falls this stored energy is used to maintain the flow until the ventricles contract again. In this way the pressure fluctuations of the blood are smoothed out to give a more continuous flow. This is known as **elastic recoil of arteries**.

The importance of valves in veins

Having squeezed through the capillary network, blood in veins is at a low pressure, yet still needs to be returned to the heart. For example, low-pressure blood in veins in your feet has to be pumped a vertical height of about a metre to reach the heart. Veins have **valves** at intervals, which prevent the backflow of blood (Figure 11.4). These valves also mean that as surrounding muscles contract and press against the veins, they force the blood from one valve to the next, so helping the return flow.

The arrangement of arteries and veins

You have already seen that mammals have a double circulation. It is the role of the right side of the heart to pump deoxygenated blood to the lungs. The arteries, veins and capillaries serving the lungs are known as the **pulmonary circulation**. The left side of the heart pumps oxygenated blood to the rest of the body. The arteries, veins and capillaries serving the body are known as the **systemic circulation**.

In the systemic circulation, organs are supplied with blood by arteries branching from the main artery known as the **aorta**. Within individual organs, the arteries branch into numerous arterioles (smaller arteries) and the smallest arterioles supply the capillary networks. Capillaries drain into venules (smaller veins), and venules join to form veins. The veins join the main vein (**vena cava**) carrying blood back to the heart. The branching sequence in the circulation is, therefore:

aorta → artery → arteriole → capillary → venule → vein → vena cava

For the pulmonary circulation this becomes:

pulmonary artery → arteriole → capillary → venule → vein → pulmonary vein

Arteries and veins are often named after the organs they serve (Figure 11.5). The blood supply to the liver is via the hepatic artery, but the liver also receives blood directly from the small intestine, via a vein called the **hepatic portal vein**. This brings much of the products of digestion, after they have been absorbed into the blood circulation in the gut.

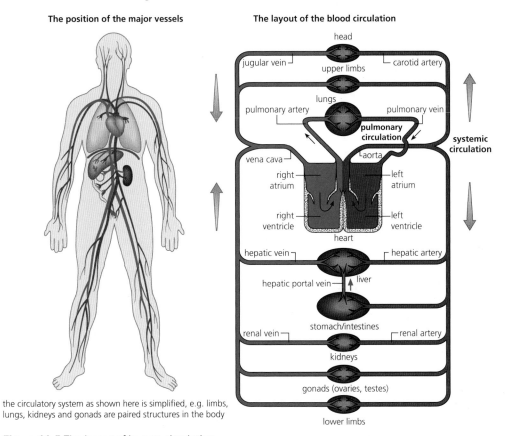

the circulatory system as shown here is simplified, e.g. limbs, lungs, kidneys and gonads are paired structures in the body

Figure 11.5 The layout of human circulation

Test yourself

4 Name the white cell that is involved in the allergic response.

5 Explain what is meant by a phagocyte.

6 Name the protein that is found in the outer layers of arteries and veins.

7 Explain why veins have much thinner walls than arteries.

8 Suggest how the flow of blood in the circulation would be different if artery walls were rigid.

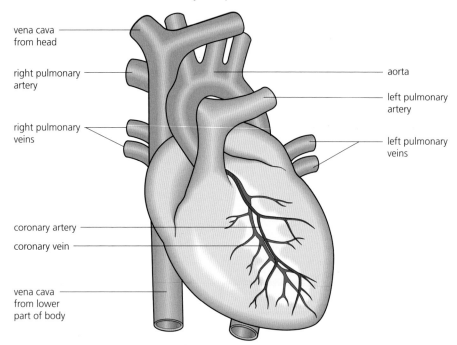

heart viewed from the front of the body with pericardium removed

vena cava from head

right pulmonary artery

right pulmonary veins

coronary artery

coronary vein

vena cava from lower part of body

aorta

left pulmonary artery

left pulmonary veins

heart in LS

vena cava from head

right pulmonary artery

right atrium

vena cava from lower part of body

tricuspid valve

right ventricle

aorta

left pulmonary artery

left pulmonary veins

left atrium

semilunar valves

bicuspid valve

left ventricle

Figure 11.6 The structure of the heart

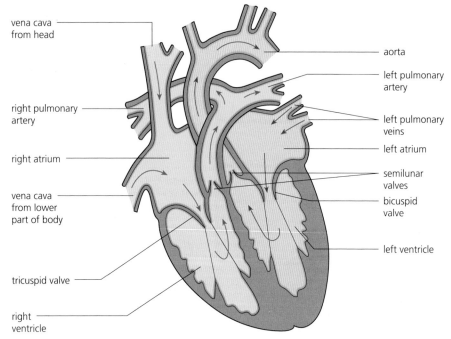

The heart as a pump

The human heart is the size of a clenched fist. It is found in the thorax between the lungs and beneath the breastbone (sternum). The heart is a hollow organ with a muscular wall, and is contained in a tightly fitting membrane, the pericardium – a strong, non-elastic sac that anchors the heart within the thorax.

The cavity of the heart is divided into four chambers, with those on the right side of the heart completely separate from the left side. The two upper chambers are thin-walled **atria** (singular: atrium). These receive blood into the heart. The two lower chambers are thick-walled **ventricles**. The ventricles pump blood out of the heart, with the muscular wall of the left ventricle being much thicker than that of the right ventricle. However, the volumes of the right and left sides (the quantities of blood they contain) are identical.

Note that the walls of the heart (the heart muscle) are supplied with oxygenated blood via **coronary arteries**. These arteries, and the capillaries they serve, deliver to the cardiac muscle fibres the **oxygen** and **nutrients** essential for the maintenance of the pumping action.

The **valves** of the heart prevent backflow of the blood, thereby maintaining the direction of flow through the heart. The **atrio-ventricular valves** are large valves, positioned to prevent backflow from ventricles to atria. The edges of these valves are supported by tendons anchored to the muscle walls of the ventricles below. These tendons do not move the valves. The opening and closing of these valves is caused by pressure differences between atria and ventricles. However, the tendons are vital, as the pressure from the ventricle is so great the flaps would simply be pushed upwards into the atrium. The tension in the tendons is equal and opposite to this pressure and so the valve flaps simply close tightly against each other.

The valves on the right and left sides of the heart are individually named: on the right side, the **tricuspid valve**; on the left, the **bicuspid** or mitral valve.

A different type of valve separates the ventricles from pulmonary artery (right side) and aorta (left side). These are pocket-like structures called **semilunar valves**, rather similar to the valves seen in veins. Once again these valves are opened and closed by pressure differences, this time between ventricles and the main arteries leaving the heart. As the ventricle relaxes the pressure drops below that in the artery and the semilunar valve 'cups' fill with blood to completely close off the artery, preventing backflow.

The action of the heart – the cardiac cycle

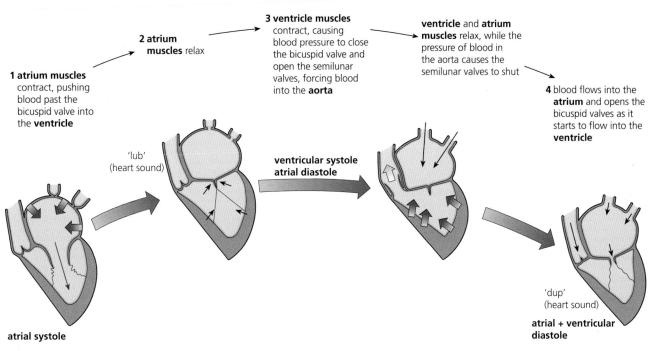

1 atrium muscles contract, pushing blood past the bicuspid valve into the **ventricle**

2 atrium muscles relax

3 ventricle muscles contract, causing blood pressure to close the bicuspid valve and open the semilunar valves, forcing blood into the **aorta**

ventricle and **atrium muscles** relax, while the pressure of blood in the aorta causes the semilunar valves to shut

4 blood flows into the **atrium** and opens the bicuspid valves as it starts to flow into the **ventricle**

'lub' (heart sound)

ventricular systole atrial diastole

'dup' (heart sound)

atrial systole

atrial + ventricular diastole

Figure 11.7 The cardiac cycle

The heart normally beats about 75 times per minute – approximately 0.8 seconds per beat. In each beat, the heart muscle contracts strongly, and this is followed by a period of relaxation. As the muscular walls of a chamber of the heart contract, the volume of that chamber decreases. This increases the pressure on the blood contained there, forcing the blood to a region where pressure is lower. Since the valves prevent blood flowing backwards, blood consistently flows on through the heart.

Look at the steps involved in contraction and relaxation, illustrated in the left side of the heart in Figure 11.7. (Both sides function together, with simultaneous contraction of the chambers.)

1 We start at the point where the atrium contracts. Blood is pushed into the ventricles (where the contents are under low pressure) by contraction of the walls of the atrium. This contraction also prevents backflow by blocking off the veins that brought the blood to the heart. This contraction step is known as **atrial systole**.

2 The atrium now relaxes. The relaxation step is called **atrial diastole**.

3 Next the ventricle contracts, and contraction of the ventricle is very forceful indeed. This step is known as **ventricular systole**. The high pressure this generates slams shut the atrio-ventricular valve and opens the semilunar valves, forcing blood into the aorta. A **pulse**, detectable in arteries all over the body, is generated.

4 This is followed by relaxation of the ventricles. Each contraction of cardiac muscle is followed by relaxation and elastic recoil. This stage is referred to as **ventricular diastole**.

Control of the cardiac cycle

If the heart is completely separated from all other connections it will continue to beat with a frequency of about 50 b.p.m. Because this rhythm originates from the heart muscle itself it is called myogenic.

The myogenic stimulation of the heart provides a perfectly coordinated sequence of activity to ensure maximum efficiency from the cardiac cycle.

The cycle begins in a specially modified structure in the wall of the right atrium called the **sino-atrial node** (SAN). The cells of the SAN depolarise to a point where a wave of excitation similar to a nerve impulse spreads rapidly across the atria, causing them to contract simultaneously (atrial systole).

The boundary between atria and ventricles is made up of connective tissue that does not conduct these impulses. To pass to the ventricles the impulses stimulate another node, the **atrio-ventricular node (AVN)**.

Key term

Myogenic Myogenic activity originates within muscles rather than through the nervous system.

P-wave	Atrial depolarisation (atrial systole)
QRS wave	Ventricular depolarisation (ventricular systole)
T wave	Ventricular repolarisation (ventricular diastole)

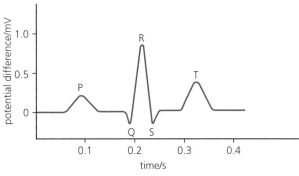

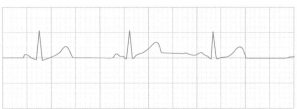

A complete ECG trace from a healthy patient

Figure 11.8 An electrocardiogram

To reach the ventricles the AVN is connected to a bundle of specially modified muscle fibres called the **Bundle of His** and then through a network of finer branching **Purkyne tissue** to the base of the ventricles. Purkyne tissue carries the impulses five times faster than the surrounding muscle to ensure that the base of the ventricles contracts first in ventricular systole, forcing blood in the right direction.

After every contraction, cardiac muscle has a period of insensitivity to stimulation known as the **refractory period**. In the heart, this is longer than most other muscles and means that the heart muscles relax to allow refilling and are less likely to suffer from fatigue.

Measuring electrical activity in the heart

Electrical activity in the heart can be measured by attaching electrodes to the thorax and recording the changing patterns of potential differences. These are displayed as an electrocardiogram (ECG) by means of a chart recorder. The patterns can be matched to the events described above and provide detailed diagnostic information about heart function (Figure 11.8).

The sequence of pressure changes during systole and diastole coupled with the electrical activity of the heart are shown in Figure 11.9.

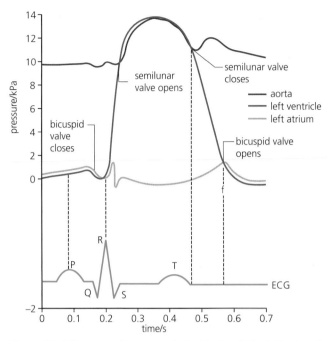

Figure 11.9 Pressure changes and electrical activity in the heart during one cardiac cycle

Test yourself

9 If a heart beats, on average, about 75 times per minute, calculate how many heartbeats will have been completed in a 70-year-old person.

10 It is essential that the total volume of the right and left ventricles is identical. Explain why.

11 Suggest a reason why the bicuspid valve of the left ventricle has only two flaps but the mitral valve has three.

12 The valve tendons are attached to small outgrowths of the ventricular muscle wall called papillary muscles. These contract at the same time as the rest of the ventricle. How might this help the valves to function?

13 State the forces that cause the opening and closing of cardiac valves.

14 Explain the reasons for the following:
    a) It is essential for the heart to have specialised tissues such as Purkyne fibres in the Bundle of His.
    b) It is essential that the Purkyne fibres reach to the base of the ventricles.

15 The electrocardiogram sequence in Figure 11.8 shows a trace taken over 4 seconds. What will be the heart rate of this patient in beats per minute?

Exchange in the tissues – tissue fluid and lymph

The formation of **tissue fluid** assists in the delivery of nutrients to cells and the removal of waste products. Tissue fluid is formed from the plasma, components of which escape from the blood and pass between the cells in most of the tissues of the body. Red cells and most of the blood proteins are retained in the capillaries.

The walls of the capillaries are selectively permeable to many components of the blood plasma, including glucose and mineral ions. Nutrients like these, in low concentration in the tissues, diffuse from the plasma into the tissue fluid (Figure 11.10). There are also tiny gaps in the capillary walls, found to vary in size in different parts of the body, which facilitate formation of tissue fluid. It is the pressure of the blood that drives fluid out (that is, the **hydrostatic pressure** generated as the heart beats). Meanwhile, the proteins and some other components are retained in the blood. These soluble substances maintain an **osmotic gradient**. The water potential of the tissue fluid is less negative than the blood, so some of the water forced out by hydrostatic pressure returns to the blood by osmosis all along the capillary. However, initially there is a net outflow because the hydrostatic pressure is greater than fluid movement due to the osmotic gradient.

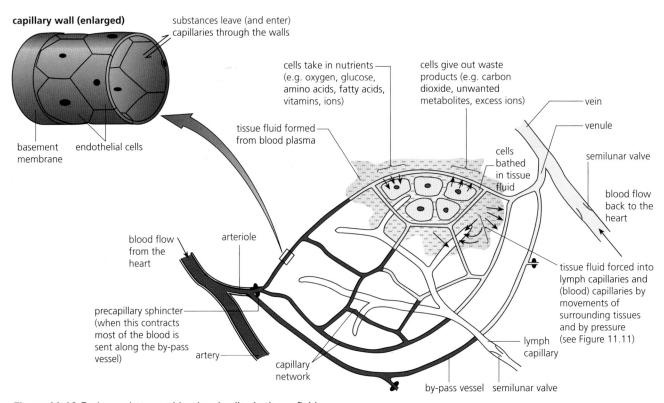

Figure 11.10 Exchange between blood and cells via tissue fluid

Return of tissue fluid to the circulation

Further along the capillary, there is a net inflow of tissue fluid to the capillary (Figure 11.11). Hydrostatic pressure has now fallen as fluid is lost from the capillaries. Water returns by osmosis, and a diffusion gradient carries unused metabolites and excretory material back into the blood.

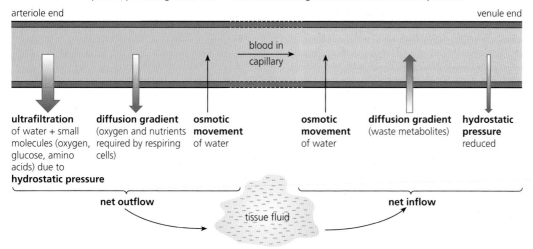

Blood proteins, particularly the albumins, cannot escape; they maintain the water potential of the plasma, preventing excess loss of water and assisting the return of fluid to the capillaries.

arteriole end   venule end

blood in capillary

ultrafiltration of water + small molecules (oxygen, glucose, amino acids) due to **hydrostatic pressure**

diffusion gradient (oxygen and nutrients required by respiring cells)

osmotic movement of water

osmotic movement of water

diffusion gradient (waste metabolites)

hydrostatic pressure reduced

net outflow   tissue fluid   **net inflow**

Figure 11.11 Forces for exchange in capillaries

Formation of lymph

Not all tissue fluid returns to the blood capillaries – some enters the **lymph capillaries**. Molecules too large to enter blood capillaries can pass into the lymph system at tiny valves in the vessel walls. Liquid is moved along these and larger lymph vessels by compression due to body movements; backflow is prevented by valves. Lymph finally drains back into the blood circulation in veins close to the heart.

Lymph nodes are also the site of production of lymphocytes, which makes the lymph system an important part of the immune system. We shall return to this in more detail in *Edexcel A level Biology 2*. The lymph system is also a pathway for lipids to be transported from intestines to the bloodstream following digestion.

Test yourself

16 State which has a higher water potential at the arterial end of a capillary network – plasma or tissue fluid.

17 Water enters and leaves capillaries by osmosis. What other force causes water to leave capillaries?

18 Name the white cells that are most abundant in the lymph system.

Blood clotting as a defence mechanism

In the event of a break in our closed blood circulation, the dangers of a loss of blood and the possibility of a fall in blood pressure arise. It is by the clotting of blood that escapes are prevented, either at small internal haemorrhages, or at cuts and other wounds. In these circumstances, a clot both stops the outflow of blood and reduces invasion opportunities for disease-causing organisms (pathogens). Subsequently, repair of the damaged tissues can get underway. Initial conditions at the wound trigger a cascade of events by which a blood clot is formed.

Firstly, **platelets** collect at the site. These are components of the blood that are formed in the bone marrow along with the red and white cells. They are circulated throughout the body, suspended in the plasma, with the blood cells. Platelets are actually cell fragments, disc-shaped and very small (only 2 μm in diameter) – too small to contain a

Key term

Cascade of events When one signal event sets off a whole sequence of reactions, leading to an important outcome. In the case of blood clotting, damage to blood vessels leads to protein fibres trapping red blood cells to form a clot, through numerous intermediate events.

nucleus. Each platelet consists of a sac of cytoplasm rich in vesicles containing enzymes, and is surrounded by a cell surface membrane. Platelets stick to the damaged tissues and clump together there. (At this point they change shape from sacs to flattened discs with tiny projections that interlock.) This action alone seals off the smallest breaks.

The collecting platelets release a **clotting factor** (a protein called thromboplastin), which is also released by damaged tissues at the site. This clotting factor, along with vitamin K and calcium ions, always present in the plasma, causes a soluble plasma protein called **prothrombin** to be converted to the active, proteolytic enzyme **thrombin**. The action of this enzyme is to convert another soluble blood protein, **fibrinogen**, to insoluble **fibrin** fibres at the site of the cut. Red cells are trapped within this mass of fibres and the blood clot is formed.

SEM of blood clot
showing meshwork of fibrin fibres and trapped blood cells

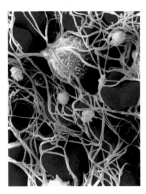

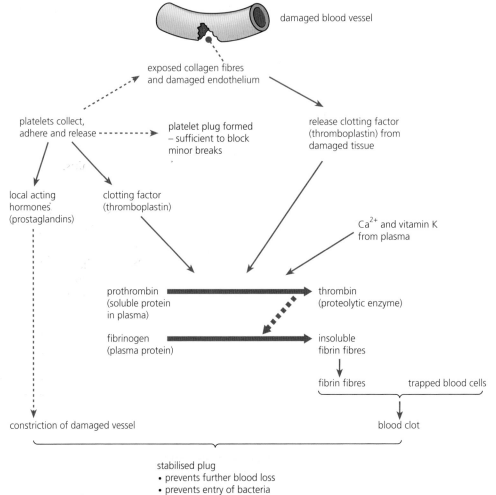

Figure 11.12 The blood clotting mechanism

Test yourself

19 Name the ions in the blood that are essential for blood clotting.

20 Name the insoluble protein used to form a network, trapping red blood cells in clotting.

21 Name one enzyme in the blood clotting cascade.

It is most fortunate that clot formation is not normally activated in the intact circulation; clotting is triggered by the abnormal conditions at the break. The complex sequence of steps involved in clotting may be seen as an essential **fail-safe mechanism**. This is necessary because a casual formation of a blood clot within the intact circulatory system immediately generates the risk of a dangerous and possibly fatal blockage in the lungs, heart muscle or brain.

Atherosclerosis – a disease of the human circulatory system

Diseases of the heart and blood vessels are known as **cardiovascular diseases (CVD)**. These are responsible for more premature deaths in the developed world than any other single cause. By premature death, we mean a death that occurs before the age of 75 years. Most of these are due to atherosclerosis. The structures of a healthy artery wall and one affected by atherosclerosis are shown in Figure 11.13.

Key term

Atherosclerosis
The progressive degeneration of artery walls.

The steps in the development of an atherosclerosis condition are as follows:

1 **Endothelial damage**. Healthy arteries have pale, smooth linings, and the walls are elastic and flexible. However, in arteries that have become unhealthy, the walls have strands of yellow fat deposited under the endothelium. This fat builds up from certain lipoproteins and from cholesterol that may be circulating in the blood. This damage causes white blood cells (macrophages) to invade the fatty streaks where they begin to take up cholesterol from low density lipoproteins and develop fibrous connective tissue forming an **atheroma** (Figure 11.13).

2 **Raised blood pressure**. These deposits start to impede blood flow and contribute to raised blood pressure. Thickening of the artery wall leads to loss of elasticity and this, too, contributes to raised blood pressure. In the special case of the arteries serving the heart muscle – the coronary arteries – progressive reduction of the blood flow impairs oxygenation of the cardiac muscle fibres, leading to chest pains. These are known as **angina**, and are usually brought on by physical exertions.

3 **Lesion formation and an inflammatory response**. Where the smooth lining actually breaks down, the circulating blood is exposed to the fatty, fibrous deposits. These lesions are known as atheromatous **plaques**. Further deposition occurs as cholesterol and triglycerides accumulate, and smooth muscle fibres and collagen fibres proliferate in the plaque. Blood platelets tend to collect at the exposed, roughened surface, and these platelets release factors that trigger a defensive response called inflammation, which includes blood clotting (see Figure 11.12). A blood clot may form within the vessel. This clot is known as a **thrombus** (Figure 11.13), at least until it breaks free and is circulated in the bloodstream, whereupon it is called an **embolus**.

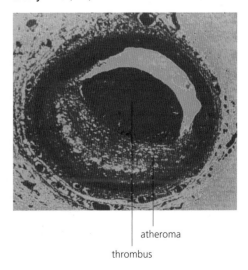

healthy

endothelium — flow of blood

diseased

blood clot = thrombus formed where atheroma has broken through the endothelium

lipid + fibre deposit = atheroma

photomicrograph of diseased human artery in TS (×20)

atheroma

thrombus

Figure 11.13 Atherosclerosis, leading to a thrombus

Myocardial infarctions, strokes and aneurysms

An embolus may be swept into a small artery or arteriole that is narrower than the diameter of the clot, causing a blockage. Immediately, the blood supply to the tissue downstream of the block is deprived of oxygen. Without oxygen, the tissue dies.

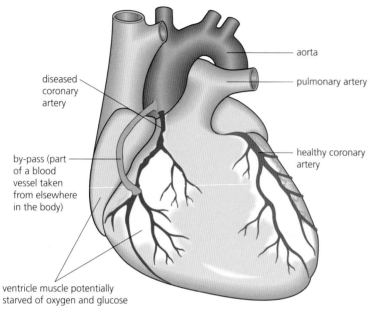

Figure 11.14 A heart by-pass

The arteries supplying the heart are the coronary arteries. These arteries are especially vulnerable, particularly those to the left ventricle. When heart muscle dies in this way, the heart may cease to be an effective pump. We say a heart attack has occurred (known as a **myocardial infarction**). Coronary arteries that have been damaged can be surgically by-passed (Figure 11.14).

When an embolus blocks an artery in the brain, a **stroke** occurs. Neurones of the brain depend on a continuous supply of blood for oxygen and glucose. Within a few minutes of the blood supply being lost, the neurones affected will die. Neurones cannot be replaced, so the result of the blockage is a loss of some body functions controlled by that region of the brain.

In arteries where the wall has been weakened by atherosclerosis, the remaining layers may be stretched and bulge under the pressure of the blood pulses. Ballooning of the wall like this is called an **aneurysm**. An aneurysm may burst at any time.

Factors affecting the incidence of coronary heart disease (CHD)

In the developed world, cardiovascular diseases remain high on the list of most serious health problems, despite recent improvements. For example, in the UK, CHD is responsible for the highest percentage of deaths before the age of 75 years in both males and females (Figure 11.15). Another feature of CHD is that the condition may go largely unnoticed for many years – often until it is too late.

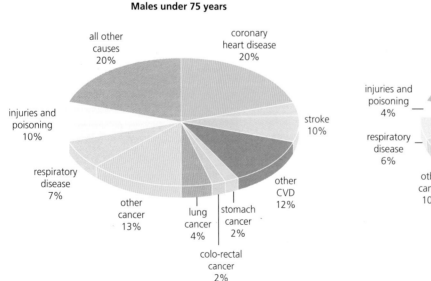
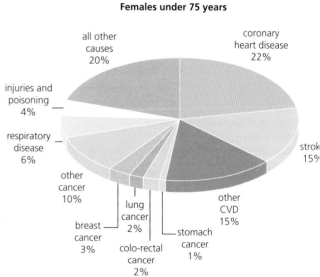

Figure 11.15 Premature death by causes in Europe, published in 2012

Major risk factors that cannot be controlled

- **Increasing age**. The risk of coronary atherosclerosis increases with age. Over 80 percent of people who die from CHD are aged 65 years or more. However, a word of caution is required here, for the evidence is that the genesis of this condition may lie much earlier in life. Post-mortem studies of soldiers killed in action have disclosed early but well-developed 'fatty plaques' in the arteries of young males of average age 22 years, suggesting that these commenced development in their bodies during their adolescence. This suggests why, for individuals brought up in the developed world, after the ages of 35 in men and 45 in women, the chance of dying from CHD increases dramatically.

- **Genetic factors** (including ethnicity). The occurrence of heart attacks at a relatively early age frequently runs in families, suggesting that there are genes that may confer vulnerability to CHD. Children of parents with heart diseases are distinctly more likely to develop the condition. This aspect will become clearer as knowledge of the human genome is developed to identify the roles of individual genes and their effects on metabolism.

Additionally, some races are more prone to CHD and strokes. This is the case with Afro-Caribbean people, for example. Another genetic factor beyond dispute is gender, for the possession of a Y chromosome (being a male) predisposes to greater risk of CHD than that carried by females. It has been suggested that the production of the hormone oestrogen by females inhibits the formation of atheroma and reduces their risk of CHD.

Major risk factors that we can control or reduce

- **Hypertension**. Persistently high blood pressure is defined as systolic pressure greater than 140 mm Hg (18.7 kPa) and diastolic pressure greater than 90 mm Hg (12 kPa). A blood pressure of 120/80 mm Hg (16/10.7 kPa) is regarded as normal. Hypertension is known as a 'silent killer' because of the damage it does to the heart by increasing its workload, causing the heart to enlarge and weaken with time. It also causes damage to blood vessels, accelerating the onset of atherosclerosis. The brain and kidneys are also damaged, although without causing noticeable discomfort. Hypertension increases the risk of strokes, too, and it makes a brain haemorrhage more likely. The following factors in this list all make the condition of hypertension more likely. However, hypertension is a condition that, once detected and regularly monitored, can be successfully treated with drugs (see below).

- **Smoking**. The habit of cigarette smoking generates the greatest risk of fatal ill-health, especially from cardiovascular diseases. It is principally the nicotine and carbon monoxide in tobacco smoke that damage the cardiovascular system. Carbon monoxide combines irreversibly with the pigment haemoglobin in red cells, reducing the ability of the blood to transport oxygen to all respiring cells, including to cardiac muscle fibres. The effects of nicotine are via its stimulation of adrenaline production. This hormone triggers an increase in heart rate and causes arteries to contract (**vasoconstriction**). The result is raised blood pressure.
  There have been huge investments in the tobacco industry for a long time, via the growth of the crop and the manufacturing of tobacco products. The enthusiastic endorsement of cigarette smoking in the developed world (manifested by the free availability of cigarettes to the troops in two World Wars, for example) has now spread to developing countries as fresh markets for tobacco products have opened up. Consequently, the dangers of smoking have not had the attention they deserve, and some people have even suggested the evidence against tobacco is equivocal.
  Yet from the earliest statistical studies there has been no room for doubt. The late Dr Richard Doll and colleagues, working at St Thomas' Hospital, London from 1947, investigated the cause of death of a sample of 3500 people admitted to hospital

for treatment. Whatever the symptoms these people had, the vast majority of those who were smokers later died from a cardiovascular disease.

This study was followed up by one of a group of 40 000 healthy, working doctors (among whom the habit of smoking was widespread at that time), which was more conclusive still. Some of Richard Doll's data from his studies with doctors are shown in Table 11.3. Today, few doctors smoke.

Table 11.3 Mortality from cardiovascular disease caused by cigarette smoking

Cause of death	Non-smokers	Continuing cigarette smokers
CHD	606	2067
stroke	245	802
aneurysm	14	136
atherosclerosis	23	111
Total	**888**	**3116**

- **High levels of cholesterol in the blood.** Lipids are a more diverse group of biochemicals than just the triglycerides we met earlier in Chaper 1, for they include steroids. Steroids occur widely in nature, and one very important steroid is **cholesterol** (Figure 11.16). The 'skeleton' of a steroid is a set of complex rings of carbon atoms. The bulk of the molecule is hydrophobic but the polar chemical −OH group is hydrophilic.

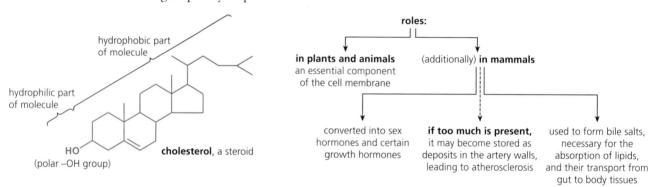

Figure 11.16 The steroid cholesterol

Lipids are absorbed into the body in the gut, and are stored as body fat. They have to be transported around the body, both in their role as respiratory substrates for the transfer of energy, and as cholesterol for use in the production of steroid hormones and the maintenance and repair of cell membranes. Since they are insoluble in water, they are carried in association with proteins, as either **low-** or **high-density lipoproteins** (LDLs or HDLs) according to the relative proportions of protein and lipid. Triglycerides combine with them. These components are introduced in Table 11.4.

Table 11.4 Low- and high-density lipoproteins

	Protein (raises density)/%	Lipid (reduces density)/%	Particle size/nm*	Known as:
Low-density lipoprotein (LDL)	10–27	5–61	20–90	'Bad cholesterol' – when combined with saturated fats.
High-density lipoprotein (HDL)	50	3	7–10	'Good cholesterol' – when combined with insaturated fats.

A nanometre is an SI-derived unit of length, and is 10^{-9} metres. So, a metre is divided into 1000 millimetres, a millimetre is divided into 1000 micrometres (μm or microns) and a micrometre is divided into 1000 nanometres (nm) – a really small unit of length!

Most cholesterol is transported as LDLs, but an excess of these in the bloodstream has been shown to block up the many receptor points in the cell membranes of cells that metabolise or store lipid, leaving even higher quantities of LDLs circulating in the blood plasma. The excess is then deposited under the endothelium of artery walls, initiating or enhancing plaque formations. However, monounsaturated fats help remove the circulating LDLs, and polyunsaturated fats are even more beneficial for they further increase the efficiency of the receptor sites at removing 'bad cholesterol' from the blood.

A note of caution is needed here. While you can (and should) avoid a diet that is excessively rich in saturated fats and cholesterol, this lipid is an essential body metabolite that is manufactured in the liver in the absence of absorbed dietary cholesterol. To some extent, your blood cholesterol levels are genetically controlled. A causal relationship is suggested by statistical studies of deaths from CHD per 1000 of the population each year, plotted against the levels of cholesterol and LDLs measured in blood (serum) (Figure 11.17). The establishment of the actual role of LDLs in triggering CHD is provided by experimental laboratory and clinical evidence that destructive plaques are created as a result of raised levels of blood serum LDLs, as described above.

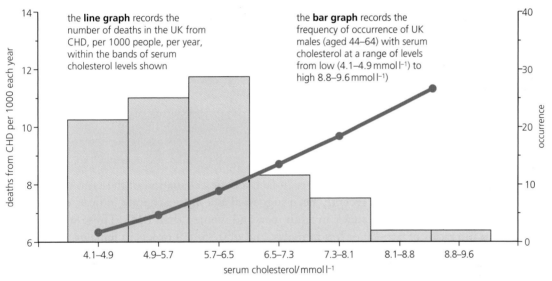

Figure 11.17 The relationship between deaths from CHD and blood serum cholesterol levels

Tip

Cause and effect – correlation isn't causation

Look carefully at Figure 11.17. It is possible to state that the straight line graph shows there is a **positive linear correlation** between deaths from CHD and blood cholesterol levels. You will see in *Edexcel A level Biology 2* how this link can also be tested statistically. But does this mean you can also state that cholesterol *causes* CHD? This would be a very dangerous conclusion to make as the history of science is littered with perfectly good correlations leading to false conclusions. In this case you can see from Table 11.4 that 'cholesterol' does not tell the whole story. In addition, this chapter lists other strong risk factors for CHD so how do you know it wasn't one of these? For example, egg yolk, liver and butter contain high levels of cholesterol. If some other ingredient, only found in egg yolk, caused CHD then people eating lots of eggs would have high levels of this ingredient but also, coincidentally, high levels of cholesterol. This is exactly what we mean by a false correlation. So the best we could say objectively is that this finding 'supports the idea that...' In other words much more evidence needs to be found before we can confidently identify the cause. There is now plenty of research to show how cholesterol has its effect but it is not the whole story.

- **Alcohol**. An excessive intake of alcohol leads to raised blood pressure, damaged heart muscle and irregular heartbeats. It also causes raised LDL levels in the blood, and is associated with certain cancers, too. This harm may arise when more than one unit of alcohol a day is imbibed by women and more than two units per day by men. In those who consistently stick to the recommended limits, the risks of CHD may be lower than in non-drinkers. But 'binge' drinking is especially dangerous. This risk factor is of growing importance.
- **Physical inactivity, obesity and generally being overweight**. A sedentary lifestyle increases the risk of CHD, but physical activity, even only moderate activity, helps prevent heart and blood-vessel disorders. Regular, vigorous physical activity is especially beneficial, not least because it helps prevent obesity.

  We need to be able to quantify the conditions of the body we believe are 'underweight', 'normal', 'overweight' and 'obese' for different people, if individuals are to be able to satisfactorily regulate their body weight.

  To accurately and consistently quantify body weight in relation to health, the **body mass index (BMI)** has been devised. We calculate our BMI according to the following formula:

  $$\frac{\text{body mass in kg}}{(\text{height in m})^2}$$

  Using our calculated BMI and Table 11.5 below, we can determine our 'weight status'.

Table 11.5 The 'boundaries' between being underweight, normal and overweight

BMI	Status
Below 18.5	Underweight
18.5–24.9	Normal
25.0–29.8	Overweight
30.0 and over	Obese

Another factor here is body shape. In fact, a basic calculation of the ratio between waist and hip size has recently been suggested as a more accurate indicator of the risk of having a heart attack in adult men and women than BMI. Those with waist to hip ratios of 0.9–1.0 or less in males, and of 0.85–0.9 or less in females, have a significantly lower risk of a cardiac infarction (heart attack). This correlation is attributed to abdominal fat cells being a major source of LDLs, and a source of metabolites that damage the insulin production system in the body.

Note that the condition of clinical obesity is defined as having a BMI of 30 and over. The incidence of obesity has substantially increased over the past 20 years. Studies by the World Health Organization, published in 2003, estimate that 300 million people are clinically obese worldwide. In the population of a developed country like the UK, 13–17 percent of men and 16–19 percent of women are clinically obese.

- **Diabetes**. This is a disease that carries a significantly raised risk of the patient developing CVD, and people with this condition require especially close monitoring of their blood pressure and blood glucose to ensure they are continuously controlled within safe parameters.

- **Other dietary issues**. These include the question of salt consumption. Sodium chloride is a widely used food preservative, traditionally popular in many cultures and used in diverse prepared food products. Today it plays a major part in the prevention of food poisoning when used as a preservative in packaged convenience foods and 'ready meals'. While it has this and other advantages, it does cause raised blood pressure after it has been absorbed into the bloodstream and prior to its excretion by the kidneys. The Food Standards Agency recommends a salt intake of no more than 6 g per day, but it is common for an individual's intake to be twice that.

Test yourself

22 Explain how atherosclerosis leads to a stroke.

23 Name the blood vessels that become blocked to cause myocardial infarction.

24 Which compound is present in higher proportion in high-density lipoproteins?

25 What is known as 'bad' cholesterol?

26 BMI is a better measure of body proportion than simply measuring mass. Explain why.

27 Define what is meant by 'hypertension'.

Exam practice questions

1 Which of the following would be found in blood plasma but not in tissue fluid?

  A a lymphocyte   **C** a calcium ion

  B a glucose molecule   **D** an erythrocyte   *(1)*

2 In the first stage of blood clotting, platelets collect at the site of damage and release which substance?

  A prothrombin   **C** thrombin

  B thromboplastin   **D** thrombus   *(1)*

3 Explain how the myogenic stimulation of cardiac muscle brings about the sequence of events in the cardiac cycle.   *(6)*

> ### Tip
>
> Question 3 is basically an AO1 question, testing your ability to show you understand a part of the specification. But take care to read it carefully and make sure that you link the electrical changes with what happens in the cardiac cycle rather than just regurgitate some facts. This could easily be a stand-alone question or attached to a longer question with some data.

4 a) Describe two ways, other than causing lung cancer, in which it is thought that smoking can damage the cardiovascular system.   *(4)*

  b) In a study of the link between smoking and coronary heart disease, the medical histories of 187 000 American men were examined. The data on smoking habits was collected from their files as recorded by their doctors following consultation.

> ### Tip
>
> Question 4 is a typical question that combines knowledge of basic specification material (part a)) for AO1 with testing your practical skills of data interpretation and experimental design (parts b) and c)) for AO3.

    The data are shown in the graph.

    i) Why were the data plotted as a percentage of non-smokers?   *(2)*

    ii) Describe two pieces of evidence from these data that support the hypothesis that smoking causes coronary heart disease.   *(2)*

  c) Describe three ways in which the data collection could be modified to improve the reliability of any conclusions made.   *(3)*

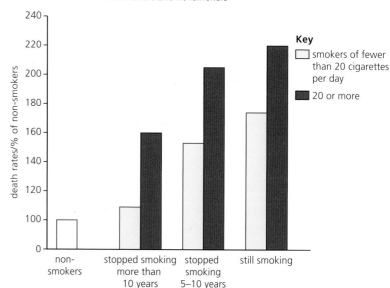

Death rates from coronary heart disease – American men aged 50–70 – smokers compared with ex-smokers and nonsmokers

Key
◻ smokers of fewer than 20 cigarettes per day
◼ 20 or more

11 Mammalian circulation

5 The diagram shows the values of hydrostatic pressure and water potential in two parts of a capillary surrounded by tissue fluid. One part of the capillary (A) was close to the arteriole supplying the capillary and the other (B) close to the venule carrying blood away.

a) i) In which direction will water flow in part A? *(1)*

  ii) Calculate the overall (net) pressure tending to cause water to flow in this direction. *(3)*

b) i) Calculate the percentage change in water potential of the blood between part A and part B. *(2)*

  ii) Explain why there is a change in water potential of the blood between part A and part B. *(2)*

Key
H = hydrostatic pressure (kPa)
W = water potential of solution (kPa)

A		B
H = +5.6		H = +3.25
W = −3.5	Blood	W = −3.7
H = +0.69	Tissue fluid	H = +0.69
W = −0.41		W = −0.41

Tip

In Question 5 you are asked to show you understand some basic concepts, including water potential from Chapter 9. This is quite common – there are synoptic questions such as this in AS papers and in A level papers. Part a) ii) asks you to apply your understanding of the formation of tissue fluid; part b) i) is a very typical percentage calculation that can appear in a wide range of questions.

Stretch and challenge

6 Statins are compounds that have been shown to reduce the level of cholesterol in the blood. In recent years medical guidelines have consistently reduced the levels of blood LDLs (low-density lipoproteins) at which statins should be prescribed and some have gone as far as recommending statins be taken by everyone over the age of 50.

a) What are statins and how do they reduce levels of LDLs in the blood?

b) Many doctors and patients are opposed to prescribing in this way. Discuss the advantages and disadvantages of mass-medication with statins as a means of reducing the incidence of CVD.

c) Several plant foods, including nuts, contain natural cholesterol-reducing compounds called sterols and stanols. These are widely advertised and added to various products.

  i) How are these thought to bring about their effect?

  ii) Why are they not recommended by doctors and the government advisory body NICE?

7 a) Describe the structure of lymph nodes and explain why they are important.

  b) i) People sitting on an aircraft for long periods can experience swelling of the ankles and feet (oedema). Explain why this might happen.

  ii) Describe how wearing elasticated stockings ('flight socks') can help alleviate this problem.

Tip

In Question 6(b) you are asked to 'discuss'. This type of question could appear in an A level exam paper. The important point is to show that you understand the arguments for and against; if the command word was 'evaluate' then you would be required, in addition, to come to a conclusion. Whichever is the case, this extension work will provide an opportunity to practise this approach.

Transport in plants

Prior knowledge

In this chapter you will need to recall that:
→ xylem tissue carries water and mineral ions in plants
→ xylem tissue plays an important part in supporting plant stems
→ phloem tissue carries the products of photosynthesis around the plant
→ transpiration is the loss of water vapour from the leaves of a plant
→ transpiration is responsible for drawing water up the xylem tissues of a plant
→ water vapour is lost from leaves though open stomata
→ light and temperature can affect the rate of transpiration
→ water enters plants through root hair cells
→ mineral ions enter plant root cells by active transport
→ plant cells have cellulose cell walls, which are freely permeable to water.

Test yourself on prior knowledge

1 State whether xylem vessels are living or dead.
2 State whether phloem sieve tubes are living or dead.
3 What is the chemical nature of the products of photosynthesis carried in phloem tissues?
4 Describe the effect that increasing temperature has on the rate of transpiration.
5 Explain why it is an advantage to the plant to close stomata at night.
6 Suggest why the cellulose structure of a plant cell wall is so permeable to water.
7 The root hair cell needs active transport to take up many important ions from soil water. Why is this necessary?

Transport tissues of a flowering plant

Key term

Vascular bundles
Groups of phloem, xylem and support tissues found in the stems and roots of plants.

Internal transport in plants requires specialised tissues just as it does in animals. There are two separate tissues. Water and mineral ions are transported in **xylem tissue**. Sugars, produced by photosynthesis, and some amino acids are transported in **phloem tissue**. Like mammals, the main transport mechanism is by mass-flow but unlike mammals there is no central pump. Xylem and phloem are collectively known as vascular tissue and they are usually found together in the vascular bundles in the stem, and in the central stele in the root. You can see the arrangement of vascular tissue in a plant in Figures 12.1 and 12.2.

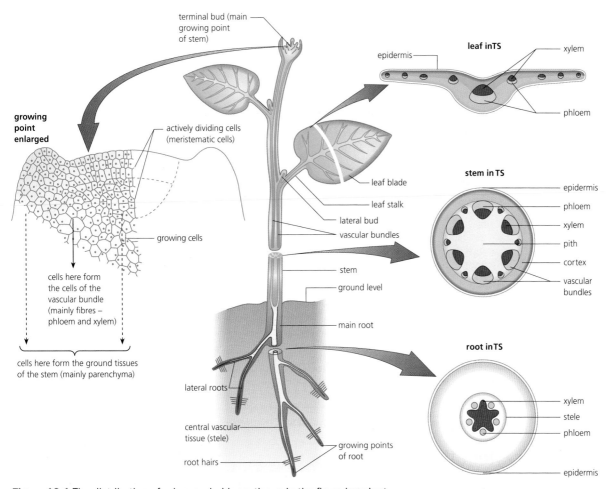

Figure 12.1 The distribution of xylem and phloem tissue in the flowering plant

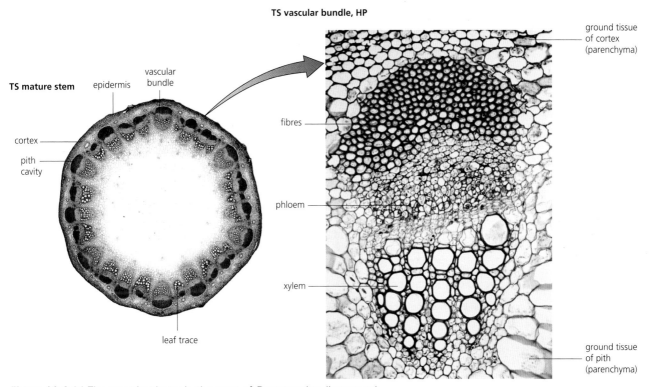

Figure 12.2 (a) The vascular tissue in the stem of *Ranunculus* (buttercup)

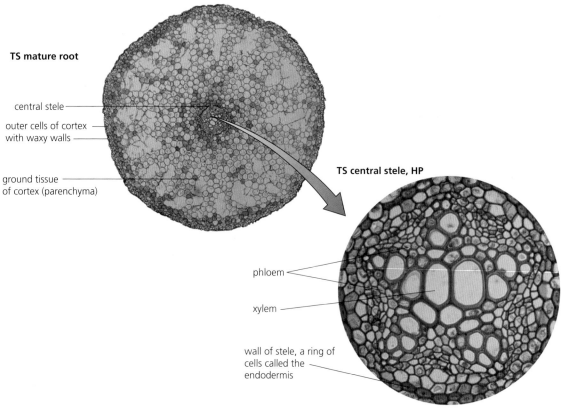

TS mature root

central stele

outer cells of cortex
with waxy walls

ground tissue
of cortex (parenchyma)

TS central stele, HP

phloem

xylem

wall of stele, a ring of
cells called the
endodermis

Figure 12.2 (b) The vascular tissue in the root of *Ranunculus* (buttercup)

Xylem tissues

Xylem vessels

Key term

Lignin A complex
polymer of phenols used
to thicken plant cell
walls.

The main water transport tissue in xylem is made up of **xylem vessels**. These cells begin life as typical elongated plant cells. As the vessels grow, a complex carbohydrate polymer called lignin is laid down in the spaces between the cellulose fibres of the cell wall. This makes them much stronger but also impermeable, so the cytoplasm dies. The end walls of each of the cells break down to form long continuous tubes, ideal for water transport. Table 12.1 summarises these adaptations. Although you may be unfamiliar with the name 'lignin', you are almost certainly very familiar with lignified xylem vessels. If you are sitting on a wooden chair or at a wooden desk, look at it carefully, as wood is nothing more than dead vessels coated with lignin. In most plants the lignification process tends to stop after vessel formation but in woody plants it continues until the vessels are blocked. Whilst this is of no use for water transport it forms an ever thicker stem for support in larger plants. New large vessels formed every Spring to restore water transport give the woody stem its characteristic **annual rings**.

Table 12.1 Structural adaptations of xylem tissues for water transport

Structure	Adaptation
Dead empty xylem vessels	Creates a wide lumen for unrestricted water flow
End walls of vessels break down	Creates a long continuous tube for water transport
Cell walls of vessels lignified	Prevents vessels collapsing when contents are under tension
Cell walls lignified with rings, spirals and in a reticulate manner	Allows vessels to be flexible, preventing breakage as the stem moves

Figures 12.3 and 12.4 show other adaptations of xylem vessels. In order to allow water and ions to be transferred to all parts of the plant, vessels have **pits**. These are areas of the cell wall where there is no lignin and water can move laterally as well as up the stem. In addition to pits, vessels have different types of lignin thickening. The transport of water inside vessels causes a slight reduction in pressure, threatening to cause them to collapse. The lignin thickening helps to overcome this problem but stems still need to be flexible if they are not to break. Therefore the pattern of thickening is often in the form of rings, spirals or networks.

Key term

Pits Areas of the cell wall that lack lignin and so allow lateral transport.

photomicrograph of xylem tissue in LS

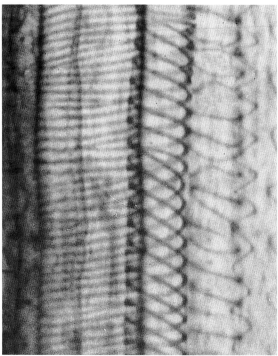

drawing of xylem vessels in TS and LS

all vessels have lignin-free 'pit' areas in their walls

TS

fibre

xylem parenchyma (the only living cells of xylem tissue)

LS

lignin thickening as spirals, rings, network or solid blocks – deposited on inside of vessel, strengthening the cellulose layers

Figure 12.3 The structure of xylem tissue

Other xylem tissues

Not all the xylem is made up of vessels. Other types of tissue can be found in different plants.

- **Xylem tracheids** are similar to vessels but are narrower and shorter with tapering ends. They are found in less-advanced species such as conifers, where they are the main water-carrying tissue. Like vessels, their sloping end walls break down to form continuous tubes as they become lignified.
- **Xylem parenchyma cells** are typical plant cells with no thickening. They are found among the vessels or tracheids and remain as living tissue.
- **Xylem fibres** are narrow, highly thickened dead cells with only a small gap (lumen) in the centre. They are very similar to the fibres found in many other parts of plants. Their structure means they cannot transport water but are used for support.

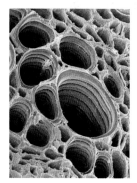

Figure 12.4 SEM of xylem vessels

Test yourself

1. Name the complex carbohydrate laid down in plant cell walls that forms wood.
2. Xylem vessels die when their walls become thickened. Explain why.
3. Explain why many xylem vessels have rings or spirals of thickening and not a solid coating.

Phloem tissues

Key terms

Sieve plates Perforated end walls of phloem sieve tubes.

Companion cells Plant cells with dense cytoplasm connected to sieve tubes in phloem tissue.

Plasmodesmata Channels in adjacent cell walls of plant cells that form cytoplasmic bridges for communication and transport.

Sieve tubes

Sieve tubes are elongated living cells with a very specialised structure. As they develop they lose their nuclei and their cytoplasm is restricted to a very thin peripheral layer with few organelles, but they still remain alive. Their end walls have many holes called **sieve pores**, which form a sieve plate between each cell. Running through the large lumen and the sieve pores between each tube element are strands of **phloem protein**. Sieve tubes are connected to their companion cells through channels in the cell walls called plasmodesmata (singular: plasmodesma).

Companion cells

Running alongside the sieve tubes are **companion cells**. These are typical plant cells but with dense cytoplasm and many organelles. Their cytoplasm is connected to the sieve tube through the plasmodesmata and it is thought that they carry out many cellular functions, enabling the sieve tubes to stay alive and transport materials, even though they have no nucleus and very little cytoplasm. You can see sieve tubes and companion cells in Figure 12.5.

Other phloem tissues

Phloem contains very similar additional tissues to xylem, including **phloem parenchyma** and **phloem fibres**. Once again these are not concerned with sugar transport. Table 12.2 summarises the main adaptations of phloem tissue to its role in solute transport.

Table 12.2 Structural adaptations of phloem tissues for solute transport

Structure	Adaptation
Sieve tubes have limited peripheral cytoplasm and organelles	Creates space for sugar transport through the cell
End walls form sieve plates with sieve pores	Form direct connections for transport from one sieve tube element to the next
Companion cells and sieve tubes are connected by plasmodesmata	Enables the sieve tube to stay alive without a nucleus and with very limited cytoplasm
Companion cells have dense cytoplasm and many organelles	Thought to be needed to support sieve tubes

companion cell and sieve tube element in LS (high power)

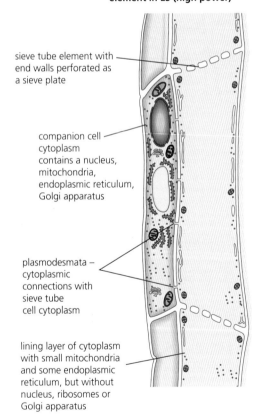

sieve tube element with end walls perforated as a sieve plate

companion cell cytoplasm contains a nucleus, mitochondria, endoplasmic reticulum, Golgi apparatus

plasmodesmata – cytoplasmic connections with sieve tube cell cytoplasm

lining layer of cytoplasm with small mitochondria and some endoplasmic reticulum, but without nucleus, ribosomes or Golgi apparatus

Figure 12.5 Phloem tissue

4 State **two** ways in which sieve tubes differ from xylem vessels.

5 Explain why sieve tubes have reduced peripheral cytoplasm.

6 How are sieve tubes connected to companion cells?

Movement of water through the plant

The three routes of water movement through plant cells and tissues are shown in Figure 12.6. They are as follows:

- **Mass flow** through the interconnected 'free' spaces between the cellulose fibres of the plant cell walls. This pathway does not pass through membranes or the living contents of the cell and is known as the apoplast pathway. The apoplast includes water-filled spaces of dead cells such as xylem vessels and is the major route of water transport.

- **Diffusion** through the cytoplasm of the cells using the channels through cell walls called plasmodesmata. This is called the symplast pathway. The cytoplasm of cells is packed with organelles and other molecules that slow diffusion, so this is a much more restricted pathway.

- **Osmosis** from vacuole to vacuole of plant cells using their partially permeable membranes (sometimes known as the **vacuolar pathway**). This is driven by a gradient of water potential. Although this is also a limited pathway for movement of large quantities of water, it is the way in which individual cells take in water. Root hair cells take up water from the soil in this way as their contents have a much more negative water potential than the very dilute solution of ions found in soil water.

Key terms

Apoplast pathway The route of water transport across plant cells that passes through the fibrous cell walls.

Symplast pathway The route of water transport across plant cells that passes from cell to cell through plasmodesmata.

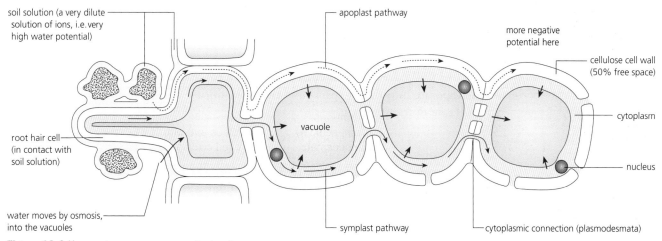

Figure 12.6 How water moves across plant cells

Movement of water up xylem vessels

Figure 12.7 shows how water entering root hair cells by osmosis can be transported across the root to the xylem vessels, which have pits, allowing the water to enter freely. At this point water must be drawn up the xylem vessels to reach the leaves. In many plants this means a long ascent. Giant redwood trees, *Sequoia sempervirens,* grow to over 100 m tall. The **cohesion–tension model** suggests the way in which this might be achieved.

The evaporation of water from the aerial parts of the plant (transpiration) means that a tension is applied to the water column in the xylem vessels. This tension is sufficient to draw up the continuous column of water from the root. However, this would not be possible if it were not for the remarkable properties of water molecules. We saw in Chapter 1 that water molecules are attracted to each other by weak **hydrogen bonds**, a property known as cohesion. Although weak, there are many billions of such bonds, giving the column of water great **tensile strength**. In simple terms this means the column of water can be drawn upwards to a great height without breaking. Water molecules are also attracted to other molecules, a property known as **adhesion**. This means they are attracted to molecules in the vessel walls, which also assists in their upward movement without the column breaking.

Key terms

Transpiration The loss of water vapour from the aerial parts of plants.

Cohesion A property of water molecules by which hydrogen bonds give columns of water great tensile strength.

Tip

Remember to use the correct terminology. 'Water vapour' is the accurate description here. Water molecules will constantly escape from liquid water at a wide range of temperatures (steam is very different). 'Water' alone can imply the liquid.

Evidence for the cohesion–tension model

- When xylem vessels are punctured air enters, demonstrating water under tension not under pressure.
- Xylem vessels have thickened lignin walls to prevent them collapsing under tension.
- Rates of water uptake are strongly linked to factors affecting transpiration from leaves.
- Fine columns of water placed under tension show sufficient tensile strength to account for transport up the highest trees.

Factors affecting the rate of water movement

The faster the rate of transpiration, the faster will be the movement of water up the xylem as this is the main driving force. The following factors affect transpiration:

- **Temperature**. Figure 12.7 shows that, in leaves, water vapour evaporates from the apoplast of the spongy mesophyll cells into the air spaces. From here it diffuses into the atmosphere through the stomata. An increase in temperature means the water molecules will have more kinetic energy. With more energy, more water molecules will evaporate into the air spaces and they will diffuse out faster, increasing the overall rate of transpiration.

- **Light**. To escape into the atmosphere, water molecules must pass through the stomatal pores. As light intensity decreases, the guard cells of the stomata begin to lose their turgidity, causing them to flatten against each other, closing the stomatal pores. As the pores close diffusion of water vapour is severely restricted.

- **Humidity**. Humidity is simply a measure of the amount of water vapour in the surrounding atmosphere. As the number of water molecules in the air increases, the diffusion gradient compared with the inside of the leaf is reduced, slowing diffusion.

- **Air movements**. In still air, water vapour diffusing out of the stomata tends to build up close to the surface of the leaf. Again, this reduces the diffusion gradient and slows transpiration. As the movement of air increases, more of the water vapour surrounding the leaf is removed, increasing the diffusion gradient and speeding up transpiration.

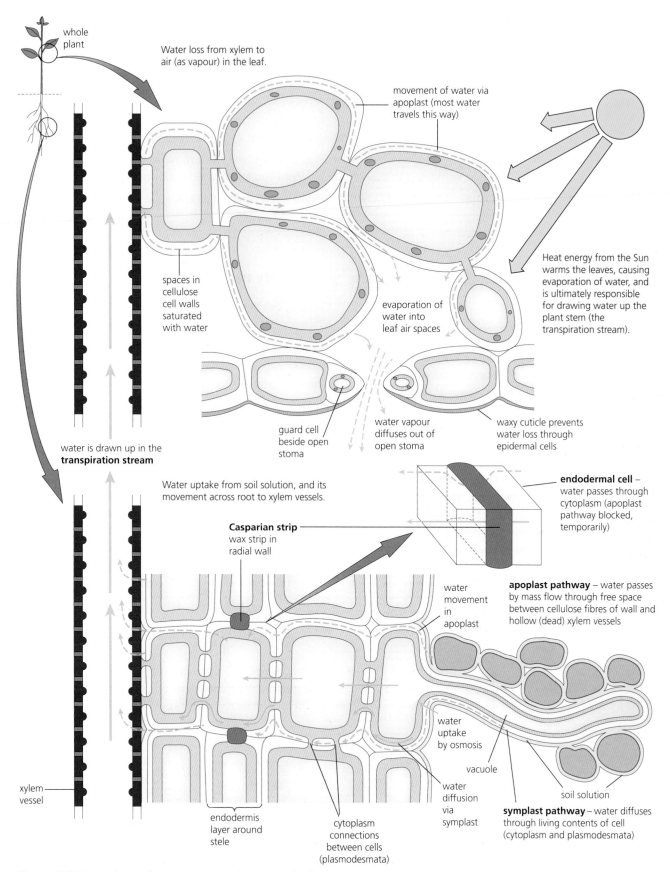

whole plant

Water loss from xylem to air (as vapour) in the leaf.

movement of water via apoplast (most water travels this way)

spaces in cellulose cell walls saturated with water

evaporation of water into leaf air spaces

Heat energy from the Sun warms the leaves, causing evaporation of water, and is ultimately responsible for drawing water up the plant stem (the transpiration stream).

water is drawn up in the **transpiration stream**

guard cell beside open stoma

water vapour diffuses out of open stoma

waxy cuticle prevents water loss through epidermal cells

Water uptake from soil solution, and its movement across root to xylem vessels.

endodermal cell – water passes through cytoplasm (apoplast pathway blocked, temporarily)

Casparian strip wax strip in radial wall

water movement in apoplast

apoplast pathway – water passes by mass flow through free space between cellulose fibres of wall and hollow (dead) xylem vessels

water uptake by osmosis

vacuole

water diffusion via symplast

soil solution

symplast pathway – water diffuses through living contents of cell (cytoplasm and plasmodesmata)

xylem vessel

endodermis layer around stele

cytoplasm connections between cells (plasmodesmata)

Figure 12.7 The pathway of water movement from soil to leaf

7 Explain why cellulose cell walls have so much 'free' space for water molecules to pass through.

8 Name the pathway for water movement that involves passage through plasmodesmata.

9 Describe the feature of water molecules that gives the liquid a high tensile strength.

10 If the lumen of a xylem vessel has a diameter of 25 μm, what will be the mass of water contained in a continuous xylem vessel of a redwood tree 100 m tall?

11 Explain why an increase in humidity will slow transpiration.

Core practical 8

Investigate factors affecting water uptake by plant shoots using a potometer

Background information

The principle of this investigation is very simple in theory but controlling variables poses interesting challenges. A **potometer** is simply a tube filled with water attached to a plant shoot. As the shoot draws water up the xylem vessels then the meniscus of the water in the tube will be drawn along a scale. The rate of water uptake can be measured by calculating the volume of the water in the tube drawn up in a fixed time. When the meniscus reaches the end of the tube it needs to be reset by refilling the tube. In the apparatus shown in Figure 12.8 this is done by using a reservoir of water. This allows the tube to be refilled simply by opening the tap and pressing the syringe, so the shoot is not disturbed. Whilst it is very convenient, this type of potometer is expensive. It is possible to collect very reliable data by using a simple capillary tube about 30 cm long attached to a shoot by a piece of rubber tubing. This can be refilled by injecting water through the rubber tubing using a syringe with a fine needle.

Carrying out the investigation

Aim: to investigate the rate of water uptake by a plant shoot

Risk assessment: There are no significant risks associated with this investigation unless the shoot is cut with a sharp knife, when care must be taken to cut downwards onto a hard surface.

1 First of all you will need to select a suitable plant shoot. To do this, check the internal diameter of the rubber tubing, which is used to attach the shoot to the potometer. To achieve a good seal the diameter of the shoot you choose needs to be slightly larger than the rubber tubing. It is also helpful to choose shoots from a woody plant with a strong stem. It is important that any shoots you use are placed in water as soon as they are removed from the parent plant as water continues to be drawn up the xylem, introducing air locks. Cutting off 1–2 cm of the shoot under water just before use helps to ensure any air locks are removed.

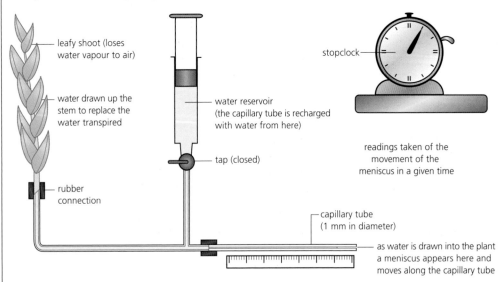

leafy shoot (loses water vapour to air)

water drawn up the stem to replace the water transpired

rubber connection

water reservoir (the capillary tube is recharged with water from here)

tap (closed)

stopclock

readings taken of the movement of the meniscus in a given time

capillary tube (1 mm in diameter)

as water is drawn into the plant a meniscus appears here and moves along the capillary tube

Figure 12.8 A potometer used to investigate water uptake by a plant shoot. There are various designs of potometer, all operating in a similar way

2 When you are ready with a suitable shoot in water close at hand, fill the capillary tube and the rubber tubing to the very brim by opening the tap and pushing down on the syringe. Close the tap and insert the shoot in the rubber tubing. Make sure the whole shoot is secured with a stand and clamps as movements may squeeze the rubber tubing, giving erratic readings.

3 The water meniscus should begin to move along the capillary. If the water leaks from the open end of the capillary then there is a poor seal and you must repeat the process above. Smearing grease around the seal can help, but make sure this does not cover the cut surface. Once the potometer is working smoothly, check how quickly it moves along the scale and decide on a suitable time interval to record the movement.

4 Now you will need to consider very carefully how to vary conditions to investigate factors affecting transpiration. Light intensity and air movements are the factors that you can manipulate more easily. For light, you will need to darken the laboratory and use a bench lamp at different distances but remember that the light intensity needs to be measured using a simple light meter. If you have a suitable 'app' on your mobile phone you can use the camera setting to give intensity readings. You will need to carry out some simple trials to decide upon suitable distances and times of recording.

Air movements can also be investigated if you have a fan with variable speed settings.

Questions

1 Why does a potometer measure rate of water uptake and not the rate of transpiration?
2 How can the volume of water taken up be measured?
3 Will it make a difference to your graph if you use just the distance moved against time rather than the volume taken up against time?
4 Why would simply covering the shoot with a black polythene bag not give you valid data about the effect of lack of light on water uptake?
5 A bench lamp placed 20 cm from the potometer gives a light meter reading of 100 AU. If the lamp is moved to 40 cm from the photometer, what will be the light intensity?

The movement of organic solutes through the plant

Scientists and models

In Chapter 7 you looked at ideas of classification, where there is scientific disagreement. Most scientific advances begin with the development of a model to try to explain something. This is not some idle speculation but is usually based upon the available evidence at the time. This model stimulates debate and allows scientists to make predictions about what might happen if the model is correct. This is the stage at which collaboration, ingenuity and the ability to think objectively need to be applied in designing reliable investigations to test the model. Very rarely can any one investigation be decisive, but as more and more predictions based on the model are shown to be correct then the model becomes widely accepted. That is, of course, until one single investigation of an important prediction proves to be incorrect!

You have seen that the cohesion–tension model of water transport has many sound features and offers a wide range of scientific explanations. It is therefore widely accepted. The story of the transport of solutes in phloem is much less clear and the mass–flow model considered here has many unanswered questions, which means that it is, at best, an incomplete explanation or there may even be much better explanations.

As a scientist, it is really important that you keep asking questions such as 'How do we know that?' or 'What other explanation might there be?' rather than expecting to look into textbooks and find the perfect answer or regard the knowledge you have as 'fact'. At A level your ability to consider or evaluate evidence will be tested in the written papers.

The movement of solutes through phloem tissue

Key terms

Translocation The name given to the transport of manufactured solutes, such as sucrose and amino acids, in the phloem.

Mass flow A method of transport in which pressure differences are used to move a fluid to carry substances in one direction.

The transport of sugars, especially sucrose, through sieve tubes can easily be demonstrated using radioactively labelled carbon dioxide and collecting their contents using aphids (greenfly), which pierce stems with their fine mouthparts to feed off phloem contents. If the feeding insects are anaesthetised and the mouthparts cut off, the contents of the sieve tubes flow out and can be analysed. Similarly, radioactively labelled sugars can be traced throughout the plant to follow their movements. Unlike water, the paths of solute movements can be in different directions, not just down from the leaves. Many plants use storage organs such as bulbs of daffodils or stem tubers of potatoes. In spring the compounds they contain must be transported to the shoots to enable them to grow and begin photosynthesis.

The sugars in sieve tubes come from photosynthesis in leaves and are transported around the plant in a process called translocation.

The mass-flow hypothesis of phloem transport

One model of transport in the phloem is called the mass-flow **hypothesis**. You saw in Chapter 11 that a mass-flow system uses a fluid to carry substances in one direction using pressure differences to move it along. This same idea has been suggested as the mechanism for movement in phloem.

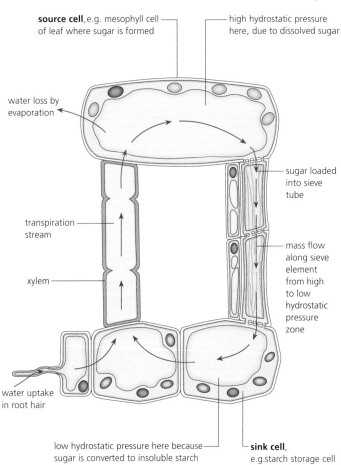

source cell, e.g. mesophyll cell of leaf where sugar is formed

high hydrostatic pressure here, due to dissolved sugar

water loss by evaporation

sugar loaded into sieve tube

transpiration stream

mass flow along sieve element from high to low hydrostatic pressure zone

xylem

water uptake in root hair

low hydrostatic pressure here because sugar is converted to insoluble starch

sink cell, e.g. starch storage cell

Figure 12.9 The mass-flow hypothesis

This model suggests that pressure differences to drive the fluid movement are generated in different parts of the plant, as shown in Figure 12.9.

In cells of the leaf, sugars are manufactured by photosynthesis. These sugars dissolve in the cytoplasm, causing a lowered (more negative) water potential. This causes water to enter the cells by osmosis and build up a high hydrostatic pressure. This is called the **source area**.

In other parts of the plant, sugars are used up rapidly in respiration or converted into starch. Starch is insoluble and forms starch grains, which have no osmotic effect. This removal of dissolved sugars raises the water potential (less negative) and therefore water tends to flow out of the cell, forming a region of lower hydrostatic pressure. This is called the **sink area**.

The model suggests that this pressure difference forces sugars into the sieve tubes at the source and induces mass-flow through the phloem towards the sink. The remaining fluid in the sink is then returned to the source through the xylem vessels.

Strengths and weaknesses of the mass-flow hypothesis

Strengths of the model:

- It is possible to measure the gradients suggested and show they are present.
- When pierced by insect mouthparts, the contents of sieve tubes flows out, showing them to be under pressure as the model predicts.
- The model links the phloem and xylem systems in a plausible way.

Weaknesses of the model:

- Organic solutes move around the plant in different directions, not just to the lowest pressure sinks.
- Sieve tubes and companion cells are living tissue and do not work if killed off. The model does not really explain why.
- Starch grains are found in many plant cells, not just sinks.
- The basic model suggests an entirely passive process but phloem has a higher metabolic rate than most other plant tissues.

Other features the model does not yet explain:

- Sugar needs to be 'loaded' into sieve tubes at the source; this is not yet fully explained.
- Why do almost all sieve tubes contain phloem protein strands?
- What is the purpose of sieve plates? They appear to be a hindrance to mass flow, not an adaptation to facilitate it.

It will be obvious that the model is far from accepted by the scientific community and much more research needs to be done to find more evidence to support it or to develop other models.

Test yourself

12 Name the process of transporting sugars and amino acids around plants.

13 Explain **two** ways in which it can be shown that phloem sieve tubes are involved in the transport of solutes.

14 Name the main sugar transported by sieve tubes.

15 Suggest where, in plants, you would expect to find the main 'source areas'.

16 Why is the presence of sieve plates a problem for the mass-flow hypothesis?

17 Describe how the fluid is moved in all mass-flow systems.

Exam practice questions

1 One feature of a mass flow system is:

   A it transports only soluble substances

   B all substances are transported in one direction

   C it involves a pump

   D it involves active transport *(1)*

2 Which of the following is an adaptation to transport in **both** xylem vessels and phloem sieve tubes?

   A presence of simple pits

   B presence of plasmodesmata

   C presence of a large lumen

   D no cytoplasm present *(1)*

3 Which of the following is an essential feature of water transport via the apoplast pathway in plants?

   A high osmotic potential in cell vacuoles

   B high proportion of free space in cellulose cell walls

   C cell membranes permeable to water molecules

   D presence of a large number of mitochondria *(1)*

4 a) Explain how the differences between the apoplast and symplast pathways in plants result in much faster water transport through the apoplast pathway. *(4)*

   b) In plant roots water is taken up by osmosis through root hair cell membranes. To enter the xylem the water must cross the endodermis surrounding the central stele. The cells of the endodermis have cell walls containing a Casparian strip of cork-like material, which is impermeable. Describe the effect that this structure will have on the flow of water into xylem vessels. *(3)*

5 A student carried out an investigation into the effect of air speed on the rate of transpiration of a woody shoot. The selected shoot was carefully attached to a potometer 1 metre in front of a slow-moving fan and the air speed measured using a flow meter. The student then measured the volume of water taken up by the shoot in 5 minutes.

   The water level of the potometer was then reset and the fan speed was increased. A flow meter was again used to measure the air speed and the volume of water taken up in 5 minutes measured.

   This was repeated four times with increased air flow each time. A graph of the results is shown below.

Tip

Question 4 is a very typical AO2 question, where you are expected to apply your knowledge of the structure to show you understand how it functions. Many questions you will meet might look a bit unfamiliar but can easily be answered by using your knowledge of the specification content.

Tip

Many questions based on the core practicals will ask you to show that you understand the principles behind them. You will also find, as in Question 5, that you will need your wider practical skills to answer them well.

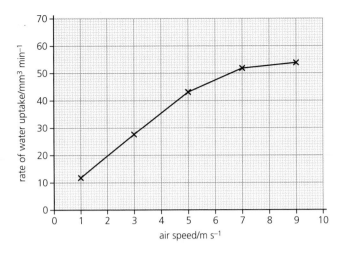

Tip

The graph in Question 5 needs you to check exactly what information is being displayed. It is really important in questions containing graphs that you pause to check exactly what the axis labels tell you — especially the units. There is more on graphs in Chapter 13.

a) What is the relationship between air speed and rate of water uptake in the range of air speed from $1-5\,m\,s^{-1}$? *(1)*

b) Analyse the data and calculate the average increase in water uptake between 1 and $5\,m\,s^{-1}$. *(3)*

c) Explain the difference in the effect on the rate of water uptake of increasing air speed from $1-5\,m\,s^{-1}$ and from $5-9\,m\,s^{-1}$. *(3)*

d) Describe two other variables that would need to be controlled to ensure that valid conclusions could be drawn from this investigation. *(2)*

e) Explain why the line graph of the results is not drawn through the origin point $(0, 0)$. *(2)*

6 a) Explain how the structure of xylem vessels is adapted for their role in water transport. *(4)*

Tip

Both parts of Question 6 are concerned with adaptations of structure to function. Take care that your answer clearly links the two together. In part a) you will find that each adaptation is given only 1 mark but requires both in the explanation.

b) Explain how the unique properties of water molecules are essential for the transport of water and ions from the roots to the shoots in flowering plants. *(4)*

Stretch and challenge

7 Plants and animals have fundamental differences. In particular, plants do not move around or maintain a high body temperature. This means plants have a much lower metabolic rate. In addition, plants manufacture their own respiratory substrates from simple molecules such as carbon dioxide and water. These differences are reflected in their internal transport systems.

a) What are the similarities and differences between transport of water from root to shoots in plants and the vascular system of a mammal?

b) How are these reflected in their structures?

Tip

Question 7 begins with recall of information but as this is from two different parts of the specification it is synoptic. Having decided on some relevant information, write your answer in extended prose to practise assembling information and writing in an accurate, concise way.

8 A key element of the mass-flow model of phloem transport is that the sucrose must be 'loaded' into sieve tubes at the source area. What theories have been proposed for this mechanism?

13 Mathematics for biology

At least 10 per cent of the total marks in the two AS examination papers and 10 per cent of the total marks available in the three A level papers will be awarded for mathematics at Level 2 or above. 'Level 2 mathematics' means the standard of the higher tier of GCSE mathematics.

This is obviously a significant part of each qualification, which will make a big difference to your final grade. If you are not studying mathematics beyond GCSE level then it is important to continue to practise your mathematical skills throughout the course. Even if you have achieved a good grade in mathematics you cannot assume that you will retain your skills 1 or 2 years later. If you are less confident with mathematics you must continue to develop your skills further. The mathematical skills required are not something that can be developed in a short period of intensive revision at the end of your course.

To help you to do this you will find specific examples of the use of mathematical skills throughout this book. They have been included in chapters where you are most likely to meet practical applications of mathematics. They will help you to understand why mathematics is relevant to a biology course and give you examples to show what might be required in written examinations.

This chapter does not repeat all of these skills but reinforces some of the most important requirements. It also explains some of the more difficult ideas, such as statistics, in a little more detail.

Basic units

All the units that you meet will be based on the SI system. This is an international standard used by all scientists. Those you will meet most often are shown in Table 13.1.

Table 13.1 SI units

Measure	Unit	Symbol	Note
Length	Metre	m	
Mass	Kilogram	kg	
Time	Second	s	
Temperature	Degree Kelvin	K	1 degree Kelvin = 1 degree Celsius but their scales are different ($0\,°C = 273\,°K$)
Amount of substance	Mole	mol	

There are many scientific reasons to use these units but scientists also use common sense when applying them. For example, it would be very awkward to keep recording in seconds when an investigation continued for several days.

Other common units are derived from these basic beginnings. The simplest way to do this is to represent fractions of the basic SI unit. Chapter 4 explains how units of length are divided to give units appropriate to a wide range of sizes. The common prefixes for decreasing values in steps of 1000 are milli-, micro- and nano-. So a millimetre is a thousandth of a metre, a micrometre is a millionth of a metre and so on.

Time is more awkward because historically 1 minute is divided into 60 seconds and there are 60 minutes in an hour. The advent of digital stopwatches means you will need to take care. A common error is to take a stopwatch reading of say 1:56 min and record this as a decimal fraction rather than 1 min 56 s.

Other units

Concentration

Controlling or measuring concentration is a very common feature of many practical investigations. The concentration is the amount of substance in a given volume. It is correctly expressed as moles per decimetre cubed ($mol\,dm^{-3}$). Here a decimetre cube (dm^3) is a derived unit of $1000\,cm^3$ or a litre. We shall consider indices in the next section but the minus sign indicates 'per'. Problems arise when we stray away from the mole as the SI unit of amount. It is much easier to make up percentage solutions where a 1% solution simply has 10 g of the substance made up in $1\,dm^3$ of the solvent. This can lead to large errors. For example, a 1% solution of glucose and a 1% solution of sucrose do not have the same concentration, simply because a sucrose molecule has twice the mass of a glucose molecule. Therefore 1 gram of the sucrose solution will only contain half the number of molecules, and the number of particles in a given volume is what we really mean by concentration.

Standard form (scientific notation)

You will find numerous examples in this book and in examination papers where numbers are written in standard form. Any number can be written in standard form but it is often used for very large or very small numbers. It will always be in the form $a \times 10^b$, where a is a number >1 and <10 and b a power of 10. For example, 1500 is written as 1.5×10^3 and 11 500 as 1.15×10^4.

If we have very small numbers of less than 1 then the power of 10 becomes negative, meaning the same as 'divide by'. For example, 0.001 5 is written as 1.5×10^{-3} and 0.000 015 as 1.5×10^{-5}.

A simple rule for calculating using standard form numbers is, if you are dividing then the powers of 10 are subtracted and if you are multiplying then the powers of 10 are added. For example, $10^5/10^2 = 10^3$ and $10^5 \times 10^2 = 10^7$.

Tip

The only way to become comfortable with units and with standard form is to practise simple examples. Using the magnification formula and a range of different sizes from electron micrographs and light micrographs is an excellent way to do this, as shown in Chapter 4.

Significant figures

A significant figure in a number adds to its precision. If you find the mass of a substance using a digital balance that has an accuracy of 0.01 g, then you would record the mass as, say, 1.00 g to indicate its precision. This means that writing 1 and 1.0 are not the same. '1' simply indicates that the value is 1 or more but not 2. '1.0' means it is 1 or more but not 1.1.

Similarly, you must be careful not to introduce additional significant figures that are not justified by your precision.

Example

You find the final mass of four results using a balance of 0.01 g accuracy and then calculate the mean.

Readings: 1.21 g, 1.33 g, 1.30 g, 1.27 g

Mean = 1.2775 (the result on your calculator)

Here we have suddenly introduced 5 significant figures when our balance can only justify three, so our mean should be limited to 1.28 g.

Graphs

Remember, graphs are pictorial representations of trends and patterns in data. When accurately drawn they can be used to make mathematical calculations and express relationships in algebraic form. The simplest relationship is a straight line graph, which can be expressed as $y = mx + c$. Here, x and y are the values on the coordinates, m is the gradient of the line and c is the value where the line cuts the y-axis.

Main graphical formats

1 Line graph

Both axes are continuous variables. In Figure 13.1 this means that it is possible to have any measurement on the temperature scale between 20 °C and 50 °C or any volume measurement up to 7 cm³.

It is advisable to join points with a straight line, as attempting to draw acceptable curves freehand is very difficult and you do not know exactly what happens between each point.

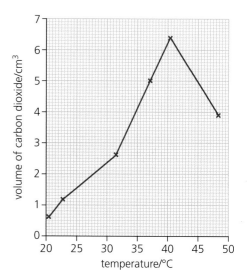

Figure 13.1 A line graph

2 Bar chart

Bar charts are used where the independent variable on the horizontal axis is not a continuous scale but is made up of distinct categories. To show this, the columns must not touch each other (Figure 13.2). A bar chart can also be the simplest way to display the means for two sets of data.

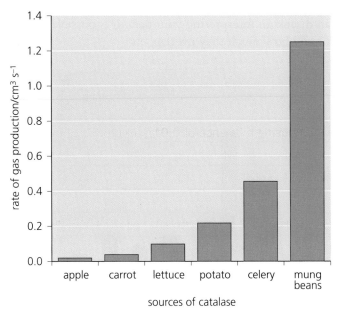

Figure 13.2 A bar chart

3 Histogram

A histogram is also drawn with columns, but the horizontal axis often shows the data from the dependent variable measurements organised into size classes. In Figure 13.3 the range of holly leaf sizes is divided into 10 size classes having 5 mm in each. You simply count how many leaves come into each size category to produce the number on the vertical axis (Figure 13.3).

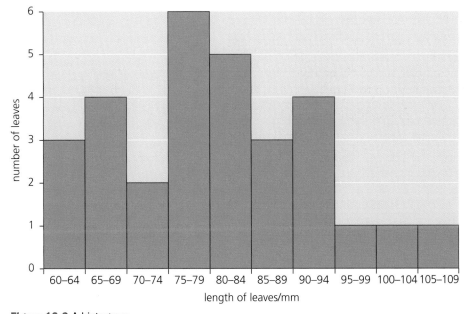

Figure 13.3 A histogram

If two sets of data are plotted on one graph it is possible to produce a comparison such as the example in Figure 13.4. Here, the sizes of dog whelks on an exposed and a sheltered shore are being compared. Note: a histogram normally has the columns touching but in this case a single column is split into two to give space for two sets of data.

A simple presentation of the data would be a two-column bar chart of the means for each shore. By working a little harder you can produce a graph that gives you much more information, which you can discuss in interpreting your data. It is possible to see how much the data are spread out and how much they overlap, etc.

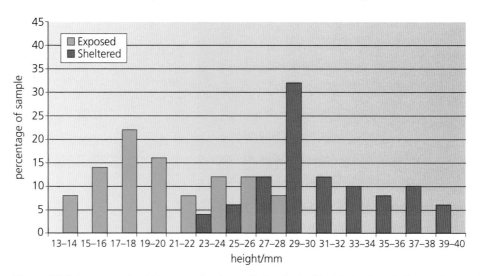

Figure 13.4 A comparative histogram showing variation in shell height in dog whelks on sheltered and exposed shores

4 Scatter graph

Scatter graphs (sometimes called scattergrams) are used when investigating a correlation between two variables. The data are simply plotted as individual points. This can often highlight possible anomalies for further investigation, such as the low diameter at pH 4.2 in Figure 13.5.

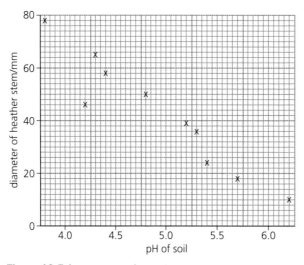

Figure 13.5 A scatter graph

The pattern of points will also indicate the general trend of any correlation present, as shown in Figure 13.6.

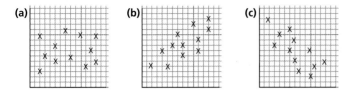

Figure 13.6 Three possible trends on a scatter graph: a) no correlation; b) positive correlation; c) negative correlation

5 'Box and whisker' plot

Box and whisker plots (Figure 13.7) are most often used to represent skewed data but they are very useful for showing lots of information about the data. Plotting two box and whiskers side by side can also be advantageous when comparing data sets.

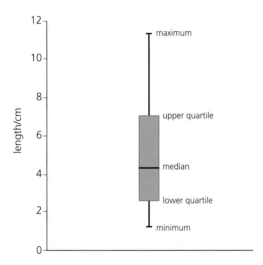

Figure 13.7 A 'box and whisker' plot

To find the values to plot, all the measurements are ranked in one list. The list is then divided into quartiles. The lower quartile boundary is the measurement that has 25 per cent of the sample. If there were 16 readings then this would mean moving up the rank order until you have 4th measurement. Similarly, the upper quartile boundary is when you have 75 per cent of the sample (the 12th measurement). The 'box' is now drawn as a column using the upper and lower quartiles on the vertical scale. The 'whiskers' are added as lines with a small cross-bar to indicate the highest and lowest measurements, and the median is drawn as a line in the box.

Graphs and examination questions

Graphs are extremely common in examination papers. They are used to test several skills but often they are linked to calculations and data interpretation.

The two graphs in Figure 13.8 show how important it is to be very clear about what data are actually plotted.

Both graphs show ion uptake into carrot root tissue. Graph A shows data recording the concentration of ions in the carrot tissue over time. Graph B shows data recording the rate of ion uptake over time.

> **Tip**
>
> Don't rush into graphically-based questions without taking a few moments to check exactly what the data on each axis actually show. What is the title? What are the units on the scales?

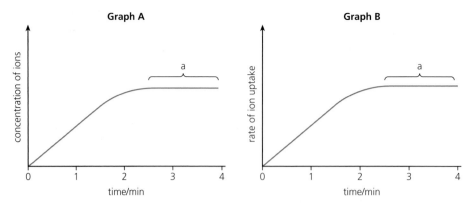

Graph A

Graph B

Figure 13.8 Different axes mean different trends

The graphs appear identical, but if we look at section 'a' on each then they mean very different things. In A, the horizontal line labelled 'a' shows no increase in concentration so transport has stopped. In B the horizontal section 'a' shows the rate has not changed but transport continues at a high level. So a lack of care in checking axes can lead to large mistakes.

Calculating rates from graphs

This is again a very common examination question so you need to practise until you are confident with the methods used. You will find the details in Chapter 2, Figure 2.13 and Chapter 9, Figure 9.7. Both of these are linked with core practicals, so you will also have the opportunity to practise with your own data. Calculations might include drawing a tangent to a curve in order to find the gradient, or taking readings from graphs and using these to find the change in value and the time taken.

Ratios and percentages

Chapter 10 introduces the concept of surface-area-to-volume ratio and demonstrates the comparative value of ratios. Knowing that a mouse might have a smaller surface area than an elephant is rather obvious, but calculating that it has a much larger surface-area-to-volume ratio leads you to some important biological conclusions.

Ratios can also be used to to give a more meaningful description of shape. The example in Figure 13.9 compares the shape of two limpets (*Patella vulgata*). Limpets are cone-shaped molluscs found attached to rocky shores. They have different shell shapes when living on exposed shores compared with those on very sheltered shores.

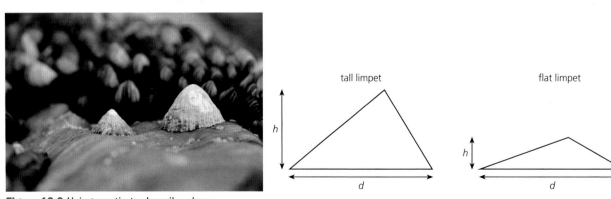

Figure 13.9 Using a ratio to describe shape

Individual measurements of height and diameter can vary, but if we use the ratio $h{:}d$ to describe the shape it is independent of these variations. Hence the 'tall' limpet will always have a much larger $h{:}d$ ratio compared with the 'flat' limpet, regardless of its overall size.

Percentage changes have very similar uses when comparing the magnitude of an overall change. A loss of mass of 1 g for a mouse could be catastrophic, but for an elephant it might be a tiny scratch on a toenail. Converting to percentage change would show this clearly, as 1 g would be 30–40 per cent loss of mass for the mouse but far less than 0.01 per cent for the elephant.

Calculating percentage change

This is a fairly simple calculation but a surprising number of students make basic errors when answering examination questions.

$$\% \text{ change} = \frac{\text{change in value}}{\text{initial value}} \times 100$$

Tip

When calculating percentage change, make sure you divide the change by the initial value not the final value, and always add a plus or minus sign to indicate the direction of the change.

Formulae

Whilst the number of equations you will need to memorise is very small, you do need to be confident in manipulating them to change the subject of the formula or to substitute values in them to calculate the final figure.

The simplest example is the magnification formula from Chapter 4. It is very useful to learn the three variations of this relationship:

$$\text{magnification} = \frac{\text{size of image}}{\text{size of specimen}}$$

$$\text{size of specimen} = \frac{\text{size of image}}{\text{magnification}}$$

$$\text{size of image} = \text{size of specimen} \times \text{magnification}$$

Some other specification formulae

In Chapter 8 you met more complex equations such as Simpson's Diversity Index (D), which is calculated using the following formula:

$$D = \frac{N(N-1)}{\Sigma n(n-1)}$$

Where N = total number of individuals of all species and n = the number of individuals of each separate species. The symbol Σ means 'sum of' so that in this case it is the sum of all the calculations of $n(n-1)$ for each separate species found in the habitat.

Statistical testing

Why do we use statistics?

Up until this point in your biology courses you may well have come across investigations and data from which you have been asked to draw conclusions. Although you will have had to apply your biological knowledge, these conclusions will have been based largely on your opinion. The trouble with opinions is that everyone has their own and this isn't good enough if we are to have reliable scientific progress.

Many investigations collect data to discover the answer to simple questions such as:

● Is there a difference between these two sets of results?
● Is there a correlation between these two variables?

Very often, the data collected do not make the answer to these questions obvious. So you need some rules on how you are to decide. These rules must be agreed before the data are collected, not selected later, as this makes it tempting to choose the rules just to fit your ideas, and they must be recognised by all other scientists.

The basic rule is all about chances or probability. Probabilities can be written in several ways (Table 13.2) but you are much more likely to meet them in decimal format.

Table 13.2 Probabilities

	Fraction	Percentage	Decimal fraction
Probability of tossing a fair coin and it falling as heads	1/2	50%	0.5

So a probability of 0 will indicate no probability and 1 or 100% will indicate total certainty.

Imagine that you are trying to investigate if there is a difference between the height of plants in two different areas. You measure the heights of random samples in each area. But is there a difference?

The problem is that, as with most living things, the height of a plant varies in each area, so it could be possible that more smaller plants than average were measured in one area and more slightly taller plants were measured in the other. This would mean you might think that the plants in the two areas had different heights when in fact they were the same.

Most of the statistical tests you might choose will calculate the probability that your results could occur purely by chance.

The rule that scientists apply to the type of data you are likely to collect is that there must be less than a probability of 0.05 (5% or 5 chances in every 100) that your results could arise simply by random sampling two areas where there was really no difference. This is called the 5 per cent significance level.

If you can show that the probability is less than this then you are entitled to claim that there is a significant difference between your two sets of data.

Tip

Always make sure you use the word 'significant' if your statistical test demonstrates this. 'Different' and 'significantly different' are not the same thing. Exactly the same argument is true for tests for a significant correlation and significant association.

Null hypotheses

In order to test a hypothesis using the 5 per cent significance level several steps are needed. In most cases this will be done by following the instructions for your chosen test but you do need to understand what is meant by a null hypothesis. These are the basic steps in hypothesis testing using the same example of plant heights as above.

1 Start by assuming that there is no difference in the height of plants in both areas. (This is the null hypothesis.)

2 Measure the height of sample plants from both areas.

3 Use the statistical test to find the probability of getting the results that you have measured if there was no difference.

4 If the probability of getting results like yours if there was no difference in heights is very low (less than $p = 0.05$) then you can reject your idea (null hypothesis) and accept that there is a significant difference (the alternative hypothesis).

A null hypothesis is usually given the symbol H_0 and the alternative hypothesis the symbol H_1.

Types of statistical test

Almost all the investigations you might undertake will be covered by three types of test (Table 13.3).

Table 13.3 Statistical tests

Type of test	Common tests	Notes
Testing for a significant difference	*t*-test Mann-Whitney U test	Only for normally distributed data Can be used for different types of data
Testing for a significant correlation	Spearman's rank test	Simple to apply and understand
Testing for an association or 'goodness of fit'	Chi-squared test	Only for categorical data not interval level measurements

You are not expected to know the more technical details of all the tests. However, do make sure you understand the principles and avoid basic errors when making your choice.

Chi-squared tests are often misused. They can only be used for categorical data. Investigations where this test is applicable are very rare at this level, so check carefully if you are thinking about using this test. You will be familiar with this type of data from genetics, for example where 'red-eye' and 'white-eye' are typical categorical counts in *Drosophila* investigations. In such cases you may form a hypothesis that there will be a fixed ratio, such as 3:1, in the results. It is rare for this to be exactly 3:1, so you can use a Chi-squared test with a 5 per cent confidence limit to test the 'goodness of fit' of your hypothesis. Are your data close enough to this predicted ratio?

An association might be tested in some cases. If you formed a hypothesis that rose bushes are more likely to suffer from black spot disease in rural areas compared with urban areas, then this could be an association. If the independent variables were rural and urban and the dependent variables were 'has black spot' and 'does not have black spot', then these would be categorical measurements. If you decided to measure the area of sample rose leaves affected by black spot then you would not have categorical data, so would need to use one of the tests for a significant difference.

Take care to use the word 'correlation' correctly. Using it inaccurately often reveals a poor understanding of what is being tested.

Accepting or rejecting a null hypothesis

For each statistical test you will need to calculate a test statistic. In most tests you will use a formula to do this. It is unlikely that you will need to learn the formula for each test as these will normally be provided, but you do need to practise calculating the test

Key terms

Correlation A relationship between two variables where it can be shown that as one increases so does the other (positive correlation) or as one increases the other decreases (negative correlation).

Categorical data Categorical data measurements are counts of numbers that fall into distinct groups or categories such as round/wrinkled, has/has not, blue eye/brown eye, etc.

Interval level measurement This type of measurement covers almost all the data you collect by measuring something, such as volume of gas, time taken, length, etc. You can always quantify any differences between interval level measurements.

statistic. Core practical 16 (see Edexcel A level Biology 2, Chapter 14) will provide you with an ideal opportunity to do this. Following this calculation you need to know the critical value of your chosen test statistic that matches your chosen level of confidence (0.05 or 5%). This involves a lot of calculations but fortunately the work will have been done for you and published in tables. If you do not have access to copies then they are freely available by typing the name of your test and 'table' into an internet search. Some are more complicated than others, so if you find one too complex, simply look up another. How you interpret the table is slightly different for each common test, and is shown in Table 13.4.

Table 13.4 Interpreting common statistical tests

Name of test	Name of test statistic	Rejecting null hypothesis
t-test	t value	Reject if your t value is higher than the critical value
Mann-Whitney U test	U1 and U2 values	Reject if the lowest U value is equal to or less than the critical value
Spearman's rank test	r_s (Spearman's correlation coefficient)	Reject if r_s is greater than or equal to the critical value
Chi-squared test	χ^2 (chi-squared)	Reject if value of χ^2 is greater than the critical value

Descriptive statistics

As the name implies, descriptive statistics are used to describe the data you collect. Several important descriptive statistics are part of the mathematical requirements of the specification. An understanding of their meaning is important when describing data.

Normal distribution

Living things often show variation. It is a vital element of natural selection. If you measure one feature of a large sample of individuals you often find the pattern shown in Figure 13.10.

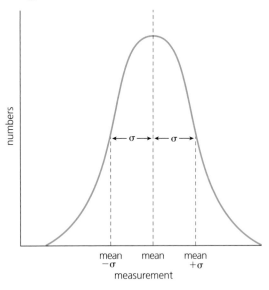

Figure 13.10 A normal distribution; the standard deviation is shown as σ

In simple terms this shows that the 'average' measurement is the one shown by most individuals and that the distribution is a symmetrical bell-shaped curve. This is what is meant by a 'normal' distribution.

The word 'average' is used in lots of different ways, so the term **mean** is better as it has a clearly defined scientific meaning.

If you look carefully at the graph in Figure 13.10 you will see that it can tell you something else about your data, namely how much the data are spread out. If all the data are in a narrow range then the curve will be tall and thin, but if the data are well spread out then the curve will be much wider and flatter. This is a really important feature if you are trying to decide if two populations are different, so it is calculated precisely. The calculation gives us a **population standard deviation**. This is given the sign σ (sigma). In a normal distribution we know that 68 per cent of all our values will lie within +/− one standard deviation of the mean and 98 per cent of our data will lie between +/− two standard deviations from the mean.

The formula for population standard deviation is:

$$\sigma = \sqrt{\frac{1}{N}\sum_{i=1}^{N}(x_i - \mu)^2}$$

This looks complicated, but if we break it down it becomes much simpler. You can use data from a number of your core practicals to practise substituting into the formula.

The standard deviation is actually the square root of the variance. So what is variance?

To calculate the variance you simply subtract each reading from the mean and square your answers. Then add up all the values and divide by the number of readings (N). In the formula above this is shown by the part of the expression under the square root sign, where x_i is the measurement and μ is the mean.

This is best illustrated by a simple example.

Key terms

Mean The average value of a set of measurements calculated by adding all the measurements and dividing by the sample number.

Population standard deviation A measure of how much the sample data are spread out from the mean value in a whole population.

Example

A student measures the mass of five seeds, in grams, as shown below.

1.2, 1.5, 1.3, 1.7, 1.7

What will be:

a) the mean

b) the variance

c) the population standard deviation of these data?

Answers

a) Mean = 1.48 (This is one more significant figure than we can justify from our original data but we tend to keep this when using it in further calculations.)

b) To calculate variance find the sum of (Σ) the squared differences from the mean:

$(-0.28)^2 + (0.02)^2 + (-0.18)^2 + (0.22)^2 + (0.22)^2$

= 0.208 (to 3 dp)

Now divide by the number in the population ($N = 5$).

So, variance = 0.0416

c) The population standard deviation shown by the formula is the square root of this = 0.204 (to 3dp).

Note: the units of standard deviation are those of the original values (g) and it is usually written as +/− 0.204 g.

Lots of statistical formulae that you will meet contain squared numbers and square roots. You may have wondered why this is the case. The answer is shown in the calculations in the example. Trying to add differences from a mean is an obvious way to check the variability of data. However, lots of data have a nice even distribution about the mean so the plus values and the minus values tend to be even and we always end up with a total of zero, which is not very helpful. Squaring a number gets rid of the minus values as minus times minus equals plus. Now we can simply add up the differences, but since they are all squared numbers we need to take the square root to get them back to their real value. In other words a neat mathematical trick to get a total!

Just to add to the complication you might also meet a sample standard deviation that is exactly the same as the population standard deviation except that, for technical reasons, the sum of the squares of the differences from the mean is divided by $N - 1$ to find the variance.

Be careful when interpreting a standard deviation. This may be a small or large number but you need to compare it with the mean. A standard deviation of 0.1 may sound very small but if the mean was 0.05 then it would represent a very large variability in your data.

Skewed data

Not every set of measurements you might take are evenly spread out like a normal distribution. Sometimes there are a lot more readings at one end of the scale. For example, if you wished to investigate the percentage cover of heather on a moorland you would quickly find that it was very common and the majority of your measurements would be in the 80–100 per cent range. This would certainly not give you a symmetrical curve. It would look more like the graph in Figure 13.11.

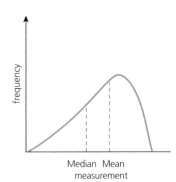

Figure 13.11 The distribution of skewed data

In this case the mean as a measure of the average value would be misleading and would not represent the data very well. Therefore for skewed data, such as this, the median value would be more useful.

Another way of expressing an 'average' value is to use the mode or **modal** value of your data. With normally distributed data the mean, median and mode are likely to be the same but for skewed data they could all be different values.

Error bars

In many ways error bars are not well-named as they do not show errors. They are used to show the variability of your data on graphs, to enable you to analyse your data in an objective way. If you are measuring some natural feature such as the area of a leaf you would expect your data to vary over a range of values. If the error bars show a lot of variability in a laboratory investigation where you intended to control many variables carefully, they may indicate a flaw in your planning.

To calculate the lengths of error bars you would normally calculate the mean of your data and then the standard deviation (to be strictly accurate this ought to be standard error but again the difference goes a little beyond A level). This could be represented as shown in Figure 13.12, using some data comparing the length of rats' tails in the wild with those kept as pets.

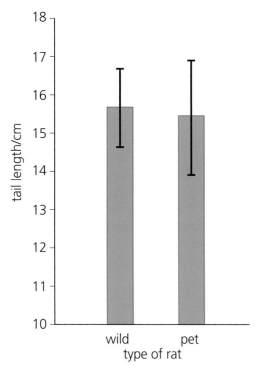

Figure 13.12 Errors bars – the mean and standard deviation of the tail lengths of wild and pet rats are shown. The error bars rise one standard deviation above and one below the mean

A much simpler way could be to use a range bar, where the bar simply represents the highest and lowest values above and below the mean.

A simple way of showing variation in skewed data is by using a 'box and whisker' plot, as shown in the Figure 13.7.

Key terms

Median A measure of 'average' value for skewed data, using the middle value of the range of readings rather than the mean.

Mode A measure of 'average' value of data, which is found by identifying the measurement that occurs most often in a set of results.

Preparing for the exams

This book covers the content of both Advanced Subsidiary (AS) Biology and the first year of Advanced level Biology for Edexcel. This chapter assumes you are preparing for the AS examination. A similar chapter in the book covering the second year of A level relates to preparation for all A level Biology examinations.

Why should I read this chapter early in my AS course?

Let's assume that, as you start your AS Biology course, your ambition is to gain a grade A. Looking at the national picture over the past few years, this means your ambition is to be in the top 20 per cent or so of candidates. How do you turn this ambition into reality?

Perhaps most important of all, you need to accept responsibility for your own learning. The best students do not simply rely on their teachers and the work they do in class to drive their learning. They know what is needed in order to achieve their ambition and ensure they develop the skills to achieve it. Accepting responsibility for your own learning is a mindset – an attitude – that you need to develop early. You cannot do it a couple of weeks before your examinations start.

Let's take an example from sport. Serena Williams is a tennis player, high in the world rankings. How has she achieved this? Clearly, she needs to know the 'rules' of tennis and to know what physical attributes successful tennis players develop. Then she needs to train. This means she recognises her own skills and practises them. She practises her best strokes to maintain them and practises her weak strokes to improve them. For her, tennis is a full-time job – you do not need to train with the same intensity she does. But *practice* is a key to her success just as it will be to your success.

Figure 14.1 Serena Williams is a tennis player high in the world rankings. How has she achieved this?

What do I need to practise?

Tip

Throughout your AS Biology course, you should continue practising your best skills as well as improving your weakest skills if you are to achieve a high grade.

Many students seem to believe that all an examination tests is memory of facts. In relation to your AS Biology examinations, this could not be further from the truth. As a consequence, preparation for your examination cannot be just a case of 'swotting facts' for a couple of weeks before the examinations. Although the AS Biology examination does test ability to recall facts (memory), it tests many more skills than that. These are the skills you need to practise.

The skills tested in each AS Biology examination are called **assessment objectives**. Table 14.1 summarises these objectives. Notice the heading of the second column in the table. It states 'the skills that will be tested'; not might, but *will*.

Table 14.1 The skills (assessment objectives) that will be tested in AS Biology examinations

Assessment objective	Description of skills that will be tested
AO1	Demonstrate knowledge and understanding of scientific ideas, processes, techniques and procedures.
AO2	Apply knowledge and understanding of scientific ideas, processes, techniques and procedures: • in a theoretical context • in a practical context • when handling qualitative data • when handling quantitative data.
AO3	Analyse, interpret and evaluate scientific information, ideas and evidence, including in relation to issues, to: • make judgements and reach conclusions • develop and refine practical design and procedures.

Table 14.2 shows the balance of marks awarded for each of the assessment objectives in the AS Biology examination papers. Notice that it is given as a percentage of the overall marks for the examinations. Notice that the balance of marks is given as a percentage of the overall marks for the examinations. The percentages are not just rough guides – examiners have to ensure the papers have this balance of marks, every year.

Table 14.2 The percentage weighting for each assessment objective in the AS Biology examinations

Paper	Percentage weighting of each assessment objective			
	AO1	AO2	AO3	All
1	17–19	20–22	10–12	50
2	17–19	20–22	10–12	50
Total for AS	**35–37**	**41–43**	**20–23**	**100**

Suppose you rely on 'swotting facts' just before your AS Biology examinations. What is the maximum mark you can expect to gain on this skill alone? Look again at Table 14.1. Recall of facts is part of AO1 – not all of it, just a part (demonstrating knowledge). Now look at Table 14.2. The maximum weighting for the whole of AO1 is 35–37 per cent of the total examination marks. Even if recall (without understanding) earns half those marks, the maximum marks for recall will be about 17–19 per cent of the total examination marks. If you are guessing that would earn you the lowest grade possible (grade U), you are correct. You need to develop *all* the skills in Table 14.1.

Tip

Because of the weighting of the assessment objectives, relying solely on memorising facts is likely to earn you a grade U in an AS Biology examination.

Do I really need to practise these skills throughout my course?

Perhaps you are hoping to pass your driving test soon, or can remember how you felt before you passed it. What did you do to prepare for that? The chances are you regularly went over the information you needed for the theory test, just to make sure you still remembered it. You might have asked someone to ask you random questions or even tried the practice test papers on the Driver and Vehicle Standards Agency website. You also probably pestered a parent, sibling or friend who had a full driving license to sit in the passenger seat, enabling you to drive the family car. If you are not learning to drive, you have probably practised a different skill recently – a performance piece on a musical instrument, a dance routine, preparing a favourite meal or reaching a new level in your favourite computer game. If so, just as with the example of learning to drive, your behaviour will have shown two key aspects of learning:

- Repetition improves performance.
- All the skills that will be tested must be repeated.

You might still be unconvinced. Look at Figure 14.2, which provides a biological explanation for the importance of repetition.

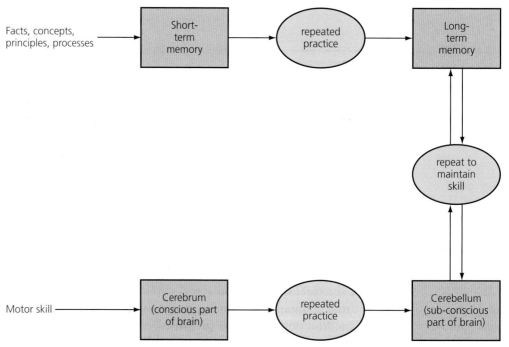

Figure 14.2 During successful learning, there is a transfer of control from one part of the brain to another

- To recall information, you need to transfer what you have just learnt from your short-term memory to your long-term memory *and* be able to retrieve this information from your long-term memory.
- For physical skills, you need to transfer skill control from the part of your brain that controls consciousness (the cerebrum) to the part of your brain that controls balance and fine motor control (the cerebellum).

In both cases, repetition helps this transfer, so you do need to practise these skills throughout your AS course.

Now look at Figure 14.3. It shows the results of an experiment performed in 1885 by a German psychologist called Hermann von Ebbinghaus. In his experiment, Ebbinghaus learnt a list of nonsensical three-letter 'words'. He then tested how well he recalled them until he could remember none of the words from the list. As you can see from curve A in Figure 14.3, he forgot the list pretty quickly. He then repeated the same experiment but this time he re-learnt his list after 2 weeks and, in a further repeat, he re-learnt his list after 2 weeks and again after 4 weeks. You can see the improvement in his ability to recall the list when he re-learnt it after 2 weeks (curve B) and even more improvement when he re-learnt it after 2 weeks and after 4 weeks (curve C). These results are generally accepted to show how we all lose the ability to recall unless we keep revising what we have learnt.

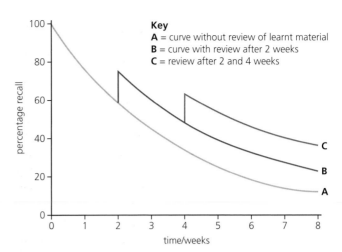

Figure 14.3 The 'Ebbinghaus forgetting curve' shows how reviewing what we learn aids memory and slows forgetting

How can I practise these skills?

Firstly, it is important that you understand what you have been taught each week. If you do not, you will find it difficult not only to recall information but also to be able to apply it. There are probably long periods of time when you are in college or school but not timetabled for classes. Use some of this time to visit the learning centre and go over your notes. If you find something you do not understand, make a note of it and plan to deal with that straight away. You could:

- ask your teacher for help
- ask others in your class for help
- use the relevant chapter in this book
- type a question into your favourite search engine. The web is full of animations, presentations and answers to students' questions that might be helpful to you.

In addition, you might choose to set aside a regular time each week to practise your skills. Research shows that an intense but short period is more effective than a prolonged period of practice. You will know best the period of time you can concentrate for; it is likely to be between 30 and 60 minutes. This is the maximum time you should plan for your practice session.

Look back to Table 14.1. It shows the skills you need to practise. Decide in advance which skills you will practise that week. Remember that the weighting of these skills, shown in Table 14.2, means you cannot afford to leave any skill as a weakness. Be prepared to plan practice sessions for skills that you find tough.

Practice is an active process – you must be doing something. Simply reading your notes is not an active process. So, your final act of preparation is to find material that you will use to practise your chosen skill.

> **Tip**
>
> You must not stay in your comfort zone by practising only those skills you find easy.

Practising AO1 skills

These are perhaps the easiest of all skills to practise. Think of how often you are alone and idle, for example, standing in a queue or sitting on a bus. How long would it take you to pick a topic and remind yourself of the key facts, principles or terminology? You could, for example, decide to recall the names and functions of the organelles of a eukaryotic cell, to recall the events in the cardiac cycle or to explain to yourself the reason why the activity of enzymes is affected by temperature. It would take minutes. And these minutes are not 'time you will never get back' – you weren't doing anything useful in the first place.

For a more structured practice session, you need test questions. Most of the 'Test yourself questions' in this book are simple tests of recall or of recall with understanding. This is also true of some of the 'Exam practice questions' in this book. For example, look back to Exam practice question 4 in Chapter 5 or to Exam practice questions 4a) and b) in Chapter 9; these test AO1. Being able to spot what skill a question is testing is an important part of preparing for your examination. The best way to do this is to look at the command word – the wording that tells you what the examiners want you to do. Table 14.3 shows you the command words used in questions that test AO1 skills (recall and recall with understanding). There is no point in writing an answer that does not do what the command word asks. Writing a description when asked for an explanation, for example, will not gain the marks you need.

> **Key term**
>
> **Command word** The verb or phrase used by examiners to tell you what is required in your answer, e.g. describe, explain, calculate.

Table 14.3 The command words used to test AO1 skills

Command word	Description of what is required
Add/Label	Requires the addition or labelling to a stimulus material given in the question, for example labelling a diagram or adding units to a table.
Complete	Requires the completion of a table/diagram.
Describe (in absence of stimulus material)	To give an account of something. Statements in the response need to be developed as they are often linked but do not need to include a justification or reason.
Draw	Produce a diagram either using a ruler or using freehand.
Explain	An explanation requires a justification/exemplification of a point. The answer must contain some element of reasoning/justification, this can include mathematical explanations.
Give/state/name	All of these command words are really synonyms. They generally all require recall of one or more pieces of information.
Plot	Produce a graph by marking points accurately on a grid from data that is provided and then drawing a line of best fit through these points. A suitable scale and appropriately labelled axes must be included if these are not provided in the question.
Sketch	Produce a freehand drawing. For a graph this would need a line and labelled axis with important features indicated, the axis are not scaled.
State what is meant by	When the meaning of a term is expected but there are different ways of how these can be described.
Write	When the questions ask for an equation.

Practising AO2 and AO3 skills

Questions testing AO2 and AO3 skills contain stimulus material, in other words text, artwork or tabulated data that you are expected to use. Consequently, you need to have questions ready for use during practice sessions for these skills. They are easy to spot because they contain lots of information that you need to use. Usually you will not have seen this information before. Do not worry; neither has anyone else. If examiners wish to test your ability to use information, it is important that they provide stimulus material that none of

> **Key term**
>
> **Stimulus material** The written, diagrammatic or tabulated information given in an examination question that you must describe, explain, analyse, interpret or evaluate.

the candidates taking the examination is likely to be familiar with or be able to recall. You will need to understand the underlying biology, but will not have seen the context before. So, in addition to the skills being tested in the question, you need to develop a further skill during your course, namely the confidence to deal with something new.

Table 14.4 shows the command words by which you can recognise questions testing AO2 and AO3 skills. Notice that, in each case, reference to one or more aspects of the stimulus material is required. If you do not refer to the stimulus material, you cannot show AO2 or AO3 skills and, consequently, will not gain many marks.

Table 14.4 The command words used to test AO2 and AO3 skills

Command word	Description of what is required
Assess	Give careful consideration to all the factors or events that apply and identify which are the most important or relevant. Make a judgement on the importance of something, and come to a conclusion where needed.
Calculate	Obtain a numerical answer, showing relevant working. If the answer has a unit, this must be included.
Comment on	Requires the synthesis of a number of variables from data/information to form a judgement.
Compare and contrast	Looking for the similarities **and** differences of two (or more) things. Should not require the drawing of a conclusion. Answer must relate to both (or all) things mentioned in the question. The answer must include at least one similarity and one difference.
Criticise	Inspect a set of data, an experimental plan or a scientific statement and consider the elements. Look at the merits and faults of the information presented and support judgements made by giving evidence.
Deduce	Draw/reach conclusion(s) from the information provided.
Describe	To give an account of something. Statements in the response need to be developed as they are often linked but do not need to include a justification or reason.
Determine	The answer must have an element which is quantitative from the stimulus provided, or must show how the answer can be reached quantitatively. To gain maximum marks there must be a quantitative element to the answer.
Devise	Plan or invent a procedure from existing principles/ideas.
Discuss	• Identify the issue/situation/problem/argument that is being assessed within the question. • Explore all aspects of an issue/situation/problem/argument. • Investigate the issue/situation etc. by reasoning or argument.
Evaluate	Review information then bring it together to form a conclusion, drawing on evidence including strengths, weaknesses, alternative actions, relevant data or information. Come to a supported judgement of a subject's qualities and relation to its context.
Explain	An explanation requires a justification/exemplification of a point. The answer must contain some element of reasoning/justification, this can include mathematical explanations.
Give a reason/reasons	When a statement has been made and the requirement is only to give the reasons why.
Identify	Usually requires some key information to be selected from a given stimulus/resource.
Justify	Give evidence to support (either the statement given in the question or an earlier answer)
Predict	Give an expected result.
Show that	Verify the statement given in the question.

What do I need to know about the examination papers?

Key term

Specification

The document from the examination board that shows the content and skills that can be tested in an examination.

You can, and should, download a copy for your own reference from the examination board's website:

http://qualifications. pearson.com/content/ demo/en/qualifications/ edexcel-a-levels/ biology-b-2015.html

There is little point in developing your skills throughout your course unless you know how you will be asked to demonstrate them in an examination paper. This involves knowing:

- what content will be tested in each paper
- the types of question that will be used in each paper, and where within that paper.

Let's deal with these in turn.

The content of each paper

Your AS examination consists of two papers, each carrying 50 per cent of the final marks. It might be helpful to you that the specification content tested in each paper is different. Table 14.5 shows this, and links the tested content to the relevant chapters in this book. Although you should practise your skills throughout your course, your final preparation should be geared towards the subject content of each examination paper.

Table 14.5 The specification content that is tested in each AS examination paper.

Paper	Exam time/ minutes	Total marks	Specification topics tested in the examination paper	Relevant chapters in this book
1	90	80	Biological molecules	1, 2 and 3
			Cells, viruses and reproduction of living things	4, 5 and 6
2	90	80	Classification and biodiversity	7 and 8
			Exchange and transport	9, 10, 11 and 12

Tip

The AS examination papers are 90 minutes long and carry 80 marks. During an exam, ask yourself, 'In the last minute, did I gain 1 mark or did I waste time?'

Types of examination question

As Table 14.5 shows, each question paper lasts for 1 hour 30 minutes and carries 80 marks. You could regard that as 1 mark per minute plus 10 minutes of thinking and reading time. It does not leave you with time to spare or time to waste. Both papers can include the following types of question:

- **multiple choice**, in which you choose the correct response from four alternatives; these usually form the first one or two parts of a structured, short-answer question
- **short open**
- **open-response**
- **calculation**, targeting mathematics at Level 2 or above
- **extended writing**, carrying up to 6 marks.

Preparing your examination strategy

Examination papers are usually designed with the following features:

- The early questions are more accessible than the later questions.
- Within each structured question, the earlier parts are more accessible than the later parts.

You will also remember from your previous experience that examinations are stressful.

Tip

Both AS Biology examination papers can contain questions testing your recall and understanding of experimental methods, so don't forget those when preparing for your examination.

Let's put these two ideas together.

- It is highly unlikely that you will obtain full marks in your examinations. What you do need to do, though, is ensure that you gain all the marks that you are capable of gaining.
- Early in the examination, you need to reduce your anxiety levels. The best way to do this is by doing something that plays to your strengths and knowing that you have gained marks.

Rather than leaving it for the day on which the examination paper lands on your desk, the best way to achieve these two objectives is to plan how you will answer the examination paper well in advance. This plan is your examination strategy. So, how do you make this plan?

Firstly, remember that the order of questions in the examination paper is not designed with you in mind. You do not need to answer them in numerical order. Secondly, ask yourself what your favourite topics are on each paper. Thirdly, ask yourself in which of your skills you have the most confidence. Finally, look through the specimen question papers and, if available, past examination papers. Is there a particular place where types of question occur? For example, if you love extended prose questions that allow you to show your AO1 skills, and gain 6 marks in the process, is there a place in the examination paper where such a question always crops up. Once you have the answers to those questions, you can start your strategy. Remember that, since it plays to your strengths, your strategy will be unique to you; don't be put off if someone else has a different approach.

Table 14.6 outlines some of the general features that your examination strategy might contain.

Table 14.6 A possible examination strategy

I plan to...	Reason(s)
• spend 5 minutes skimming through the examination paper	• I need to ensure that I have seen every question, so that I don't miss any that I can answer. • I need to find those that are easy for me to ensure I gain marks by answering them and don't waste valuable time by getting stuck on a harder question. • I need to plan the order in which I answer the questions to ensure that, early on, I get all the marks that are 'easy' for me.
• pick one or two questions that play to my strengths and answer those first	• I will gain valuable marks early in the examination. • Knowing I have been successful will give me confidence and reduce my stress levels. This will help me to feel in control.
• answer the other questions by picking those parts I find straightforward and, when a question is not easy, move quickly to the next question or part of a structured question	• I am now 'harvesting' marks, ensuring I gain all the ones that are easy to get and avoiding wasting time on those I find harder. I can come back to those once I have gained my 'easy marks'.
• go back through the paper, filling in the gaps I have left, again without wasting too much time on any part of a question	• Having built up my confidence by gaining many marks, I am now ready to tackle the parts of questions I found really hard on the first run-through.
• make an intelligent 'guess' rather than leave a gap	• There is no point in leaving gaps in the paper since I know that a gap gains no marks.
• go back through the paper to check my answers and correct any errors	• Under the pressure of an examination, it is easy to make a silly mistake – an exam howler. I will just check that I haven't made any and correct any that I find.

Sources you might find useful

In addition to the materials you gain in class and this textbook, there are many other helpful sources that are readily available. They include:

- the examination board's website for the specification and specimen examination papers:

http://qualifications.pearson.com/content/demo/en/qualifications/edexcel-a-levels/biology-b-2015.html

- the intranet in your college or school
- your favourite search engine – key in your question, topic area, or 'biology animations'
- Edexcel A level Biology Teaching and Learning Resources
- *Biological Sciences Review* for articles of interest, including structured examination-style questions about some of them:

http://www.hoddereducation.co.uk/magazines/Print-Magazines

- your classmates (Figure 14.4).

Figure 14.4 Whilst the examination is a solo effort, solving problems or revising in a group can be rewarding

Index

Free online resources

Answers for the following features found in this book are available online:

- Test yourself questions
- Activities

You'll also find an Extended glossary for each chapter to help you learn the key terms and formulae you'll need in your exam.

Scan the QR codes below for each chapter.

Alternatively, you can browse through all resources at:
www.hoddereducation.co.uk/EdexABiology1

How to use the QR codes

To use the QR codes you will need a QR code reader for your smartphone/tablet. There are many free readers available, depending on the smartphone/tablet you are using. We have supplied some suggestions below, but this is not an exhaustive list and you should only download software compatible with your device and operating system. We do not endorse any of the third-party products listed below and downloading them is at your own risk.

- for iPhone/iPad, search the App store for Qrafter
- for Android, search the Play store for QR Droid
- for Blackberry, search Blackberry World for QR Scanner Pro
- for Windows/Symbian, search the Store for Upcode

Once you have downloaded a QR code reader, simply open the reader app and use it to take a photo of the code. You will then see a menu of the free resources available for that topic.

1 Introducing the chemistry of life

3 Nucleic acids and protein synthesis

2 Proteins and enzymes

4 Cell structure and viruses

5 The eukaryotic cell cycle and cell division

10 Gas exchange and transport

6 Sexual reproduction in mammals and plants

11 Mammalian circulation

7 Classification

12 Transport in plants

8 Natural selection and biodiversity

9 Cell transport mechanisms

Acknowledgements

The Publisher would like to thank the following for permission to reproduce copyright material.

Photo credits:

p.7 © enskanto – Fotolia; **p.11** *both* © Andrew Lambert Photography/Science Photo Library; **p.14** *both* © Andrew Lambert Photography/Science Photo Library; **p.15** © Biophoto Associates/Science Photo Library; **p.16** © Biophoto Associates/Science Photo Library; **p.19** © molekuul.be / Alamy; **p.21** © Steve Gschmeissner/Science Photo Library; **p.31** © Steve Gschmeissner/Science Photo Library; **p.70** *l* © J.C. Revy, Ism/Science Photo Library, *r* © Gene Cox; **p.72** © Gene Cox; **p.74** © Power And Syred/Science Photo Library; **p.75** *l* © George Chapman/Visuals Unlimited/Getty Images, *r* © G. Wanner/Sciencefoto.de; **p.76** © Sinclair Stammers/Science Photo Library; **p.77** *t* © Dr. Kevin S. Mackenzie, Institute of Medical Sciences, Aberdeen University, *r* © Eye Of Science/Science Photo Library; **p.79** *l* © Steve Gschmeissner/Science Photo Library, *r* © Dr. Jeremy Burgess/Science Photo Library; **p.80** © Biophoto Associates/Science Photo Library; **p.81** *t* © CNRI/Science Photo Library, *r* © Dr. Kari Lounatmaa/Getty Images; **p.82** *t* © Medimage/Science Photo Library, *b* © Omikron/Science Photo Library; **p.83** © Carolina Biological Supply Co/Visuals Unlimited, Inc. /Science Photo Library; **p.84** © Dr. Kevin S. Mackenzie, Institute of Medical Sciences, Aberdeen University; **p.85** © Dr. Jeremy Burgess/Science Photo Library; **p.86** © Kwangshin Kim/Science Photo Library; **p.96** *l* © James Cavallini/Science Photo Library, *r* © BSIP SA / Alamy; **p.99** *all* © Michael Abbey/Science Photo Library; **p.102** © Wim Van Egmond/ Visuals Unlimited, Inc. /Science Photo Library; **p.111** © Pr. Philippe Vago, Ism/Science Photo Library; **p.114** © Science VU/B. John, Visuals Unlimited /Science Photo Library; **p.121** © Jean-Claude Revy-A. Goujeon, Ism/Science Photo Library; **p.122** © Gene Cox; **p.124** © Edelmann/Science Photo Library; **p.128** © Gerry Cambridge/ NHPA/Photoshot; **p.130** © Dr. Keith Wheeler/Science Photo Library; **p.133** © CNRI/Science Photo Library; **p.139** *l* © idp wildlife collection / Alamy, *m* © Lip Kee / http://www.flickr.com/photos/ lipkee/5657636385/sizes/o/in/pool-42637302@N00/ http://creativecommons.org/licenses/ by-sa/2.0, *r* © FLPA / Alamy; **p.148** *l* © Springfield Gallery – Fotolia, *r* © C. J. Clegg; **p.164** © Sally A. Morgan/Ecoscene/Corbis; **p.168** © Tony Watson / Alamy; **p.174** © kjorgen/iStock/ Thinkstock; **p.179** © Robert Bird / Alamy; **p.182** © Frans Lanting Studio / Alamy; **p.186** *both* © Don W. Fawcett/Science Photo Library; **p.194** © *both* Biophoto Associates/Science Photo Library; **p.207** © Biodisc/Visuals Unlimited, Inc./Science Photo Library; **p.214** © PHOTOTAKE Inc. / Alamy; **p.215** © Ned Therrien/Visuals Unlimited, Inc./Science Photo Library; **p.216** © Gene Cox; **p.227** © PHOTOTAKE Inc. / Alamy; **p.236** © CNRI/Science Photo Library; **p.237** © BSIP/Vem/Science Photo Library; **p.247** *l* © Dr. Keith Wheeler/ Science Photo Library, *r* © Ed Reschke/Getty Images; **p.248** *both* © Gene Cox; **p.249** *t* © Gene Cox, *b* © Dr. David Furness, Keele University/Science Photo Library; **p.266** © Stiggg/ iStock/Thinkstock; **p.274** © Actionplus / TopFoto; **p.282** © Rido – Fotolia.

t = top, *b* = bottom, *l* = left, *r* = right, *m* = middle

Every effort has been made to trace all copyright holders, but if any have been inadvertently overlooked, the Publisher will be pleased to make the necessary arrangements at the first opportunity.

New College Nottingham
Learning Centres